Springer Proceedings in Complexity

Springer Complexity

Springer Complexity is an interdisciplinary program publishing the best research and academic-level teaching on both fundamental and applied aspects of complex systems—cutting across all traditional disciplines of the natural and life sciences, engineering, economics, medicine, neuroscience, social, and computer science.

Complex Systems are systems that comprise many interacting parts with the ability to generate a new quality of macroscopic collective behavior the manifestations of which are the spontaneous formation of distinctive temporal, spatial, or functional structures. Models of such systems can be successfully mapped onto quite diverse "real-life" situations like the climate, the coherent emission of light from lasers, chemical reaction-diffusion systems, biological cellular networks, the dynamics of stock markets and of the Internet, earthquake statistics and prediction, freeway traffic, the human brain, or the formation of opinions in social systems, to name just some of the popular applications.

Although their scope and methodologies overlap somewhat, one can distinguish the following main concepts and tools: self-organization, nonlinear dynamics, synergetics, turbulence, dynamical systems, catastrophes, instabilities, stochastic processes, chaos, graphs and networks, cellular automata, adaptive systems, genetic algorithms, and computational intelligence.

The three major book publication platforms of the Springer Complexity program are the monograph series "Understanding Complex Systems" focusing on the various applications of complexity, the "Springer Series in Synergetics", which is devoted to the quantitative theoretical and methodological foundations, and the "SpringerBriefs in Complexity" which are concise and topical working reports, case-studies, surveys, essays, and lecture notes of relevance to the field. In addition to the books in these two core series, the program also incorporates individual titles ranging from textbooks to major reference works.

More information about this series at http://www.springer.com/series/11637

Paul Bourgine • Pierre Collet • Pierre Parrend
Editors

First Complex Systems Digital Campus World E-Conference 2015

Springer

Editors
Paul Bourgine
Paris, France

Pierre Collet
Univ de Strasbourg
Strasbourg, France

Pierre Parrend
ECAM Strasbourg-Europe
Schiltigheim, France

ISSN 2213-8684 ISSN 2213-8692 (electronic)
Springer Proceedings in Complexity
ISBN 978-3-319-45900-4 ISBN 978-3-319-45901-1 (eBook)
DOI 10.1007/978-3-319-45901-1

Library of Congress Control Number: 2016958135

Printed on acid-free paper

This Springer imprint is published by Springer Nature
The registered company is Springer International Publishing AG
The registered company address is: Gewerbestrasse 11, 6330 Cham, Switzerland

Contents

The original version of this book was revised.
An erratum can be found at DOI 10.1007/978-3-319-45901-1_44

From Individual to Social Cognition: Piaget, Jung, and Commons

Francisco Antonio Pereira Fialho

1 Introduction

Butterflies won't exist if life didn't go through a silent and lonely metamorphosis processes (Rubem Alves)

Education is a metamorphosis process. Assimilation, accommodation, as defined by Piaget, and transcendence, the way toward the "self," as stated by Jung, are stages of this growing up process. The final goal is the realization of the promise present in all young birds: the possibility of flying.

Individuation must be the education goal. We are fragmented beings, thousands subpersonalities that must be united. We live inside conversation networks that are both external and internal. We are never alone. Personas are the masks we use to establish relations with the world outside. Animas are the subpersonalities existing inside us. Education in the past was concerned in strengthening the "ego" in order to avoid "schizophrenia." As pointed out by Deleuze and Guattari[1] we need to empower the thousands of voices that share a living inside us. A flexible "ego" is a goal to be achieved. We believe like Fernando Pessoa (poem "After All," using the Anima of Álvaro de Campos) that:

The more I feel, the more I feel like many people,
The more personality I have,
The more intensely, stridently I have them,
The more simultaneously to feel with all of them,

[1]*Anti-Oedipus: Capitalism and Schizophrenia* is a 1972 book by philosopher Gilles Deleuze and psychoanalyst Félix Guattari. *A Thousand Plateaus* is a book of 1989.

F.A.P. Fialho (✉)
Santa Catarina Federal University, Florianópolis, Santa Catarina, Brazil
e-mail: fapfialho@gmail.com

P. Bourgine et al. (eds.), *First Complex Systems Digital Campus World E-Conference 2015*, Springer Proceedings in Complexity, DOI 10.1007/978-3-319-45901-1_1

The more unified diverse, sparsely attentive,
Are, feel, live, be,
Longer I'll get the total existence of the universe,
More complete I'll be by the entire space outside.
More analogous to God I will be, whoever could he be,
Because, whoever could he be, surely is Everything,
And outside Him there's only Him, and All for him is little (less).

As pointed out by Professor Roeris Gonzáles Sivilla in revising this paper, the first statement "There is no individual Cognition" is too "strong." The social historic cultural theory of Lev Vygotsky adopted by the scientific community's majority states that learning has two moments: An interpsychological and an intrapsychological moment. Even in this theory he defends that learning processes are social, he acknowledges that a stage of the learning of the individual develops internally and not only in his interactions with others.

Besides human beings are not only social beings. We learn in our interaction with the nature around us. How can social connections influence the sensation to please that we experiment when we took a bath in a river, seeing the dawn from a hill or the sunset in front of the sea? Can we deny that these experiences bring us an important ethical learning?

I agreed both with Dr. Roeris and Vygotsky. I see no contradiction in this. In accordance with Jung there is no "individual." We are a social community dominated by what is called "Ego." Descartes is wrong, says Freud: "We are where we do not think." We also understand "social" as including people, animals, and things, as actants (agents) in accordance with Algirdas Julien Greimas semiotics. Things and animal live also inside us as subpersonalities.

2 Designing Education as a Flow Process

There are schools that are cages and there are schools that are wings. Schools like cages are for the birds to unlearn the art of flight. Birds are caged birds under control. Caged, its owner can take them wherever he wants. Caged birds always have an owner. Because the essence of birds is flying, schools like wings do not deal with caged birds. What they love are birds in flight. They exist to give courage to birds to fly. Teaching flight is impossible, because the flight is born within the birds. The flight cannot be taught. It can only be encouraged. (Rubem Alves)

Csikszentmihalyi [1] stated that: "Flow is characterized by one or more of the following, but it is not yet clear which of these elements must be present so that there is flow."

(a) Challenge-Skill Balance

Children are naturally curious. Teaching is like giving the feather to Dumbo, the magical elephant of Disney, encouraging him to do what he was born to: fly.

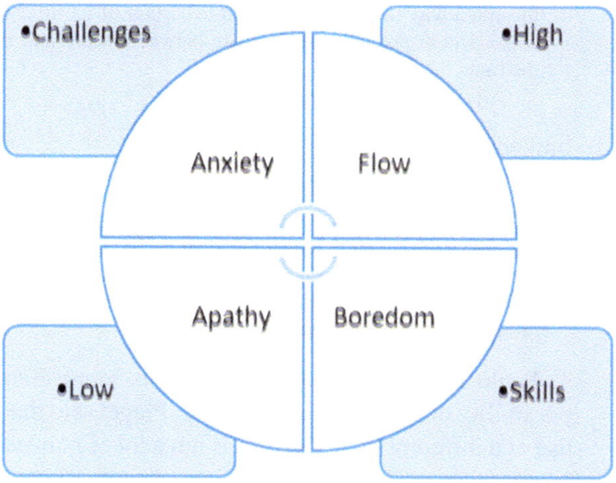

Fig. 1 Flow characteristics (Source: Adapted from Csikszentmihalyi [2, 3])

Plato believes that learning is remembering. The path of curiosity guides the student in the process of remembering. Challenges are invitations to adventure. While playing a wonderful game we record old forgotten skills.

Professors are gardeners. They prepare the soil, enriching it with the best chemicals and natural fertilizer existing. Watch the small seeds grow and flourish. Figure 1 suggests that we must adequate the level of challenge to student's skills.

(b) Merging Action and Attention

> Life cannot be saved for tomorrow. Life happens always in the present. (Rubem Alves)

The path toward illumination demands reflection and action. Making mistakes is necessary. Piaget thinks that we learn more from our own errors then when we are right. But the action that guides us to an error must be followed by the act of reflection.

The professor must be a "Magister Ludi" like Joseph Knecht, the Castalia Rector in Hermann Hess "The Glass Bead Game." The pleasure of playing games awakes in the student the passion for learning.

(c) Clear Objectives

The objectives must not be "curriculum." It is not clear for us what we really want. This is the work to be done, to help the student to make clear what he really wants. The rest is magic, is the encounter of something you are looking for, something you are in love with.

The professor here must be a "Dreams Interpreter." If a dream can be interpreted you will be able to learn (construct) what to expect in the future. Ergonomics is the art of transforming a "*ponein*" (painful) life in "*ergon*" (something wonderful).

> You must not give way to desires which you don't believe in.
>
> I know what you desire. You should, however, either be capable of renouncing these desires or feel wholly justified in having them. Once you are able to make your request

> in such a way that you will be quite certain of its fulfillment, then the fulfillment will come. But at present you alternate between desire and renunciation and are afraid all the time.
>
> All that must be overcome.[2]

(d) Immediate Feedback

> If I were to teach a child the beauty of music I would not start with scores, notes and guidelines. Together we would hear the hottest tunes and learn about the instruments that make the music. Then, enchanted by the beauty of music, he asks me who taught him the mystery of those written black polka dots on five lines. Because black polka dots and the five lines are just tools for the production of musical beauty, the experience of beauty has to come first. (Rubem Alves)

Paulo Freire states that "dialogic is more than dialectics.. Dialogic is the possibility of several interpretations. Piaget said that there is no mistake, only the use of a different logic. A class is not a lot of non-meaningful knowledge that the teacher deposit inside their students heads. Rubem Alves states that inside the classroom there are three movements. Teacher teaches students; students teach teachers; and students teach students. The last one in accordance with Rubem Alver is the most powerful.

From a successful feedback, it becomes possible to develop an optimal experience, one able to provide a flow state. There are many challenges to achieve this, since individuals are different and may perceive the experience in their own way.

Seminar comes from "semen." A good seminar must eroticize the participants. The indicator here is the intensity of the orgasm. In a seminar all actors must expose themselves sharing their knowledge.

We have here the Professor as a "Love Master," teaching their students to make love with words and ideas. The feedback is measured by the intensity of the orgasm reached.

(e) Intense Concentration on the Task

> All the words literally taken are false. The truth lives in the silence that exists around the words. Pay attention to what has not been said, read between the lines. Attention floats: play the words without being haunted by them. Beware the lure of clarity! Beware of the deception of the obvious! (Rubem Alves)

Design Thinking suggests that the first "insight" is not (usually) the best. One must say that Fernando Pessoa "Sheep Keeper" contradicts this assertive.[3] Creative Thinking demands a wandering thought. Diverge and then converge, is the trick.

Neuroscience talks about the importance of meditation. Focused Meditation prevents age. More than this, it helps students to concentrate. Mindfulness is the key to innovation, creative thought.

---

[2]Hermann Hesse. Demian (1919).

[3]Pessoa wakes up in a morning and wrote the entire poem at once. After a year he was not able to change one line. It came complete.

The professor here acts like a "Meditation Master," teaching their students the art of meditation. Several schools around the world discovered the importance of meditation for education.

(f) Absolute Actions Control

> I was given the freedom to discover my own inclination and talents, to fashion my inmost pleasures and sorrows myself and to regard the future not as an alien higher power but as the hope and product of my own strength.[4]

The locus of control must remain in the student's hands. The professor is a follower, a fellow traveler, nourishing the student adventure desires.

Professors as "Fellow Travelers" must encourage their student's adventures.

(g) Loss of Self-Consciousness

> The sacred sense of beyond, of timelessness, of a world which had an eternal value and the substance of which was divine had been given back to me today by this friend of mine who taught me dancing.[5]

To be in flow is like dancing a waltz, playing jazz, wave surfing, gliding through strange spaces, and a total immersion in the pleasure of learning.

The professor here is the "Game Animator," the one who provides the wave and the air, the surfboard and the parachutes. He is also the partner you are dancing with, and also the Master of Ceremony who brings to your attention other partners. Lose yourself in the experience of dancing.

The loss of self-consciousness refers to the loss of representation that the person has of herself and the lack of concern you have with your own personality when engaged in activities that you like. The individual begins to forget who he is being involved by the pleasure provided by the experience.

As pointed out by Professor Roeris Gonzáles Sivilla "lost of self-consciousness" could make the individual vulnerable to manipulation by others. It conflicts with (f) "absolute actions control." In psychology we talk about "transference" and its risks. The professor here must be a therapist. In other to perform his job he demands "power." This power must be freely given by the student (f). It allows the professor to "manipulate" him. He sacrifices his free will because he has confidence in the ability of his teacher to show him the way.

(h) Loss of Sense of Time

> No wonder that Adelia Prado said "erotic is the soul". They are wrong who think that erotic is the body. The body is only erotic through the worlds walking inside it. Erotic do not walk according to the directions of the flesh. She lives in the interstices of words. There is no love that resists face a body empty of fantasies. A body empty of fantasies is a silent instrument, which does not leave any melody. Therefore, Nietzsche said that there is only one question to be asked when you want to marry, "I will continue to take pleasure in talking to this person 30 years from now?" (Rubem Alves)

---

[4]Hermann Hesse. Gertrude (1910, p. 4).

[5]Hermann Hesse. The Steppenwolf (1927, p. 154).

The loss of sense of time is another important aspect to be analyzed in an experiment in culinary (education) flow (and complements the previous since the loss of self-consciousness leads to the notion of time loss), since it clearly demonstrates satisfaction of a customer (student), or at least the desire to remain in a certain experience. It is clear that a customer (student) of a restaurant (classroom) is enjoying the experience when it stays there for a long time.

(i) Autotelic Experience

> What is essential is that which, if we were robbed, we would die. What cannot be forgotten. Substance of our body and our soul ... Poets are those who, in the midst of ten thousand things that distract us are able to see the essential and call it by name. When this happens, the heart smiles and feels at peace ...
>
> (The chronicle entitled: The Essential Things—Rubem Alves, from the book: The Return and Suit)

"When experience is intrinsically rewarding, life is justified in the present, instead of being held hostage by a hypothetical future gain."

The autotelic experience occurs when "being there" is enough to make you feel great. That is, the moment that stands on its own, without the need to seek a goal beyond, or without creating expectations the next time.

According to Flusser [4, p. 182], the very word design "takes place in a context of trickery and fraud. The designer is therefore a malicious conspiratorial dedicated to engender traps." A good instructional designer concentrates not in "solving" but in "creating" problems, students must solve. Like a "gymkhana" full of hurdles to be transposed or a "game" to be played. Something fun that turns knowledge gathering in a wonderful adventure. Each obstacle results in a "learning by discovering" experience.

In accordance with Brown [5] we live in an "economy of experience." Consumers participate actively, and no longer can be considered as passive elements. This type of economy seeks a change going from basically functional for emotional basically. Thus, an experiment must meet the primary need that motivated his quest (kill hunger) and also go beyond satisfying emotional.

Teachers are storytellers. The rule is not to be true, but to charm, impress, calls the attention. Through enchantment we learn. A good instructional designer must think of himself at least as a demigod capable to create enchantment worlds, full of glamor, imaginary, and ephemeral dimensions, where alumni goes in search for light (the word alumni in Latin means without light). In this sense "Alumni" is not an adequate word. Our students hide inside then the fires of heaven. Teachers must perform like Michelangelo Buonarroti not "designing" but revealing what is hidden.

Professors are (Fig. 2):

Fig. 2 Professor Metaphors (Source: The author)

3 Education as a Common

> Siddhartha ... had begun to suspect that his worthy father and his other teachers, the wise Brahmins, had already passed on to him the bulk and best of their wisdom, that they had already poured the sum total of their knowledge into his waiting vessel; and the vessel was not full, his intellect was not satisfied.[6]

In accordance with Hess and Ostrom [6], Commons are resources shared by a group of individuals subject to social conflicts. Knowledge Commons consists of shared knowledge and can be organized around intellectual and cultural practices [7].

The tragedy of the commons is a type of social trap, usually economic, that involves a conflict between public and individual interests in the use of finite resources. It states that free access and unrestricted demand for a finite resource ends up structurally condemning this recourse because of their over-exploitation.

This notion is not merely an abstraction, but its consequences manifest themselves literally on practical issues, such as the Pau-Brazil, where exploration has made the Common no longer used due to the risk of extinction.

Associative Commons are those controlled by a group that involves rules, regulations, and restrictions and can publicly articulate a comprehensive set of values. Religious congregations, universities, scientific organizations, and professional groups are examples of Associative Commons and vary in its rules and structures, but they often have the function to protect or improve something.

Free Commons are those that anyone has the right to use and contribute. By being open, it is difficult to regulate, even though the vast majority of participants feel that specific rules should be imposed. Squares and sites linked to nature as beaches, lagoons, and also the culture of a place are examples of this type of Commons [8].

The concept of "community of practice" was coined by Etienne Wenger and, in short, can be explained as a group of individuals who meet periodically, because they have a common interest in learning and applying what has been learned.

Education is a "common." Usually there are two models for exploring this human need for completeness called education: private enterprises or treating it as a public politics issue. Nevertheless, most of the examples of success, mainly in fundamental and middle schools, come from a third possibility: "auto governance."

[6]Hermann Hesse. Siddhartha (1971, p. 5).

A "community of practice" understanding knowledge as a "Free Commons," through an auto organization process, which includes open innovation and design thinking find a "new and wonderful approach of how to provide the best possible education." All actors here must cooperate: citizens, authorities, and private organizations.

4 Design Thinking, Cuisine, and Education

> I suspect that our schools teach very accurately science to buy the tickets and packing. But I have serious doubts that they teach students the art of seeing while traveling. (Rubem Alves)

Analyzing the use of the Design Thinking approach to education, we can see the possibility of applying some of the concepts proposed by Brown [5]. These include empathy, prototyping, and experiences design.

Empathy means becoming the other. Tim Brown says that we can build bridges insights through empathy, trying to see the world through the eyes of others, of other people's experiences and emotions [5].

Savarin [9] has said that the discovery of a new recipe brings more happiness than the discovery of a new star. This quote is relevant and refers to one of education goals: pleasure. The pleasure of learning captures stars, recipes, dreams, and expectations in a magical place called memory. "We do not see what we see, we see what we are. Only see the beauties of the world, those who have beauty within" (Rubem Alves).

In 1923 Célestin Freinet reinvented education. In 1915 he was recruited into the French army and was wounded in the lung. Since his lung injury made it difficult for him to talk for long periods he purchased a printing press, originally to assist with his teaching. It was with this press that he printed free texts and class newspapers for his students. Anticipating co-working and design thinking ideas, the children would compose their own works on the press and would discuss and edit them as a group before presenting them as a team effort. Designers must leave their comfort zones in order to innovate. Freinet's students would regularly leave the classroom to conduct field trips.

Brown [5] states that an experiment should be designed in the same way as any other product, and that the components of a product must fit to create an excellent experience. The newspapers were exchanged with those from other schools. Gradually the group texts replaced conventional school books.

In education, new opportunities need to be created so that people have access to knowledge. In Manaus, a city placed within Amazon Forest, we found a school floating above the Rio Negro (Black River). In Florianopolis (Santa Catarina—Brazil) children living in "Costa da Lagoa" go and return to school in a boat. A good example of design alternatives is the Boat School in Amapá (Brazil, near French Guyana) (Fig. 3).

Fig. 3 Boat School

Anísio Teixeira was the inventor of public school in Brazil. A school for all that must be a full-time school for teachers and students, as the Park School he founded in 1950 in Salvador, who later would inspire the Integrated Centers for Public Education (CIEPS) of Rio de Janeiro. It is a shame that the idea of the CIEPS, created by Darcy Ribeiro[7] and designed by Oscar Niemeyer[8] was abandoned.

We think education as a COMMON. We don't believe in the state or in the private economy for providing good education. The example of the CIEPS is clear. Implemented initially in the state of Rio de Janeiro, Brazil, during 1983–1987 and 1991–1994, aimed to provide quality public education, the idea was abandoned and not copied in other Brazilian states.

The class schedule stretched from 8 to 17 h, offering in addition to the regular curriculum, cultural activities, directed studies, and physical education. The CIEPs provided full meals to their students, as well as medical and dental care. The average capacity of each unit was about a 1000 students.

A concept of design thinking that is present in gastronomy (not in education, unfortunately) is prototyping. We can say that gastronomy, when practical, is purely prototyping. Brown [5] states that prototyping is the culmination of a design thinking project. In most post-graduation programs we have "special topics" disciplines. Some of them stand for one period; others become permanent, living a little longer. No new discipline is created without going through this prototyping phase.

All schools should be "Experimental Schools," learning by doing. Innovation must become cultural. The world is changing and we need not just follow these

[7]Darcy Ribeiro was a Brazilian anthropologist, author, and politician. His ideas of Latin American identity have influenced several later scholars of Latin American studies. As Minister of Education of Brazil he carried out profound reforms which led him to be invited to participate in university reforms in Chile, Mexico, and others after leaving Brazil due to the 1964 coup d'état.

[8]Oscar Niemeyer was a Brazilian architect who is considered to be one of the key figures in the development of modern architecture.

changes but be the cause of these changes. As pointed by Peter Drucker the best way to foreseen the future is to construct it.

We need an "International Education Design Society," a network of people, in passion with education, contributing with better ideas for transforming this world into a better one through education. There are good examples of this, like Riane Eisler's[9] partnership education and others.

5 The Schools of the Future

> It is a mistake to try to look too far ahead. The chain of destiny can only be grasped one link at a time. (Winston S. Churchill)

It is perfectly plausible to suppose that the first places that most attracted populations for a permanent settlement necessarily had to have water. Most of the medieval cities have a mile of diameter with a water pool in the center. A mile was the distance allowing transportation of water in a comfortable way.

Imagine the world mapped as a cooperative of medieval cities with a mile of diameter. In the center of each of these we place a school: Water for the body and knowledge for the soul.

Paulo Freire says that "worst then no school at all is this school we have." We are not talking about "just another school," but a Walt Disney designed school similar to the Tivoli Gardens in the center of Copenhagen (Fig. 4). Amusement Park School where children within the mile perimeter goes each day by bike (no cars at all, we need to educate for a sustainable environment) to have fun. Imagine children walking toward these Gardens—Amusement Park—Restaurant like schools in order to play, to have fun, to socialize, and to learn. Each day lived as an adventure to be remembered. Emotions driving student's minds toward a significant knowledge inside a space designed for creation, for innovation, and for auto knowledge. We need to reinvent the school and their users each morning.

Learning with "cooperatives" and "communities of practice" we will never build something "big" (like the CIEPs for 1000 students), so big that forgets its builders and acquires a new and ferocious identity just trying to get bigger and bigger even with the prejudice of their founders.

Learning with inter-cooperatives we imagine middle school as attending at most eight of fundamental schools living and breathing around then. Universities will be created in the same scale. Amusement Parks.

All schools must be virtually connected. One school in a Florianópolis (Brazil) neighborhood will be linked with other similar schools situated in "sister cities" like

[9]http://www.mediaed.org/cgi-bin/commerce.cgi?display=home.

Fig. 4 Tivoli Gardens in Copenhagen. A school designed for a wonderful learning experience

Wonju (Korea); Kisumu (Kenya); Pskov (Russia); Opole (Poland); Lijiang (China); Saint-Lô (France), and Roanoke Valley (USA). Virtual activities will be performed in order to create a family spirit between children from all these places.

6 Final Considerations

> I want to unlearn to learn again. Scrape the paint that painted me. Emotions must be unpacked, sensations must be retrieved. (Rubem Alves)

Good education requires teachers that acted like cookers and cookers that performed like teachers. Prepare a good class demand a design thinking approach. We propose a "co-design" process enrolling all stakeholders. The instructional designer creates problems that the students must solve. A balance between challenge and skills must be reached.

We must remember that: "Anyone who tries to help a butterfly emerging from cocoons will kill it. Who tries to help a sprout out of the seed destroys it. There are certain things that cannot be adjusted. It has to happen from the inside out" (Rubem Alves).

Anísio Teixeira said: "You want democracy. Teach democracy." Although what we are proposing here seems an expensive utopia it is much cheaper than war, violence, and corruption: "While happy society does not come, there is at least future fragments where laughter is served as a sacrament, so that children learn that the world can be different. The school, itself, is a fragment of the future..." (Rubem Alves).

In accordance with Jung there will be no peace outside us if we do not conquer the peace inside us: "What people want most is someone to listen in a calm and quiet manner. In silence. Without giving advice. Unknown to say: If I were you . . . " (Rubem Alves).

This article does not have a conclusion. The author only asks for some time to inhale, breath, and continue to attack their windmills. For this, the author has the words. May the reader receive them fondly. Because:

(Weaving the Morning)
A rooster alone did not weave a morning:
he will always need other roosters.
One that take his cry
and bid it to another; another rooster
to pick up the cry of a rooster before him
and bid to another; and other roosters
than with many other roosters crossing
the sun rays of all rooster' s cries,
so that a new morning, from a fine web,
goes weaving itself between all the roosters.
(João Cabral de Melo Neto)

7 Scientific Validation

This paper has been unanimously validated in a collaborative review mode with the following reviewers:

- Reviewer 1 Roeris Gonzalez Sivilla, Universidad de Camagüey "Ignacio Agramonte Loynaz." Educación de la Ciencias Naturales, Cuba
- Reviewer 2 Angel Ric, Instituto Nacional de Educación Física de Cataluña, Grupo de Investigación Sistemas Complejos y Deporte, Universitat de Barcelona, Spain.

References

1. Csikszentmihalyi M (1991) Flow: the psychology of optimal experience. HarperCollins, New York
2. Csikszentmihalyi M (1975) Beyond boredom and anxiety. Jossey-Bass, San Francisco
3. Csikszentmihalyi M (1990) Flow: the psychology of optimal experience. Harper & Row, New York
4. Flusser V (2007) O mundo codificado: por uma filosofia do design e da comunicação. Org. Rafael Cardoso. Trad. Raquel Abi-Sâmara. Cosac Naify, São Paulo
5. Brown T (2010) Design thinking: uma metodologia poderosa para decretar o fim das velhas ideias. Campus, Rio de Janeiro
6. Hess C, Ostrom E (2007) Understanding knowledge as a commons, from theory to practice. MIT Press, London

7. Ouma M (2014) The policy context for a commons-based approach to traditional knowledge in Kenya. In: de Beer J et al (eds) Innovation and intellectual property: collaborative dynamics in Africa. UCT Press, Claremont, pp 132–151
8. Levine P (2007) Collective action, civic engagement and the knowledge commons. In: Hess C, Ostrom E (eds) Understanding knowledge as a commons, from theory to practice. MIT Press, London
9. Savarin B (1995) A Fisiologia do gosto. Companhia das Letras, São Paulo

POEM-COPA Collaborative Open Peer Assessment

Pierre Collet, Raaj Seereekissoon, Isaac Abotsi, Marie Michaud-Maret, Anna Scius-Bertrand, Emma Tillich, and Pierre Parrend

1 Introduction

A long time ago, education was provided by personal tutors who were paid by rich families to take care of 3 or 4 children. This type of education was of very high quality, as the tutor could adapt his tuition to the capabilities and inclinations of each student, therefore providing personalized education. Unfortunately, this was also very expensive, meaning that very few people were educated. The advent of schools allowed for many more people to learn how to read, write and count but this was only possible through mass education, with classrooms of 30 pupils and national education programmes that are identical for all. This means that slow students are often left behind whilst bright students have to wait for the others. This is even more so with the advent of Massive Open Online Courses (MOOC) [7], that are currently invading the world of e-education. With MOOCs, a single course can be followed by thousands of students.

The aim of the Personalised Open Education for the Masses (POEM) platform is to use complex systems to create an intelligent Learning Management System that

P. Collet (✉) • R. Seereekissoon • I. Abotsi • E. Tillich • P. Parrend
CSTB Team, ICUBE Laboratory, Strasbourg University, Strasbourg, France

e-laboratory on Education, CS-DC UNESCO UniTwin, Paris, France
e-mail: pierre.collet@unistra.fr

M. Michaud-Maret
Institut de Formation des Métiers de Santé du Nord Franche-Comté, Belfort, France

A. Scius-Bertrand
e-laboratory on Education, CS-DC UNESCO UniTwin, Paris, France
École Pratique des Hautes Études, Paris, France

P. Bourgine et al. (eds.), *First Complex Systems Digital Campus World E-Conference 2015*, Springer Proceedings in Complexity, DOI 10.1007/978-3-319-45901-1_2

is able to educate hundreds of thousands of students along personalized trajectories, depending on their previous knowledge, skills and experience, as with the personal tutor of ancient times.

> One could think that massive education and personalized education are antagonistic objectives but on the contrary, they are in synergy.

A long time ago, personal tutors would use their teaching experience to find the best series of exercises on topics adapted to each child they were in charge of. Then, if they detected a particular skill or interest in literature, science or the arts in their pupil, the tutors (who were often multifaceted) would adapt their teaching to nurture and develop this inclination for a more personalized education.

Their pedagogic experience increased with the number of children they taught, as trial and error improved their tuition skills. Having the opportunity to statistically study large numbers of educational trajectories, modern Intelligent Tutoring Systems can draw conclusions on previous successes and failures to improve their interactions with students online, and are in the position of predicting the best future for a student.

More accurate predictions require assimilating data of a massive number of such trajectories (which once more is what good professors do as their experience increases). For this reason, the participation of everyone to such an educational ecosystem is extremely desirable. Not only will it improve the system, but anyone wanting to resume their education will be quickly learning things they do not already know. Such synergy between massive and personalized education is only possible within a social intelligent ICT platform. The aim of POEM is therefore to implement an educational ecosystem responding to the objectives of 4P-education, i.e.:

- **P**articipative (to collect data and reinforce the "experience" of the system),
- **P**redictive (to guide the student using the elaborated experience),
- **P**reventive (to avoid failure) and
- **P**ersonalized, thanks to multi-level Quality Measurements allowing for an experience tailored to each student.

Many functionalities must be developed in order to develop such a comprehensive platform, typically:

- **Constructing and visualizing dynamic Knowledge Maps** of domains, to help students determine their objectives.
- **Developing individual MOOC and curricula trajectories.** POEM conjectures that, given an individual profile, the best next incremental step is determined in probability by the distribution of the choices of previous learners with similar profiles. This conjecture is developed in the Personalized Educational Man-Hill Problem, because of the similarity with ants' collective behaviour, which is known to quickly find optimal paths towards food sources [2–4, 6].
- **Providing inter-tutoring** between students, which is needed if direct Student-Teacher interaction is impossible due to the very large number of students. In such cases, POEM provides each student with a tutor, who is also a student but

more advanced in the same curriculum. The student can ask questions to his tutor. If the tutor cannot provide an answer to a difficult question, he himself can forward the question to his own tutor and so on, until there is no tutor anymore and the question reaches a professor.
- **Offering an automatic skill-level assessment system** depending on success/failure along the personalized trajectories of students. This is implemented through Elo-points as in chess or Tennis ranking.
- **Offering a high-quality assessment of open answers to open questions**. This is proposed through peer assessment, which is recognized to be of mutual benefit.
- **Provide crowdsourcing** by letting students bring new questions and new content as part of their evaluations. Good questions and content will find their way into participatively evolved trajectories, while poor content will eventually get discarded.

This paper will focus on the two last points, implemented in the COPA (Collaborative Open Peer Assessment) module of the POEM[1] educational ecosystem of the CS-DC (*Complex Systems Digital Campus*) UNESCO UniTwin.[2]

2 Peer Evaluation of Open Answers to Open Questions

Peer evaluation has been studied for a long time [16] but was only recently experimented extensively, in distance learning with platforms such as Spark (Self and Peer Assessment Resource Kit) or more recently in different MOOCs [9, 15]. It is quite frequently used for operational teaching, such as management or software development [10] but remains a top down approach, from the teacher to the learner, as topics are not dynamic and imposed to students. In the evaluation of these technical teachings, peer assessment shows a limited deviation that can be lower than 3 % [11], which means that it is quite accurate (at least in computer science, in which this experiment was published). It is also efficient for the assessment of complex tasks such as composition, typically when associated with appropriate coaching [12], which is more difficult to put into practice within MOOCs.

Peer assessment is often better accepted by students than an evaluation performed by the teachers. It enables them to improve the content of the course as well as their ability to evaluate others.

One of the current limitations of available solutions is the static character of the set of questions, which increases the risk of cheating. By including in the assessment the requirement to ask a question (that will be posed to other students), COPA provides a solution to this problem by turning learners into producers of new knowledge and analyses, that will feed the system to create a virtuous circle.

[1]http://poem.unistra.fr.

[2]http://cs-dc.org.

COPA therefore inserts into xMOOCs the transmission of specific knowledge [13], some dynamic elements of connectivist cMOOCs [14] based on individual experience and interaction between learners.

3 Principles and Implementation of COPA

Implementation of a COPA evaluation needs several databases:

- A database containing the courses (videos, Powerpoint or PREZI presentations, PDF or Word documents, etc.),
- A database containing questions,
- A database containing answers,
- A database containing students and professors.

Then, COPA evaluation uses three stages: a participative stage where students must ask a question on the course, a more passive stage where they should answer three questions and then, a third stage where they should evaluate the answers given by other students.

Before the COPA evaluation takes place, students should have followed a course (face-to-face or video) on which they are to be evaluated. If digital contents are associated with the course, they can be stored in the courses database, so that students can access them for reference during all stages of the COPA evaluation.

Interestingly enough, because COPA allows open questions and answers, topics are not limited to hard sciences where one can expect exact answers to precise questions. COPA can also be used to evaluate knowledge in social sciences, skills, literature, arts, a.s.o.

3.1 COPA Phase 1

Write a question on the course and
Provide a model answer.

When opening phase one, students have access to the course but rather than being asked to answer some questions, they are asked to pose a question on the corresponding course. This activity is much more demanding than answering a question because the students must be creative in order to ask relevant and interesting questions. Indeed, the quality of their questions will be rated (by other students) against questions from the pedagogic team, so if the question they imagine is less interesting than the teacher's questions, they will not get a good grade on the exercise.

Then, after they have been asked to pose a question, an even more challenging task is asked from them: they must provide a model answer for the question they have imagined.

Here again, the quality of the provided model answer will be evaluated by other students.

4 COPA Phase 2

Answer (and evaluate the quality of) three questions.

This phase has several purposes:

1. assessing if the students have understood the content of the course,
2. evaluating the quality of the questions posed by students in Phase 1 and
3. improving the quality of the database of student questions.

The contents of this phase are inspired by CAPTCHAs [1] and Re-CAPTCHAs [7]. CAPTCHA is an acronym for "Completely Automated Public Turing test to tell Computers and Humans Apart". Indeed, during 1950 Turing came out with a very famous test [5] to allow humans to determine if an unseen interlocutor is a human or a machine, a CAPTCHA can be seen as a reverse Turing test, created to allow computers to determine whether their interlocutor is another computer or a human.

Re-CAPTCHAs make use of the time and energy given by humans to pass the CAPTCHA test in a constructive way, i.e. to solve a problem for the computer (cf. Fig. 1.) Because POEM-COPA is run by a computer, the algorithms used in POEM-COPA are computer-oriented, not human-oriented, and therefore CAPTCHA-like. As in re-CAPTCHAs, COPA is not only asking the students to answer the questions, but also asking them to rate the relevance, originality and quality of the formulation of the questions they are asked.

Fig. 1 Re-CAPTCHA: the computer not only tests whether its interlocutor is human or not (by asking him to decipher the twisted text on the right) but also asks him what is the word on the left (that it could not recognize *via* Optical Character Recognition because it was badly printed)

However, because the aim of this phase is to evaluate how much of the course has been understood by the student, it is important that most of the questions come from the pedagogic team.

4.1 Answering 3 Questions

Re-CAPTCHAs are typically divided into two parts: a part to tell if the user is human or not, and another "crowdsourcing" part where the collaboration of users is sought.

In COPA, 2 questions come from a database of questions validated by the pedagogical team (providing for an approved evaluation), and one question comes from another student (crowd-sourcing part, where the participation of the user is sought) but of course, the user does not know which question among the 3 is by the student.

This 2/3–1/3 proportion means that students are mostly evaluated on validated questions (with model answers provided by the pedagogical team).

Each provided answer will be anonymously evaluated by 3 other students (peers), following the current practice in scientific journals/conferences where the quality of submitted research is also evaluated by peers.

4.2 Constrained Evaluation of Questions

In this collaborative part, students are asked to evaluate the relevance, originality and formulation of the 3 questions they are asked, with the aim of evaluating the quality of the student question that is posed along with 2 questions from the pedagogical team.

The risk of self-assessment between students is to observe some bias induced by the type of training undergone by the students: in competitive training (ending with a competitive exam), it is in the interest of the students to give bad marks to the others, in the hope of obtaining better relative marks. On the contrary, in courses where all students above a certain grade pass the exam, there is no competitive pressure so students may decide mark other students more generously than they otherwise would.

In order to reduce such bias, students are not asked to give grades to the questions but to rank them *via* a constrained rating, meaning that in effect, the question asked by a student will be compared to the questions asked by the pedagogical team.

Students must give 0–5 points to each of the 3 questions but they only have exactly 10 points that they must distribute entirely.

This method imposes that only the cases of Table 1 can be encountered. One can see in this table that:

Table 1 Rating combinations and marking

Teacher Q1	Teacher Q2	Student Q	Teacher average	Difference	Grade	Mark/5
0	5	5	2.5	2.5	7.5	5
1	4	5	2.5	2.5	7.5	5
2	3	5	2.5	2.5	7.5	5
3	2	5	2.5	2.5	7.5	5
4	1	5	2.5	2.5	7.5	5
5	0	5	2.5	2.5	7.5	5
1	5	4	3.0	1.0	6	4
2	4	4	3.0	1.0	6	4
3	3	4	3.0	1.0	6	4
4	2	4	3.0	1.0	6	4
5	1	4	3.0	1.0	6	4
2	5	3	3.5	−0.5	4.5	3
3	4	3	3.5	−0.5	4.5	3
4	3	3	3.5	−0.5	4.5	3
5	2	3	3.5	−0.5	4.5	3
3	5	2	4.0	−2.0	3	2
4	4	2	4.0	−2.0	3	2
5	3	2	4.0	−2.0	3	2
4	5	1	4.5	−3.5	1.5	1
5	4	1	4.5	−3.5	1.5	1
5	5	0	5.0	−5.0	0	0

Col 3 and 4: The quality of the student question can never be equal to the average quality of the 2 questions of the pedagogical team.

Col 5: In 11 cases, the student question gets better grades than the average of the teacher questions, *vs* 10 cases when it is deemed worse.

Col 6: If one adds 5 points so that the student gets grades between 0 and 7.5 and if

Col 7: one multiplies by 5/7.5 in order to get the grades back into a [1,5] interval,

the student question gets a mark (Col 7) that is identical to the grade given to him by the student undergoing the COPA evaluation.

Students whose question have been estimated as being of slightly lower quality than the average of the teacher's questions will get 3/5, which is fine as it can be considered difficult for students to only obtain points if their question is better than the questions of the teacher.

If the question of the student is used several times, it will get marked several times. COPA will use the average mark, weighed by the number of evaluations.

4.3 *Improving the Quality of the Questions in the Database*

Because there are obvious questions for all courses, there is a great risk that many students will ask the same questions. Then, some of the obvious questions may also be asked by the pedagogic team.

This could cause a problem as it allows for the possibility that the selected student question be semantically identical to the question provided by the pedagogic team, in which case the student undergoing the COPA evaluation will be faced with two identical questions (one from a student and another one from the pedagogic team).

If this is the case, the student has the possibility to indicate that 2 questions are identical. If the student clicks on the "similar questions" button, the student question will be replaced by a question from the pedagogical team, therefore guaranteeing that all 3 questions are different.

Then, the student question that was noted as being similar to one of the teacher's questions will be flagged and not be selected along with the incriminated teacher question in the future.

Another button is available for students to signal if the contents of a question are inappropriate. Inappropriate questions are removed from the students question database and an email is sent to the pedagogic team containing the question and the email of the student who submitted it.

5 COPA Phase 3

Evaluate 9 answers from other students and
Evaluate the quality of model answers.

The current practice in science is that new research submitted to a conference or journal is evaluated by 3 experts on the domain. Because no "super-scientist" exists, the "experts" cannot be anyone else other than scientists (i.e. peers) working in the same field as the scientist who submitted the work, so basically, research is driven by anonymous peer reviewing.

We pose that (even if double blind peer reviewing has its pros and cons) what is currently good enough for research evaluation could also be used for student evaluation.

Each of the three answers given in phase 2 will be evaluated by 3 peers, to determine if the answer is correct or not. Such student peer review is very desirable because:

1. The online evaluation scheme is not limited to Multiple Choice Questions: students have brains that can be used to analyse the provided answers. Using students as evaluators allows COPA to use human brains to evaluate the quality of *open responses to open questions*.

2. Evaluating somebody else's answers is not a waste of time for the evaluators: all teachers know that studying someone else's point of view has many pedagogic virtues.
3. Students take their task seriously, as they know that the mark they give could have some influence. Asking them to be evaluators *involves* participating students.

Practically speaking, because in phase 2, students must answer 3 questions and because each of the three questions must be reviewed by 3 other students, phase 2 creates the need for 9 reviews (3 × 3).

5.1 Evaluation of 9 Answers

Phase 3 therefore consists of 9 evaluations of other student's answers. In order to help the evaluating student in his task, he/she is presented with:

1. the question that was posed,
2. the model answer that was proposed by the person who wrote the question (teacher or student) and
3. the answer to be evaluated.

Because the evaluating student is presented with the question and its model answer, this is an occasion for him to see 9 more questions and answers than those he had to answer in phase 2. Then, he has to fully understand the question and the proposed model answer in order to form a judgement on the quality of the student answer he must evaluate. Doing this is not a waste of time for the evaluating student: because he is involved (his accurate evaluation will have an influence on another student's mark) the time he spends on this task is of high pedagogic quality.

Here again, a constrained evaluation scheme is given to the evaluating student so that he/she is not tempted to underrate or overrate the other student's answers.

Because the added value of POEM is more to have students revise and understand their lessons (formative evaluation) rather to evaluate how the lessons have been understood (summative evaluation) or evaluate how well students compare with peers (normative evaluation), in the first versions of POEM, the students had to distribute *exactly* 30 points over the 9 answers to be evaluated. 30 was chosen so that the evaluating student had to give an average of 3.22 points to the 9 answers, in order to be encouraging for students.

However, even though this was meant as an encouragement to students to get involved in POEM as it made it difficult for them to participate and to get bad marks, this hard constraint of 30 points was considered as too strong by many students.

Indeed, Table 2 shows the probabilities for each grade to be given: if an evaluator decided to give 0 to someone, he would have 30 points to distribute over 8 questions only, meaning giving an average of 3.75, which is a lot. Then, if an evaluator gave 3 × 0 because he evaluated that 3 answers were plainly wrong, he/she had no other choice than giving 5 to all the other answers.

Table 2 Possible occurrences of the different grades and probabilities

Grade	0	1	2	3	4	5
Grade occurrences	16,808	25,488	36,688	50,288	65,808	82,384
Probabilities	6.06 %	9.19 %	13.22 %	18.12 %	23.72 %	29.69 %

Table 3 Possible occurrences of the different grades and probabilities

Grade	0	1	2	3	4	5
Grade occurrences	615,553	732,033	842,499	939,099	1,014,399	1,062,279
Probabilities	11.83 %	14.06 %	16.19 %	18.04 %	19.48 %	20.40 %

In the current version, more leeway was given to evaluators: they could distribute between 22 (average of 2.44/5) and 30 (average of 3.33/5) to distribute over the 9 answers. This increased the number of possible grades combinations from 277,464 to a whopping 5,205,862 and the number of possible combinations with no 0 or 5 jumped from 2598 to 159,436.

The probabilities for each grade to be given was more even (cf. Table 3). Evaluators could still not give bad grades to all answers (impossible to give a lower average than 2.44, which is pretty close to 2.5/5) while still being able to give greater than average marks to everyone (encouraging for participating students).

More importantly, this made it easier to distribute grades 0 and 5. Previously, giving 2×0 meant you had to give at least 2×5 and therefore 5×4: $0 + 0 + 4 + 4 + 4 + 4 + 4 + 5 + 5 = 30$.

With the less constrained notation, evaluators had more choice to give 0s or 5s.

5.1.1 Grade Interpretation to Create Marks

Because evaluators have more leeway in the way they can distribute grades, it is considered that *grades from 0 to 5* are interpreted as *marks from 0 to 3* the following way:

Grades 0 and 5: are given by *choice* rather than by constraint. It is therefore desirable to respect the evaluator's choices when they are extreme. *Grade 0* gives **mark 0** and *grade 5* gives **mark 3**.

Grades 1 and 2: if grades 0 and 5 are given by choice, more constraints apply on grades 1, 2, 3, 4 in order to stay within the 22/30 points limit. Therefore, *grades 1 and 2* give **mark 1**.

Grades 3 and 4: Similarly, *grades 3 and 4* give **mark 2**.

Thanks to this relaxation of constraints, what was previously ranks can now be considered as a summative evaluation. For an evaluated student, each of the 3 questions he answered on phase 2 is evaluated three times between 0 and 3. The marks are summed and then multiplied by 5/27 in order to obtain a global summative mark within $[0, 5]$.

5.2 Evaluation of Model Answers

As said above, in order to help the evaluator in his task, he is presented with:

- the question that was posed,
- the model answer that was proposed by the writer of the question,
- the answer to be evaluated.

Because the writer of the question may be a student, it is possible that the student posed a good question, but has not thoroughly understood it and therefore, did not provide a good or complete enough model answer.

Therefore, whenever shown a model answer, the evaluator is also asked to evaluate the quality of the model answer proposed by the writer of the question, by giving it a grade between 0 and 5.

Because COPA is thought as a collaborative platform, if the given grade is 2 or less, this means that the evaluator clearly thinks that the model answer could be improved. A pop-up window therefore opens, asking him to provide a better model answer. In the future, both model answers will be shown to future evaluators, who will then have to evaluate two model answers.

Because this is a participative grade, no constraint is given on the grades that can be given. If the model answer is evaluated several times, the average mark is given weighted by the number of evaluations.

5.3 Evaluation of the Quality of the Marking

The conscientiousness of the evaluator can also be evaluated. Indeed, if all 3 evaluators do their job correctly, they should give equivalent marks to a same answer. One can then determine how different the marks of an evaluator are with reference to the marks given by his/her two colleagues, on the same question. To this effect, one can compute the difference it makes on the average mark to include his mark or not.

Supposing the evaluator gives the maximum grade (5, i.e. mark 3) on an answer where the other two gave 0. The average when including his mark is 1, while the average without his mark is 0. It is possible to sum all the differences made by this user (maximum 9) and evaluate the quality of the conscientiousness of evaluator e as:

$$\mathcal{C}_e = 9 - \sum(\overline{m} - \overline{(m - m_e)})$$

with $\overline{(m - m_i)}$ the average of the marks of the other evaluators than e.

The mark is multiplied by 5/9 in order to bring it into the $[0, 5]$ interval.

6 Global Marking

For each course, POEM-COPA requires many interactions involving the student, from writing a question to writing a model answer, to answering 3 questions (while evaluating their quality) to evaluating 9 answers (and evaluating 9 model answers).

COPA provides the professor with two marks: one corresponding to a summative evaluation (within [0, 5]) and with participation marks coming from:

- the evaluation of the quality of the provided question ([0, 5] weighted),
- the evaluation of the quality of the provided model answer ([0, 5] weighted),
- the evaluation of the consciousness of the student as an evaluator ([0, 5]).

The POEM-COPA interface asks over how many points the student should be graded (usualy 20 in France, but could be anything). Then, a selector is provided for the teacher to tell if he wants to combine (or not) the summative and participative evaluations, and offers two parameters to weigh the marks.

Finally, each of the three participative marks (PQ, PMA and C) can also be weighted depending on the importance the teacher wants to associate with each mark.

7 Conclusion

POEM-COPA has been mainly used along face-to-face classes, for more than 10 different courses up to now, at Strasbourg University (in computer science and English courses), French National Engineering School for Advanced Techniques (ENSTA), Institut de Formation des Métiers de Santé du Nord Franche-Comté, in a training course for nursing auxiliaries and others.

Students always have been quite enthusiastic about using the COPA collaborative peer to peer formative evaluation. In one of the feedback polls, 97 % of the students stated the need to open their courses and research content over the Internet in order to fulfill the evaluation. 72 % found that COPA allowed them to be better prepared for the final exam and 71 % said they would like COPA to be used in other courses.

8 Scientific Validation

This paper has been unanimously validated in a collaborative review mode with the following reviewers:

- Christophe Schnitzler, Strasbourg University, Sport Sciences Faculty
- Roeris Gonzalez Sivilla, from University of Camaguey “Ignacio Agramonte Loynaz”.

Acknowledgements We thank the Strasbourg University IDEX Investissements d'Avenir Program for funding the edX-POEM project.

References

1. Ahn LV, Blum M, Hopper NJ, Langford J (2003) Captcha: using hard ai problems for security. In: Proceedings of the 22nd international conference on theory and applications of cryptographic techniques. Springer, Berlin, pp 294–311
2. Johnson J, Shum SB, Willis A, Bishop S, Zamenopoulos T, Swithenby S, MacKay R, Merali Y, Lorincz A, Costea C et al (2012) The FuturICT education accelerator. Eur Phys J Spec Top 214(1):215–243
3. Louca J, Johnson J, Bourgine P, Portelli PCT, Scius-Bertrand A, Lenhard W, Escalona M, Taramasco C, Kohlhase M, Cointet J, Collet P (2013) Poem platform for massive personalized education. In: Proceedings of the European Conference on Complex Systems (2013)
4. Semet Y, Lutton E, Collet P (2003) Ant colony optimisation for e-learning: observing the emergence of pedagogic suggestions. In: 2003 IEEE swarm intelligence symposium, SIS 2003, Indianapolis, IN, 24–26 April, pp 46–52. http://dx.doi.org/10.1109/SIS.2003.1202246
5. Turing AM (1950) Comput Mach Intell 59:433–460
6. Valigiani G, Biojout R, Jamont Y, Lutton E, Republique CB, Collet P (2005) Experimenting with a real-size man-hill to optimize pedagogical paths. In: Proceedings of the 2005 ACM symposium on applied computing (SAC). ACM, New York, pp 4–8. http://doi.acm.org/10.1145/1066677.1066683
7. von Ahn L, Maurer B, McMillen C, Abraham D, Blum M (2008) Recaptcha: human-based character recognition via web security measures. Science 321(5895):1465–1468. http://science.sciencemag.org/content/321/5895/1465

Ecological Dynamics: A Theoretical Framework for Understanding Sport Performance, Physical Education and Physical Activity

Ludovic Seifert and Keith Davids

1 Introduction

The *Ecological dynamics* framework sustains a scientific approach to studying the behaviours of neurobiological systems, especially processes of action, perception and cognition. Kugler and Turvey [1] considered that "Ecological science . . . is a multidisciplinary approach to the study of living systems, their environments and the reciprocity that has evolved between the two . . . Ecological Psychology . . . [emphasizes] the study of information transactions between living systems and their environments, especially as they pertain to perceiving situations of significance to planning and executing of purposes activated in an environment".

In line with this philosophical direction, ecological dynamics has provided a systems orientation for the theoretical analysis of behaviours of athletes and sports teams over different timescales (performance, learning and development), conceptualised as complex adaptive systems [2]. In this research endeavour, we have sought to integrate key concepts from ecological psychology and nonlinear dynamics, in order to enhance understanding of performance, learning and expertise acquisition in sport. As we note later in this paper, the latter two processes are developed through the design of practice task simulations which exploit the transfer (from practice to competition) of adaptive human behaviours [3–5].

A major aspect of ecological dynamics is James Gibson's theory of ecological psychology, at the heart of which is "systems thinking" [6], emphasising continuous organism–environment interactions as the relevant scale of analysis for understanding human behaviours in performance environments (e.g. an athlete–performance

L. Seifert (✉)
CETAPS - EA 3832, Faculty of Sport Sciences, University of Rouen, Rouen, France
e-mail: ludovic.seifert@univ-rouen.fr

K. Davids
Centre for Sports Engineering Research, Sheffield Hallam University, Sheffield, UK

P. Bourgine et al. (eds.), *First Complex Systems Digital Campus World E-Conference 2015*, Springer Proceedings in Complexity, DOI 10.1007/978-3-319-45901-1_3

environment relationship). The central ideas of Gibsonian theorising, with respect to the directness of perception, have been linked back to the philosophical influence of Aristotle. In ecological psychology the continuous regulation of human behaviour is predicated on the role of information (e.g. optical, proprioceptive and haptic) that emerges from the individual–environment system to continuously guide activities like dribbling with a ball or intercepting a moving target. This conceptualisation of information differs greatly from traditional psychological connotations, which regarded that sensory systems yielded information from an environment, which was ambiguous and impoverished, needing interpretation with cognitive processes such as memory, attention and anticipation. In contrast, ecological psychology emphasises that *knowledge of* an environment underpins actions and is directly gained through perception without mediation by internalised representations [7]. Gibson's theory of direct perception proposes that perception is an act, emphasising that it is an active process in which perceptual and action systems function in a highly integrated and cyclical manner [6]. A key concept of ecological psychology, which we elucidate further, is *affordances,* which are opportunities or invitations for action that performers are surrounded by in a performance environment, becoming attuned to them through learning and experience [6,8]. These ideas from ecological psychology have been interpreted to suggest that with extensive practice and experience, an individual and a performance environment become ever more integrated as dynamical, self-organising systems coupled by information [5,9]. In these complementary systems intentional behaviours emerge from a process of exploration and learning before being stabilised into functional action patterns [9]. Therefore, in a performance environment, the most significant information sources that constrain athlete behaviours are *affordances*, which provide invitations or opportunities for action offered by each individual's perception of functional relations with a performance environment [6,8].

This theory of ecological psychology has been enhanced with the integration of tools and concepts from nonlinear dynamics to explain how information is cyclically related to the dynamics of a performance environment [10]. Dynamical systems theory provides a conceptual framework for understanding the emergence of coordination tendencies that exist between, and within, system components and levels of complex neurobiological systems (e.g. in human beings and sports teams) at multiple levels (e.g. from behaviour to brain) [10–12] with roots in thermodynamics [13] and synergetics [14]. Physical principles and concepts from nonlinear, dissipative, self-organising systems explain coordination dynamics as a natural process of pattern formation in neurobiological systems. In particular, stability in neurobiological complex systems is ensured by a constant flow of energy, consisting of the exchanges of information between the system and its environment [14]. More broadly, the constant exchange of matter, energy or information through continual and circular interactions between a system and its environment revealed self-organisation, i.e., assemblage and implementation of the components by the system itself without prescription from a higher-order system [14,15]. Therefore, those systems are called dissipative dynamical systems [15]. From there, dynamical systems theory seeks to explain and predict how patterns of coordination emerge, adapt, persist and change in humans considered as integrated complex systems [10]. Organisation in neurobiological systems is both facilitated and bounded by interacting *constraints* which shape the dynamics of emergent behaviours [16]. Constraints on behaviours include task, performer and environmental factors. The unpredictable nature of the environment and the many individual differences that exist in personal characteristics and functions of athletes signify that indeterminate solutions exist for many tasks in sport. This idea emphasises the circular causality of the relationship between each individual and a performance environment, mediated by brain and behaviour, and processes of perception and action [10]. The causality between brain and behaviour, and between processes of perception and action, is not linear but cyclical [6], as the individual continuously constructs goal-directed interactions with a performance environment. From this perspective, expertise is the continuous functional adaptation of behaviours to a set of interacting constraints in order to exploit them to the fullest in achieving specific intended performance goals [17].

Therefore by articulating Ecological psychology theory and Dynamical Systems theory, *Ecological dynamics* presents a viable theoretical framework for studying sport performance because its systems oriented perspective on the relationship between each individual and a performance environment addresses the weaknesses of traditional approaches to expert performance in sport, which separately focus on the performer and the environment. From there this paper presents the ecological dynamics theoretical rationale to identify key properties of expertise in sport, predicated on the performer–environment relationship as the appropriate scale of analysis. Indeed, Seifert et al. [17] previously observed that, with this theoretical orientation, key properties of expert movement systems that are worthy of further investigation include *multi-* and *meta-stability* [18], *adaptive variability according to interacting constraints* [16,19,20], *redundancy*, *degeneracy* [21–24] and the *attunement to affordances* [6,25]. From there, our presentation is organised in four sections and discusses: (1) the functional role of movement variability, which could be exhibited by multi- and meta-stability of movement patterns, reflecting neurobiological degeneracy properties; (2) the interest of representative learning and training design to guarantee ecological validity, generalisability and skill transfer; (3) the affordances picking-up and the attunement and calibration to specifying information for action and (4) all together, these properties are the basis for a nonlinear pedagogy. Indeed, this last section about nonlinear pedagogy [26] argues the idea that skill acquisition does not emerge from the internal representation of declarative and procedural knowledge, or the imitation of expert behaviours to linearly reduce a perceived "gap" separating movements of beginners and a putative expert model. Rather skill acquisition and expert performance correspond with the on-going co-adaptation of an individual's behaviours to dynamically changing, interacting constraints, individually perceived and encountered [17,27].

2 The Functional Role of Movement Variability

Developing knowledge of the *functional role of movement variability* is essential to understanding expert performance in many different sports (involving individuals and teams; ball games and outdoor activities; land and aquatic environments) [19,28,29]. Studying the functional role of movement variability involves assessing how adaptive human behaviour is by analysing the balance between movement pattern stability (i.e. persistent behaviour) and flexibility (i.e. variable behaviour) relative to a performance context [9,17]. Specifically, our work has sought to explore how individuals adapt their motor behaviours in various performance contexts, i.e., when the environmental properties are stable or in unstable or transitional regions where movement patterns co-exist (technically known as a meta-stable region; [18]). According to Kelso [18], meta-stability is the "simultaneous realisation of two competing tendencies: the tendency of the components to couple together and the tendency for the components to express their intrinsic independent behaviour" (p. 186). In a meta-stable performance region, component tendencies of independence co-exist, explaining how rich and varied movement patterns can spontaneously emerge in dynamic sport environments as an individual adapts his/her motor behaviours to achieve particular performance goals. These continuous changes of motor behaviours could merely result in an adaptive pattern refinement or the adoption of new form through a nonlinear phase transition. There exists some evidence of meta-stability in skilled sport performance, such in boxing [30] and cricket [31]. For example, Hristovski et al. [30] investigated how boxers' striking patterns were adapted when they punched a heavy-bag at various scaled distances to the target. At greater distances from the boxing bag, "jab" movement pattern emerged, whereas at closer distances, "uppercuts" or "hooks" patterns were observed. At a critical intermediated distance, the novice boxers explored a rich, varied and creative range of movement patterns involving "uppercuts", "hooks" and "jabs", which characterised meta-stable performance region [30].

In sum, key properties of complex, dynamical systems concern the continuous formation of stable behavioural patterns (i.e. *attractors*); *nonlinear transitions* from one attractor to another attractor (e.g. system transitions or bifurcations, hysteresis and critical slowing down); *multi-stability* (i.e. ability to transit between multiple states of organisation under given constraints); *meta-stability* (i.e. ability to exploit co-existing coordination tendencies in a transition or unstable region) and *variability* (exploiting critical fluctuations to enable adaptive behavioural transitions) [10,17]. In our research programme, we have investigated how athletes perform different types of movement and/or to adopt one of a number of co-existing modes of coordination between system components, in order to achieve the same functional performance outcomes, predicated on the neurobiological system property of *degeneracy*. Edelman and Gally [21] defined *degeneracy* as the capacity of system components that differ in structure to achieve the same function or performance output. This structural property in humans indicates the availability of an abundance of motor system degrees of freedom, which can take on different roles when assembling functional actions during sport performance. These key properties signify that, in sport performance, although basic movement patterns need to be acquired by developing athletes, there exists no ideal movement template towards which all learners should aspire, since relatively unique functional movement solutions emerge from the interaction of key *constraints* [16].

Newell [16] defined three types of constraints: environmental, task and organismic (personal). *Personal* constraints are structural or functional and refer to characteristics of an individual such as genes, anthropometric properties, cognition, motivation and emotions. In swimming, personal constraints may include passive drag and flotation parameters (hydrostatic lift, sinking force acting at the ankle) that could be artificially modified by wearing a wet suit and influence movement and coordination organisation [32]. Other body characteristics like strength, endurance and laterality (handedness and the preferred breathing side) can also influence movement and coordination organisation [32]. *Environmental* constraints are external to an individual and can be physical, reflecting the environmental conditions of the task. For instance, in climbing, temperature and altitude can make the ascent harder. In particular, it is well known that a high ambient temperature can cause swelling in the feet, which means that climbing becomes uncomfortable. Humidity can cause a climber's hands to become sweaty, which challenges the graspability of holds. In swimming, environmental constraints refer to water properties such as the temperature and density and viscosity of the fluid, the direction and type of water flow (laminar vs. turbulent, quantified with Reynolds number), underwater visibility and waves on the surface of the water (e.g. assessed with Froude number) [32]. *Task* constraints include the goal of the task, the rules, boundary locations, instructions or equipment specifying a response. For instance, among these task constraints, Vilar et al. [33] showed that the cooperative (within team) or competitive (between teams) nature of the interpersonal coordination in team sport influences the way of how individuals interact to reach the task goal.

3 Representative Learning and Training Design: A Condition for Skill Transfer

In an ecological dynamics framework, the concept of *representative design* underpins the organisation of experimental and learning environments so that observations and acquired skills can be linked to emergent functional behaviours in a specific performance context [34]. The representative design framework provides guidance for the development of ecological constraints that best reflect continuous performance interactions of athletes and performance environments, and the role of variability during these interactions. This is an important feature of *skill transfer* in complex adaptive systems, since it ensures that cognitions, perceptions and actions used to regulate behaviour in one performance context (e.g. a practice environment) can be expected to generalise and underpin performance in another context (competitive performance environments) [35]. Skill transfer can be characterised as positive (i.e. performance is improved under different constraints than would otherwise be the case without learning), neutral, or negative (i.e. performance decreases under different constraints than would otherwise have been the case without learning) [36,37]. Positive transfer can occur when the existing intrinsic dynamics (i.e. performance disposition or tendencies) of an individual cooperate with the dynamics of a new task to be learned, facilitating successful performance behaviours [38]. Negative transfer occurs when intrinsic and task dynamics compete [38]. Cooperation and competition phenomena are influenced by the intrinsic dynamics of the individual which are modifiable in terms of structural and functional adaptations, many of which that are specific to, and can be shaped by, constraints on experience. An important research challenge is to effectively characterise different performance ecologies along each axis of transfer (near and far), in order to predict how processes of skill transfer might support performance (positive, neutral or negative adaptation of the intrinsic dynamics). Near domain transfer refers to when skills are generalised to a new set of performance constraints that, although different, maintain interactions among key system components [36,37]. Far transfer refers to an across-domain transfer (i.e. domain scaled) where separate subsystems become important in different performance environments (e.g. when involving qualitatively different decision making processes, physiological or psychological performance aspects). In sum, the relationship between two performance environments is captured by numerous, dynamic and interacting constraints: environmental, task and personal (both structural, e.g., strength, flexibility, height and functional, e.g., decision making, attunement and calibration to specifying information for action) [3,4]. This makes it difficult to predict transferability of movement behaviours between two performance domains. In our programme of research we have sought to undertake necessary theoretical and empirical developments to enhance understanding of this important process.

According to our analysis of key ideas in ecological dynamics, transferability that enhances functional skill adaptations might relate to *specificity and generality* of transfer processes induced by the selection and/or design of a performance environment. This is because the design of some practice task constraints simulate the properties of performance environment more specifically than others [39]. An ecological dynamics framework suggests that the level of specificity between performance environments relates to the *ecological niche* of an individual [6], providing a *landscape of available affordances* that supports emergence of functional adaptive behaviours [40]. The issue of generality or specificity of transfer particularly relates to the capability of an individual to pick up and use these affordances under different performance conditions. Bruineberg and Rietveld [40] defined the landscape of affordances as the affordances available in an ecological niche, i.e., the whole spectrum of abilities available in our socio-cultural practices. The richness of the landscape of affordances raises questions over the nature of learning designs that can help an individual to pick up only relevant affordances in a given performance environment, i.e., through attunement and calibration to specifying information for adaptive behaviours [25].

4 Attunement to Affordances

An important concern is the information that guides behaviours of complex systems, emphasised through the role of *affordances* as a key property of performer–environment interactions [6]. Affordances are particular properties of a performance environment, which are perceived in "animal-relevant" terms, i.e., what they offer, invite or demand of an organism in terms of actions [8]. The concept of affordances provides a powerful way of understanding how processes of perception and action function in complex adaptive systems, since "within the theory of affordances, perception is an invitation to act, and action is an essential component of perception" (p. 46) [6]. Affordances are defined by the complementary relations between an individual and an environment. They are both objective and subjective to each performer, since they are ecological properties of the environment picked up relative to an individual [41,42]. Affordances for action in an environment are specified within a unique frame of reference for each individual performer, whether learner or expert, adult or child. Descriptions of the state of the environment are "frame dependent" because affordances for action are perceived relative to relevant properties of an individual including the scale of key body dimensions (e.g. height, limb sizes). For example, when dribbling in team games, affordances might include angles and distances to and distances between approaching defenders [43,44]. Ecological dynamics rationalises that such distance or angular values are not perceived in arbitrary metric scales, such as different units of measurement (e.g. yards or metres). Rather, gaps, distances, angles and heading orientation afford different actions from different performers based on their personal constraints [16]. Such personal constraints can be structural (e.g. height and weight and limb length of dribblers) or functional (speed, movement time, power and technical skill in dribbling a ball). In order to negotiate the environment with tools, as extensions of their bodies, during functional behaviours, research has verified theoretical ideas that individual performers perceive different affordances with respect to their own morphological and biomechanical properties, as well as those properties in other performers (e.g. defenders). Athletes need to be able to learn how to solve particular performance problems by picking up relevant body-scaled environmental properties, such as whether their speed in dribbling with the ball can affords passing through gaps of varying dimensions on field, with respect to their relevant physical organismic constraints.

From a Gibsonian philosophical orientation, learning in neurobiology involves the pick up of affordances. Gibson [6] proposed that all organisms perceived affordances, which can constrain their actions. Skilled individuals become progressively attuned available affordances as their expertise levels increase [25]. Affordances are action opportunities invited from an individual, which are predicated on knowledge of a performance environment (rather than knowledge about) [6,7]. This type of knowledge highlights the importance of adopting a person-environment scale of analysis in an ecological dynamics orientation. Basically, it is argued that the environment is composed of physical properties (such as light amplitudes or surface hardness) and the individual is made up of measurable action capabilities. The relationship between the physical properties of the environment and the individual's action capabilities constitutes an affordance [8]. In climbing, affordances refer to "climbing opportunities" [45], i.e., environmental properties that invite hold reach-ability, grasp-ability and climb-ability. In particular, in rock climbing, Boschker et al. [45] demonstrated how experts recalled more information specifying the *functional* properties of a climbing wall (knowledge of a performance environment), neglecting to perceive its *structural* features (knowledge about a performance environment). Conversely, novices were not able to recall such functional properties of the wall to support their actions and they tended to report almost exclusively the structural (less functional) features of the holds [45]. For instance, if a rock climber grasps a surface hold because of its large size, instead of its shape or its orientation, he/she may be using the wrong structural feature (e.g. hold size instead of hold shape or hold orientation) to decide which hold to grasp and how to grasp it [46]. An affordance in rock climbing specifies what a hold is and what a hold means, not separately, but unified in one perceiving-acting process. Perceiving opportunities for specific actions requires perceptual attunement to and calibration of relevant informational variables, meaning that individuals need to pick up a range of perceptual variables from different system modalities (haptic, kinesthesis, auditory and visual) that specify a relevant property of a performance environment [25,47]. The term "relevant" signifies functionality, as this property enables an individual performer to achieve a specific task goal with efficacy. An important characteristic of experts is their perceptual attunement to relevant informational variables, revealing that experts are better at perceiving affordances than beginners.

5 Nonlinear Pedagogy

The previous key theoretical concepts in ecological dynamics provide a powerful underlying principled philosophical basis for the implementation of a *nonlinear pedagogy* [26]: a pedagogical framework for enhancing learning in athletes and sports teams as complex adaptive systems. Chow et al. [26] proposed several principles for a nonlinear pedagogy in sport and physical education: the representativeness of the situation, focus of attention, functional variability through exploratory behaviours, manipulation of constraints and ensuring relevant information-movement couplings (e.g. favouring perception of affordances). Through these principles, nonlinear pedagogy emphasises the non-linearity and non-proportionality between the amount of practice undertaken and the skill acquisition (i.e. the level of expertise) [26]. This philosophical orientation to practice contrasts completely with ideas from the theory of deliberate practice, which are beginning to be discredited in the field of expertise acquisition [48,49]. In contrast to such a linear pedagogical orientation that prescribes the movement to be learned, giving numerous verbal instructions, nonlinear pedagogy respects and even enables the use of the degenerate nature of neurobiological systems, encouraging behavioural exploration, interaction with task and environmental constraints [50,51]. This theoretical approach to learning design requires a respect for processes like task simplification rather than part-task decomposition, in the sense that learning and training designs have to preserve the complexity of an activity to maintain the coupling between information and movement [4]. In nonlinear pedagogy, error-reduction towards a specific model is thus not relevant. The challenge for coaches, instructors and physical education teachers is instead to create conditions that facilitate the exploratory process for the performer, rather than merely providing a precise description and prescription of a movement pattern or a team tactical pattern to follow. For instance, key aspect of nonlinear pedagogy is to use external focus of attention instructions to enhance a more subconscious control of movement with attention paid to the effects of a movement on the environment, rather than the precise form of the movement, which seems to encourage the acquisition of important affordable in skill learning.

6 Conclusion

In summary, an ecological dynamics framework provides a unique philosophical orientation, which can address current challenges in understanding sport performance and skill acquisition in athletes and sports teams considered as complex adaptive systems. Our research programme has shown that addressing these challenges can enhance the effectiveness of current understanding and practice in sport science, pedagogical practice and performance analysis. In our work, our specific aims have included:

- To identify key properties of expertise and skill acquisition in athletes,
- To enhance understanding of how to design affordances into practice environments in sport which simulate competitive performance environments,
- To understand the functional and adaptive role of movement variability (within and between individuals),
- To help to design representative and interactive task constraints in sports development programmes, facilitating an individualised approach,
- Engineering technology and equipment to enhance skill acquisition during practice and training, predicated on complexity and neurobiological principles,
- To apply principles of a nonlinear pedagogy to re-shape current physical education and coaching practices.

7 Scientific Validation

This paper has been unanimously validated in a collaborative review mode with the following reviewers:

• Ana Sousa, University of Porto, Faculty of Sport, Porto, Portugal.

- María Carmen Lemos, University of Sevilla, Department of Condensed Matter Physics, Sevilla, Spain.
- Carlota Torrents Martin, National Institute of Physical Education (INEFC), Barcelona, Spain.

Acknowledgments This project received the funding of the CPER/GRR Logistic Mobility Numeric XTerm.

References

1. Kugler PN, Kelso JAS, Turvey MT (1980) On the concept of coordinative structures as dissipative structures: I. Theoretical lines of convergence. Adv Psychol 1:3–47
2. Davids K, Hristovski R, Araújo D, Balague Serre N, Button C, Passos P (2014) Complex systems in sport. Routledge (Taylor and Francis Group), London
3. Davids K, Araújo D, Hristovski R, Passos P, Chow JY (2012) Ecological dynamics and motor learning design in sport. In: Hodges NJ, Williams AM (eds) Skill acquisition in sport: research, theory and practice. Routledge (Taylor and Francis Group), New York, pp 112–130
4. Davids K, Button C, Bennett SJ (2008) Dynamics of skill acquisition: a constraints-led approach. Human Kinetics, Champaign
5. Araújo D, Davids K, Hristovski R (2006) The ecological dynamics of decision making in sport. Psychol Sport Exerc 7:653–676
6. Gibson J (1979) The ecological approach to visual perception. Houghton Mifflin, Boston
7. Davids K, Araújo D (2010) The concept of 'organismic asymmetry' in sport science. J Sports Sci 13:633–640
8. Withagen R, de Poel HJ, Araújo D, Pepping G-J (2012) Affordances can invite behavior: reconsidering the relationship between affordances and agency. New Ideas Psychol 30:250–258
9. Warren WH (2006) The dynamics of perception and action. Psychol Rev 113:358–389
10. Kelso JAS (1995) Dynamic patterns: the self-organization of brain and behavior. MIT, Cambridge
11. Jantzen KJ, Oullier O, Kelso JAS (2008) Neuroimaging coordination dynamics in the sport sciences. Methods 45:325–335
12. Zanone PG, Kostrubiec V (2004) Searching for (dynamic) principles of learning. In: Jirsa VK, Kelso JAS (eds) Coordination dynamics: issues and trends. Springer, Berlin, pp 58–89
13. Prigogine I, Stengers I (1979) La nouvelle alliance [the new alliance]. Gallimard, Paris
14. Haken H (1983) Synergetics: an introduction: nonequilibrium phase transition and self-organization in physics, chemistry, and biology. Springer, Berlin
15. Prigogine I (1994) Les lois du chaos. Flammarion, Paris
16. Newell K (1986) Constraints on the development of coordination. In: Wade M, Whiting H (eds) Motor development in children: aspects of coordination. Martinus Nijhoff, Amsterdam, pp 341–361
17. Seifert L, Button C, Davids K (2013) Key properties of expert movement systems in sport: an ecological dynamics perspective. Sports Med 43:167–178
18. Kelso JAS (2012) Multistability and metastability: understanding dynamic coordination in the brain. Philos Trans R Soc Lond B Biol Sci 367:906–918
19. Davids K, Bennett SJ, Newell KM (2006) Movement system variability. Human Kinetics, Champaign
20. Newell KM, Corcos DM (1993) Variability and motor control. Human Kinetics, Champaign
21. Edelman GM, Gally JA (2001) Degeneracy and complexity in biological systems. Proc Natl Acad Sci. U S A 98:13763–13768
22. Mason PH (2010) Degeneracy at multiple levels of complexity. Biol Theory 5:277–288

23. Whitacre JM (2010) Degeneracy: a link between evolvability, robustness and complexity in biological systems. Theor Biol Med Modell 7:6
24. Price CJ, Friston KJ (2002) Degeneracy and cognitive anatomy. Trends Cognit Sci 6:416–421
25. Fajen BR, Riley MR, Turvey MT (2009) Information, affordances, and the control of action in sport. Int J Sport Psychol 40:79–107
26. Chow JY, Davids K, Hristovski R, Araújo D, Passos P (2011) Nonlinear pedagogy: learning design for self-organizing neurobiological systems. New Ideas Psychol 29:189–200
27. Seifert L, Davids K (2012) Intentions perceptions and actions constrain functional intra- and inter-individual variability in the acquisition of expertise in individual sports. Open Sports Sci J 5:68–75
28. Davids K, Glazier PS, Araújo D, Bartlett RM (2003) Movement systems as dynamical systems: the functional role of variability and its implications for sports medicine. Sports Med 33:245–260
29. Glazier PS, Davids K (2009) On analysing and interpreting variability in motor output. J Sci Med Sport 12:e2
30. Hristovski R, Davids K, Araújo D, Button C (2006) Hox boxers decide to punch a target: emergent behaviour in nonlinear dynamical movement systems. J Sci Med Sport 5:60–73
31. Pinder RA, Davids K, Renshaw I (2012) Metastability and emergent performance of dynamic interceptive actions. J Sci Med Sport 15(5):437–443
32. Seifert L, Komar J, Barbosa T, Toussaint H, Millet G, Davids K (2014) Coordination pattern variability provides functional adaptations to constraints in swimming performance. Sports Med 44:1333–1345
33. Vilar L, Araújo D, Davids K, Button C (2012) The role of ecological dynamics in analysing performance in team sports. Sports Med 42:1–10
34. Brunswik E (1956) Perception and the representative design of psychological experiments. University of California, Berkeley
35. Pinder RA, Davids K, Renshaw I, Araújo D (2011) Representative learning design and functionality of research and practice in sport. J Sport Exerc Psychol 33:146–155
36. Issurin V (2013) Training transfer: scientific background and insights for practical application. Sports Med 43:675–694
37. Rosalie S, Müller S (2012) A model for the transfer of perceptual-motor skill learning in human behaviors. Res Q Exerc Sport 83:413–421
38. Zanone PG, Kelso JAS (1997) Coordination dynamics of learning and transfer: collective and component levels. J Exp Psychol Hum Percept Perform 23:1454–1480
39. Seifert L, Wattebled L, L'Hermette M, Bideault G, Herault R, Davids K (2013) Skill transfer, affordances and dexterity in different climbing environments. Hum Mov Sci 32:1339–1352
40. Bruineberg J, Rietveld E (2014) Self-organization, free energy minimization, and optimal grip on a field of affordances. Front Hum Neurosci 8:599
41. Scarantino A (2003) Affordances explained. Philos Sci 70:949–961
42. Turvey M, Shaw R (1999) Ecological foundations of cognition I: symmetry and specificity of animal–environment systems. J Conscious Stud 65:95–110
43. Passos P, Araújo D, Davids K, Gouveia L, Serpa S, Milho J, Fonseca S (2009) Interpersonal pattern dynamics and adaptive behavior in multiagent neurobiological systems: conceptual model and data. J Mot Behav 41:445–459
44. Passos P, Milho J, Fonseca S, Borges J, Araújo D, Davids K (2011) Interpersonal distance regulates functional grouping tendencies of agents in team sports. J Mot Behav 43:155–163
45. Boschker MSJ, Bakker FC, Michaels CF (2002) Memory for the functional characteristics of climbing walls: perceiving affordances. J Mot Behav 34:25–36
46. Seifert L, Orth D, Hérault R, Davids K (2013) Affordances and grasping patterns variability during rock climbing. In: Davis T, Passos P, Dicks M, Weast-Knapp J (eds) Studies in perception and action XII: seventeenth international conference on perception and action. Psychology Press, Taylor & Francis, Estoril, Portugal, pp 114–118
47. Jacobs DM, Michaels CF (2007) Direct learning. Ecol Psychol 19:321–349

48. Tucker R, Collins M (2012) What makes champions? A review of the relative contribution of genes and training to sporting success. Br J Sports Med 46:555–561
49. Hambrick D, Oswald F, Altmann E, Meinz E, Gobet F, Campitelli G (2014) Deliberate practice: is that all it takes to become an expert? Intelligence 45:34–45
50. Chow JY, Davids K, Button C, Shuttleworth R, Renshaw I, Araújo D (2007) The role of nonlinear pedagogy in physical education. Rev Educ Res 77:251–278
51. Chow JY, Davids K, Button C, Shuttleworth R, Renshaw I, Araújo D (2006) Nonlinear pedagogy: a constraints-led framework for understanding emergence of game play and movement skills. Nonlinear Dyn Psychol Life Sci 10:71–103

A Multi-agent System Approach to Load-Balancing and Resource Allocation for Distributed Computing

Soumya Banerjee and Joshua P. Hecker

1 Introduction

Recent years have seen a trend in moving large computational tasks to collections of inexpensive, commercial off-the-shelf (COTS) computers that are geographically distributed. This has contributed significantly to the advancement of science by providing access to large-scale shared computing resources on which to solve computationally expensive problems. Some common examples are SETI@home [1] which runs tasks on millions of computers worldwide and Google MapReduce [9] which distributes calculation of web crawled metrics among thousands of computers. This move towards distributed computing has created a need for efficient task allocation and scheduling algorithms. Such algorithms should be very scalable since these systems typically have thousands to millions of computers. They should also be robust to single-point failures and be adaptive to task demand. Recent research on grid resource allocation has focused on volunteer resource allocation, agreement-based resource allocation, and economic resource allocation [17]. Multi-agent decentralized systems offer an exciting approach to distributed resource allocation. They have emergent global properties which arise from local interactions and have been previously used to model biological phenomena [2–7, 16, 19] and

S. Banerjee (✉)
The Broad Institute of MIT and Harvard, Cambridge, MA, USA

Complex Biological Systems Alliance, North Andover, MA, USA

Ronin Institute, Montclair, NJ, USA

University of Oxford, Mathematical Institute, Oxford, UK
e-mail: neel.soumya@gmail.com

J.P. Hecker
Department of Computer Science, University of New Mexico, Albuquerque, NM, USA
e-mail: jhecker@cs.unm.edu

P. Bourgine et al. (eds.), *First Complex Systems Digital Campus World E-Conference 2015*, Springer Proceedings in Complexity, DOI 10.1007/978-3-319-45901-1_4

solve real-world problems [11–15]. Here we use such a decentralized computing approach to allocate and schedule tasks on a grid. The remainder of the paper is organized as follows: Sect. 2 formalizes the problems and states the assumptions, Sect. 3 briefly reviews decentralized computing and the advantages it can afford to a distributed allocation problem, Sect. 4 introduces multi-agent systems, Sect. 5 introduces the simulator used for the experiments in this paper, Sect. 6 discusses the dRAP algorithm, Sect. 7 deals with analysis of the cost of searching through the global queue, Sect. 8 discusses some dRAP optimization techniques influenced by the immune system, Sect. 9 deals with experiments and results, Sect. 10 discusses related work in this area, and Sect. 11 presents concluding remarks and outlines future work.

2 Statement of Problem and Assumptions

Assume there is a queue Q of processes waiting to be allocated to processors. Each process is required to declare a priori its resource requirements viz. the number of threads into which it can be parallelized (TH_n) and the number of system resources it requires (the number of CPUs is assumed to be equal to the number of threads which can be run in parallel, CPU_{req}). Our system departs from traditional resource allocation techniques in that there is no centralized dispatcher. Instead, we dynamically organize a system of geographically distributed computers into clusters to service each process in Q. Over time, clusters of computers are dynamically created, dissociated, and created again in order to serve the resource requirements of the processes in Q. We define a *cluster* as a network of computers which together can completely service the resource requirements of a single process. Clusters of computers are created so as to be proximal to each other in order to reduce latency and communication costs.

We acknowledge the following assumptions in our system:

1. Distributed computers can communicate with each other.
2. There are advantages to computing with geographically proximal computers due to network latency and bandwidth limitations.
3. A new process P_1 that comes in the system will declare a priori the number of threads that it can be parallelized into and its resource requirements (e.g., the number of CPUs it will require, I/O devices required, amount of memory, etc.).
4. The approach will become viable in the asymptotic region of millions or billions of geographically dispersed computers, when there will be expected benefits from a decentralized computing approach that exploits geographical proximity and reduces latency costs, as opposed to a centralized monitor.

3 Decentralized Computing

The extreme size of the computing grid and an ever-increasing demand for computational power places exacting demands on any scheduling, allocation, and load-balancing algorithm. Here we argue that a decentralized computing paradigm presents an ideal solution to the bottlenecks and single-point failures inherent to a centralized monitor tasked with allocating resources and balancing loads in the grid:

1. The workload assigned to a centralized monitor increases as computers are added to the computing grid. A decentralized approach can alleviate the computing load on monitors. In this approach, each individual computer, or cluster of computers, will do some computation.
2. A centralized monitor makes the system susceptible to single-point failures. Distributing load-balancing and resource allocation tasks to individual computers will increase system robustness.
3. Individual computing nodes are naturally aware of their own workloads. As a result, the decentralized paradigm can achieve application-level resource management with significantly less communication overhead than a centralized monitor.
4. A decentralized system uses peer-to-peer networking to scale communication as the system grows, whereas a centralized monitor has to communicate with an increasing number of nodes.
5. A decentralized system is more robust to single node disruptions and failures, whether malicious or benign.
6. A decentralized system may be able to better respond to fluctuations in process requirements, e.g., in a scenario where the scheduler has to "forget" past process requirements and completely rebuild new clusters after servicing one process, i.e., there is no locality in process requirements.

4 Multi-agent Systems

Multi-agent systems use distributed agents to either model or solve a problem. An agent is an entity which matches some real-world object. It could be a biological cell, a virus particle, an ant, or in our case an individual computer. A computer program encodes simple rules or behaviors for interacting with other agents. The agents move about in space and interact with other agents in their neighborhood according to the encoded rules. Thus the behavior of low-level entities is specified and high-level behaviors evolve as simulation time progresses. Multi-agent systems emphasize local interactions based on first principles, and these interactions give rise to the complex high-level *emergent properties* of interest. Such systems have been used to model biological phenomenon such as the human immune system [16], as well as solve real-world problems like communication between distributed radar transmitters [13] and efficient resource collection in swarms of foraging robots [11, 12, 14, 15].

There is no centralized dispatcher to facilitate the formation and dissociation of clusters in the proposed dRAP algorithm. Instead, the algorithm relies on the *self-emergent* properties of a multi-agent system. A *multi-agent* or *agent-based system* is an architecture in which the global properties of the system emerge from local interactions.

The concept of a decentralized system presents a powerful counterpoint to the more common centralized control model often seen in business, government, and military organizations. Decentralization provides a number of important advantages over closed systems, such as robustness, adaptability, flexibility, innovation, and distributed intelligence. The key to this compelling architecture is the impressive ability of a decentralized system to react, mutate, or grow in response to challenging situations.

In any such decentralized system, the agent represents the base unit of computing power for the system. It behaves according to very simple rules. At each unit of model time (or *time step*), the agent senses its immediate local environment and takes actions based on its encoded rules. One rule might instruct the agent to divide if the number of neighbors is greater than 3, while another would cause it to die and be removed from the simulation if the number of neighbors is less than 2. These two examples are rules in the "Game of Life" [10], a paradigmatic system where complex patterns arise from local interactions and simple rules.

If we recast each agent's local sensing functionality as a peer-to-peer communication protocol with other nearby agents, then we can define a new set of rules for each agent that induces actions based on the state of these other, neighboring agents. Using this localized communication scheme, such rule-action pairs can be viewed as instructions for individual agents that produce *decentralized computation* across the system. There is no centralized monitor and yet this system is capable of performing complex computations. In fact, the computational power of such a system of distributed agents acting on simple rules has been proven to be Turing-complete [8].

We use such an agent-based system to dynamically create and dissociate clusters based on the resource requirements of each process. A snapshot of this system is shown in Figs. 1 and 2.

5 Software Platform

For this project we utilize the multi-agent simulation toolkit MASON [18]. MASON consists of a fairly small and portable set of Java library files that provide for design of both model (the "algorithm" component) and visualization (the "graphical user interface" component).

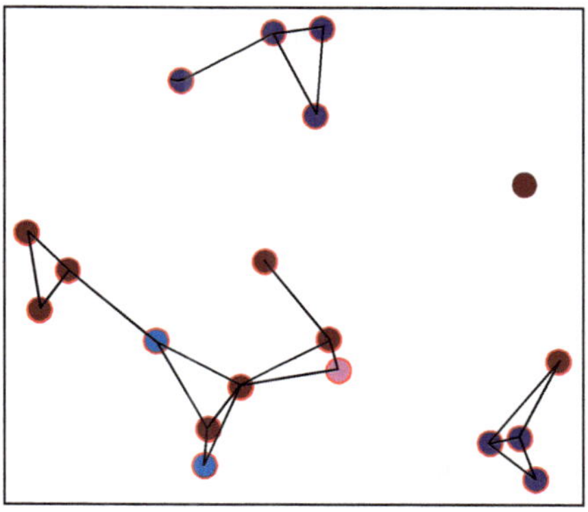

Fig. 1 Agents in large clusters; 1 free agent

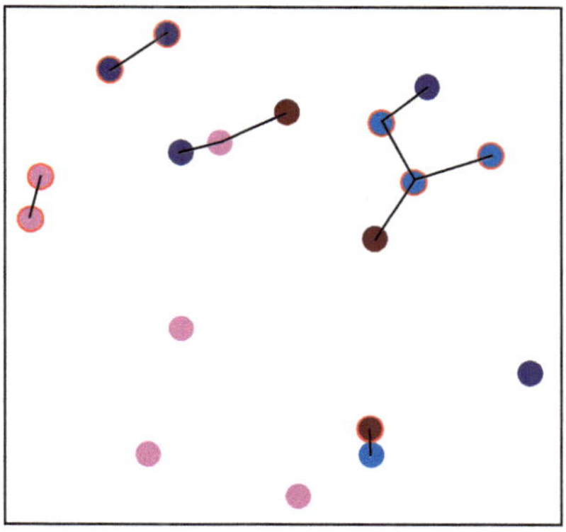

Fig. 2 Agents in several smaller clusters

The agent, the base component of computation in MASON (as in any multi-agent system), is coded in the familiar object-oriented programming format: the class "Agent" that contains all generalized methods and parameters needed for the object "agent" that is simply an instantiation of the Agent class. Following this format, each instantiated agent may contain a unique set of parameters, thereby allowing for minor variation in the replicated objects.

Agents are allowed to make decisions (and even communicate with one another) in a randomized batch lock-step. That is, the MASON scheduler moves through the (randomized) queue of all agents at each time step of the simulation. Scheduling of agents continues as long as the simulation itself is running, although the user may interrupt at any point by pausing or stopping the model.

MASON in particular was selected because of its all-in-one toolkit approach, making multi-agent simulation much easier than if done from scratch, as well as the authors' familiarity and experience with the MASON system.

6 dRAP Algorithm

The distributed resource allocation protocol (dRAP) is described below and some intended optimizations are suggested for future work. An agent in our system is simply a computer. Each agent has a vector containing the time remaining to finish executing its current process ($time_{rem}$) and the number of CPUs in its current cluster ($CPU_{cluster}$). Each agent (or node) is guaranteed to be in exactly 1 of 4 modes (or states) during the simulation:

Mode 1: An agent/node that is currently not part of a cluster and has no task assigned to it

1. The agent scans the queue Q, considers the resource requirements CPU_{req} of unallocated tasks, and takes on the task which minimizes the equation $|CPU_{req} - 1|$.

Mode 2: An agent/node that is currently not part of a cluster and has a task assigned to it

1. The agent continues executing the task and updates its information vector ($time_{rem}$, $CPU_{cluster}$).
2. If the task requirements are not completely satisfied (i.e., if $CPU_{req} > 1$), the agent will query its neighbors and attempt to form a cluster such that $CPU_{req} = CPU_{cluster}$.
3. When the agent finishes executing the task, it returns to **Mode 1**.

Mode 3: An agent/node that is currently part of a cluster and has no task assigned to it

1. The agent scans the queue Q, considers the unallocated tasks, and takes on the task which minimizes the equation $|CPU_{req} - CPU_{cluster}|$.

Mode 4: An agent/node that is currently part of a cluster and has a task assigned to it

1. The agent continues executing the task and updates its information vector ($time_{rem}$, $CPU_{cluster}$).
2. When the task completes, the agent dissociates from the cluster and returns to **Mode 1**.

A key feature of our algorithm is that nodes query their neighbors (other nodes that are close to them physically) in order to form clusters. This has the effect of reducing latency and communication costs. One optimization to consider would be to delay cluster dissociation in **Mode 4**. This would lead to *learning* or *memory* in the system where the scheduler would be able to remember past process requirements.

7 Analysis of Queue Cost

The dRAP algorithm requires a traversal through the global task queue in **Mode 1** and **Mode 3**. The algorithmic complexity is given by $\sum (n-i)m = O(n^2 m)$ where $m =$ the number of tasks in the global task queue, and $n =$ the average number of clusters. At a given timestep, the worst case can be approximated as $O(nm)$.

8 Optimizations Inspired by the Immune System

The immune system is able to find rare spatially localized pathogens and eliminate them in a timely manner [5, 6]. Similar to how in our system clusters of computers find processes, the immune system uses specialized cells to find pathogens in anatomical regions called lymph nodes. In previous work we showed how a sub-modular arrangement of lymph nodes could lead to fast elimination of pathogens in the immune system and also faster search for solutions in immune inspired distributed systems of computers [5, 6, 19]. Let an artificial lymph node be composed of a number of clusters and a process queue. Also let there be a number of such artificial lymph nodes that have the capability of communicating with each other. An "artificial lymph node" is supposed to be a computer in charge of a number of clusters. This computer will store the process queue and also will have some memory and CPU to communicate with other "lymph nodes."

We are interested in making the system sub-modular so that we can minimize the total time to find a cluster. There is a tradeoff between the local cost and the global cost; the local cost is $O(n^2)$ and the global cost $O(N/n)$. The total cost of traversing through the queue in a lymph node and the cost of communicating with other lymph nodes can be summed up as:

$$t_{\text{total}} = t_{\text{local}} + t_{\text{global}} \quad (1)$$

$$t_{\text{total}} = O(n^2) + O(N/n) \quad (2)$$

where n is the number of clusters in a single lymph node and N is the total number of clusters in the complete system. We assume that the global cost of finding another cluster in another lymph node that can service some process requirement is proportional to the number of lymph nodes (where N/n is the number of lymph nodes in the system).

Minimizing the total time cost, we get $2n - N/n^2 = 0$

$$n = O(N^{1/3}) \quad (3)$$

This implies that in larger systems (more computers, more clusters, and more lymph nodes), the number of clusters within a single lymph node should grow larger but only sub-linearly in the number of total clusters in the system. This would balance local costs of queue traversal and global costs of finding another lymph node with another cluster that can service the process. The key point here is that the number of clusters in a lymph node should scale sub-linearly with the size of the whole system, i.e., if a system of networked artificial lymph nodes were to grow a 1000 times bigger (1000 times more clusters), then the number of clusters within a lymph node need only increase by a factor of 10. Such sub-modular systems inspired by the immune system have been proposed previously for mobile ad hoc networks, control of mobile robots, intrusion detection systems, and peer-to-peer networks [5, 6, 19].

More generally, if the local and global communication costs scale with exponents α and β, we have

$$t_{\text{total}} = O(n^{\alpha}) + O(N^{\gamma}/n^{\beta}) \tag{4}$$

Minimizing the expression with respect to N, we get

$$n = O\left(N^{\frac{\gamma}{\alpha+\beta}}\right) \tag{5}$$

1. If $\gamma < \alpha + \beta$ we have sub-linear scaling.
2. If $\gamma > \alpha + \beta$ we have super-linear scaling.
3. If $\gamma = \alpha + \beta$ we have linear scaling.
4. If $\gamma/(\alpha + \beta) = 0$ we have no scaling (constant).
5. If $\gamma/(\alpha + \beta) < 0$ we have negative scaling.

9 Experiments

We conduct several experiments that compare our dRAP algorithm to a null model, i.e., a first-in first-out (FIFO) scheduling system. Additionally, we measure the effective computational complexity of queue traversals and examine the scaling properties of our system by varying the number of nodes and measuring the effect on performance. We define two timing metrics on which our system performance will be judged: T_{complete} is the time required to complete all tasks in the queue, and T_{wait} is the average wait time for a task added to the queue. Unless otherwise noted, system parameters are defined as such: number of nodes $=$ 100, number of tasks $=$ 1000, tasks are randomly selected from a normal distribution s.t. CPU_{req} varies from 1 to 5, with initial time_{rem} varying from 25 to 125 in increments of 25. That is, a task t_i with $\text{CPU}_{\text{req}} = 1$ has an initial $\text{time}_{\text{rem}} = 25$, and a task t_j with $\text{CPU}_{\text{req}} = 5$ has an initial $\text{time}_{\text{rem}} = 125$. All averages are computed across ten trials.

Table 1 Average timing comparison of dRAP and FIFO scheduling algorithms with 95 % confidence intervals

	T_{complete}	T_{wait}
dRAP	845.60 (861.94, 829.26)	342.54 (349.30,335.78)
FIFO	1071.20 (1088.99, 1053.41)	475.31 (485.79,464.82)

Table 2 Average cluster utilization of dRAP and FIFO scheduling algorithms with 95 % confidence intervals

	μ_{cluster}
dRAP	100 %
FIFO	56 % (54 %,58 %)

9.1 Comparison to Null Model

Here we present three separate experiments which compare the dRAP and FIFO algorithms. The first is a simple timing comparison that looks at T_{complete} and T_{wait} for each case. Values are presented in Table 1 (including 95 % confidence intervals).

We observe an approximate 20 % reduction in T_{complete} and an approximate 25 % reduction in T_{wait} when comparing dRAP to FIFO.

Our second experiment comparing dRAP and FIFO involves average cluster utilization. Because dRAP assigns tasks s.t. $\text{CPU}_{\text{cluster}} == \text{CPU}_{\text{req}}$, this ensures that all nodes in the cluster will be utilized. However, the FIFO scheduling system hands out tasks to the first available cluster, meaning it allows for the possibility that $\text{CPU}_{\text{cluster}} > \text{CPU}_{\text{req}}$. For example, a task with $\text{CPU}_{\text{req}} = 2$ that is assigned to a cluster with $\text{CPU}_{\text{cluster}} = 5$ will leave three unused nodes. Thus, we present an analysis of cluster utilization using the metric in Eq. (6):

$$\mu_{\text{cluster}} = \frac{\text{CPU}_{\text{req}}}{\text{CPU}_{\text{cluster}}} \tag{6}$$

If $\text{CPU}_{\text{cluster}} \leq \text{CPU}_{\text{req}}$, we simply set $\mu_{\text{cluster}} = 1$. Values are presented as percentages in Table 2 (note that dRAP's μ_{cluster} is always 100 % by definition).

Finally, our third experiment is designed to measure global node utilization over the time of the simulation. Here we simply document the number of nodes that do computation on a given timestep and normalize by the total number of nodes in the system. Results are displayed in Fig. 3 (taken from a single simulation run).

We observe that the dRAP algorithm utilizes approximately 90–95 % of the nodes for the majority of the simulation, while FIFO utilizes approximately 70–75 %.

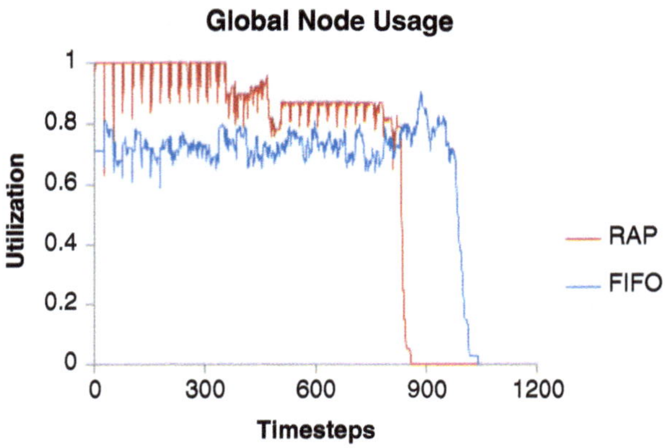

Fig. 3 Global utilization of nodes throughout simulation

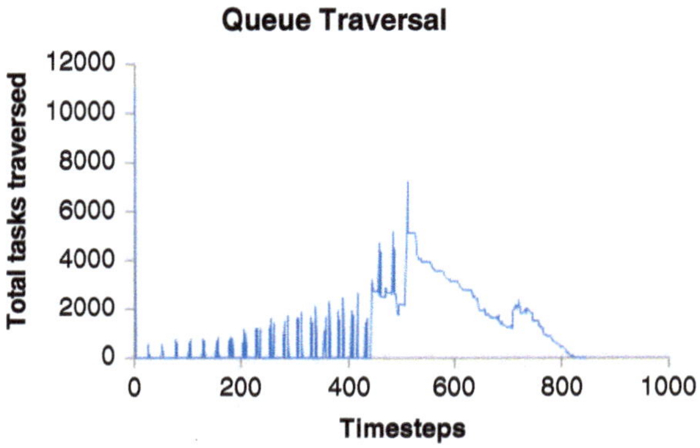

Fig. 4 Total task traversals by all clusters per timestep

9.2 Effective Complexity

For this experiment, we estimate the "effective" computational complexity of the dRAP algorithm. That is, in comparison to the qualitative $O(nm)$ worst case runtime per timestep, we are interested in how much of the task queue must be traversed in order to properly fit the $CPU_{cluster} == CPU_{req}$ requirement. Total tasks traversed per timestep from one selected simulation run are presented in Fig. 4. Note that the initial traversal (timestep "0"), although difficult to see, is approximately 11,000.

"Worse case" here, as addressed above, is $O(nm)$, or 100,000 tasks traversed per timestep if n = number of clusters = number of nodes = 100 and m = number of tasks = 1000. From this plot (plus additional runs not included here), we can conclude that effective computational complexity is no more than approximately 10 % of the worst case runtime $O(nm)$.

9.3 Scaling

For our last experiment, we are interested in collecting information on the scaling ability of our algorithm. Our goal in this test is to increase the number of nodes (in intervals of 50), while also maintaining an equal number of neighbors for each node. That is, we ensure that the neighborhood size parameter defined in the simulation scales inversely with the number of nodes s.t. a given node has approximately the same number of neighbors regardless of the total nodes in the system. Results are presented for our two timing metrics: $T_{complete}$ scaling in Fig. 5 and T_{wait} scaling in Fig. 6. Data in both figures are $\log_2$-transformed in order to correlate doubling of nodes with halving of the timing metrics.

We note a near-perfect scaling for both timing metrics, as shown in the fitted power law equations inset into each plot (Figs. 5 and 6). Note that the T_{wait} exponent above 1 is most likely a result of inexact tuning of the neighborhood size with increasing nodes, and this issue will rectified in future work.

Fig. 5 Scaling of $T_{complete}$

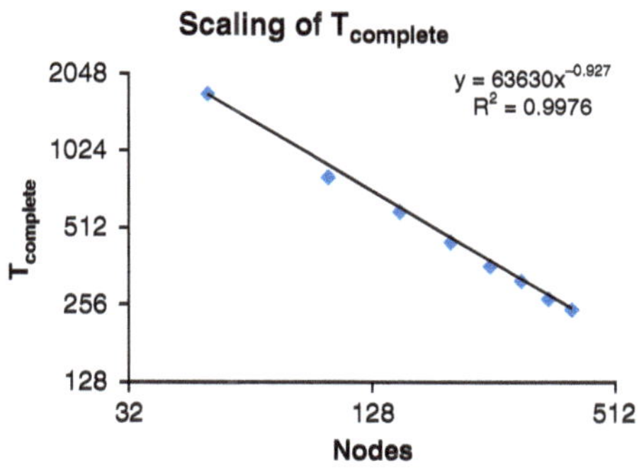

Fig. 6 Scaling of T_{wait}

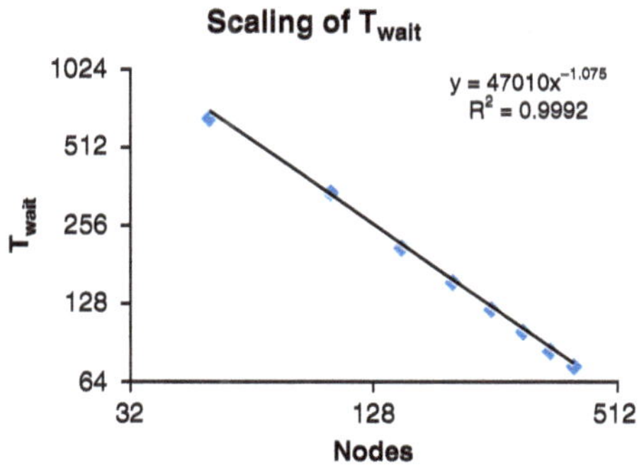

10 Related Work

Resource allocation for grid computing is an active area of research. For example, SLURM [20] is a configurable Linux utility for cluster allocation that uses static allocation of nodes to clusters, called partitions, in contrast to the dynamic cluster allocation presented in this paper. LSF [21] is another proprietary cluster management facility; however, details of its allocation algorithm are not publicly available.

11 Conclusions and Future Work

In this paper we have presented an algorithm for allocating, scheduling, and load-balancing processes on a massively distributed system. This is very relevant to current research in operating systems, especially with a trend of moving computation tasks onto inexpensive, distributed hardware. The proposed decentralized algorithm draws inspiration from biology, adaptively creating, and dissociating clusters from nodes to match task demand. Decentralization enables scalability, robustness, alleviation of computing load on monitor, better response and adaptability to process queue fluctuations, and learning about process requirements. The dRAP algorithm outperforms a FIFO scheduling algorithm on time to complete all tasks, average waiting time, and CPU utilization. The scheduling is also shown to be robust to a malicious adversary that might permute the order of the tasks such that high demand tasks would be queued first followed by low demand tasks. A key feature of our procedure is that nodes communicate with neighboring computers in order to dynamically form clusters. Hence our algorithm also holds promise in areas where it is advantageous to communicate with immediate neighbors due to network latency, e.g., Google MapReduce uses a locality optimization to reduce latency due to network communication [9]. The comparison of this algorithm to other scheduling algorithms like Shortest Remaining Time First (SRTF) on other metrics like response time, as well as collection of data on the exact distribution of process demand in a queue in a real-world scenario, will be the subject of future investigation.

12 Scientific Validation

This paper has been unanimously validated in a collaborative review mode with the following reviewers:

- Cyrille Bertelle, from Université du Havre (France)
- Pierre Collet, from Strasbourg University
- Carlos Jaime Barrios Hernandez.

Acknowledgements We thank Dr. Dorian Arnold for fruitful discussions.

References

1. Anderson D, Cobb J, Korpela E, Lebofsky M, Werthimer D (2002) SETI@ home: an experiment in public-resource computing. Commun ACM 45(11):61
2. Banerjee S (2009) An immune system inspired approach to automated program verification. arXiv preprint arXiv:0905.2649. http://arxiv.org/abs/0905.2649
3. Banerjee S (2013) Scaling in the immune system. Ph.D. thesis, University of New, Mexico
4. Banerjee S, Moses M (2009) A hybrid agent based and differential equation model of body size effects on pathogen replication and immune system response. In: Timmis J (ed) The 8th international conference on artificial immune systems (ICARIS). Lecture notes in computer science. Springer, Berlin, pp 14–18. http://www.springerlink.com/content/b786g874642q2j37/
5. Banerjee S, Moses M (2010) Modular radar: an immune system inspired search and response strategy for distributed systems. In: Artificial immune systems. Springer, Berlin, pp 116–129
6. Banerjee S, Moses M (2010) Scale invariance of immune system response rates and times: perspectives on immune system architecture and implications for artificial immune systems. Swarm Intell 4(4):301–318. http://www.springerlink.com/content/w67714j72448633l/
7. Banerjee S, Levin D, Moses M, Koster F, Forrest S (2011) The value of inflammatory signals in adaptive immune responses. In: Artificial immune systems. Springer, Berlin, pp 1–14. http://www.springerlink.com/index/U634HJ83W62W5383.pdf
8. Cook M (2004) Universality in elementary cellular automata. Complex Syst 15(1):1–40
9. Dean J, Sanjay G (2008) MapReduce: simplified data processing on large clusters. Commun ACM 51(1):107–113
10. Gardner M (1970) Mathematical games: the fantastic combinations of John Conway's new solitaire game 'Life'. Sci Am 223(4):120–123
11. Hecker JP, Moses ME (2013) An evolutionary approach for robust adaptation of robot behavior to sensor error. In: Proceedings of the 15th annual conference companion on genetic and evolutionary computation (GECCO '13 companion). ACM, New York, NY, pp 1437–1444. http://doi.acm.org/10.1145/2464576.2482724
12. Hecker JP, Moses ME (2015) Beyond pheromones: evolving error-tolerant, flexible, and scalable ant-inspired robot swarms. Swarm Intell 9(1):43–70
13. Hecker J, Wu A, Herweg J, Sciortino J Jr (2008) Team-based resource allocation using a decentralized social decision-making paradigm. In: Proceedings of SPIE, vol 6964, p 696409
14. Hecker JP, Letendre K, Stolleis K, Washington D, Moses ME (2012) Formica ex machina: ant swarm foraging from physical to virtual and back again. In: Swarm intelligence: 8th international conference, ANTS 2012. Springer, Berlin, pp 252–259
15. Hecker JP, Stolleis K, Swenson B, Letendre K, Moses ME (2013) Evolving error tolerance in biologically-inspired iAnt robots. In: Proceedings of the twelfth European conference on the synthesis and simulation of living systems (Advances in artificial life, ECAL 2013). MIT Press, Cambridge, MA, pp 1025–1032
16. Jacob C, Litorco J, Lee L (2004) Immunity through swarms: agent-based simulations of the human immune system. In: Artificial immune systems. Lecture notes in computer science. Springer, Berlin, pp 400–412
17. Krawczyk S, Bubendorfer K (2008) Grid resource allocation: allocation mechanisms and utilisation patterns. In: Proceedings of the sixth Australasian workshop on grid computing and e-research-Volume 82. Australian Computer Society Inc, Darlinghurst, pp 73–81
18. Luke S, Cioffi-Revilla C, Panait L, Sullivan K (2004) Mason: a new multi-agent simulation toolkit. In: Proceedings of the 2004 SwarmFest workshop, vol 8

19. Moses M, Banerjee S (2011) Biologically inspired design principles for scalable, robust, adaptive, decentralized search and automated response (radar). In: 2011 IEEE symposium on artificial life (ALIFE), pp 30–37
20. Yoo A, Jette M, Grondona M (2003) SLURM: simple linux utility for resource management. In: Job scheduling strategies for parallel processing. Lecture notes in computer science. Springer, Berlin, pp 44–60
21. Zhou S (1992) LSF: load sharing in large-scale heterogeneous distributed systems. In: Proceedings of workshop on cluster computing, pp 1995–1996

Is the Lean Organisation a Complex System?

Pierre Masai, Pierre Parrend, Nicolas Toussaint, and Pierre Collet

1 Introduction

Lean is a holistic management method. As such, it embeds emergent decision processes, feedback loops through training, coaching for the human actors [17], or kanban for production flow control [19], and a high connectivity through a highly formalised communication mechanism [4]. This shows intuitively that lots of its core characteristics are tightly bound with complex system behaviours: our hypothesis is that this assumption is true, and that the characterisation of these complex system properties has a high potential impact for identifying the success factors of Lean organisations, and is thus a strong enabler for their efficient implementation.

P. Masai (✉)
ICube Laboratory, Université de Strasbourg, Strasbourg, France

Toyota Motor Europe, Bruxelles, Belgium

Complex System Digital Campus (UNESCO Unitwin), Paris, France
e-mail: pierre.masai@etu.unistra.fr

P. Parrend
ICube Laboratory, Université de Strasbourg, Strasbourg, France

ECAM Strasbourg-Europe, Schiltigheim, France

Complex System Digital Campus (UNESCO Unitwin), Paris, France
e-mail: pierre.parrend@ecam-strasbourg.eu

N. Toussaint
ICube Laboratory, Université de Strasbourg, Strasbourg, France

P. Collet
ICube Laboratory, Université de Strasbourg, Strasbourg, France

Complex System Digital Campus (UNESCO Unitwin), Paris, France
e-mail: pierre.collet@unistra.fr

P. Bourgine et al. (eds.), *First Complex Systems Digital Campus World E-Conference 2015*, Springer Proceedings in Complexity, DOI 10.1007/978-3-319-45901-1_5

In this article, we summarise what Lean is and what the main properties of a Complex System are. We develop our hypothesis on why the Lean organisation exhibits the properties of a complex system, and propose two models showing the behaviour of Lean processes as complex systems: (1) the Hoshin Kanri process (or management of the organisation objectives) which involves different levels of agents (Top Management, Management, and employees), spiralling top down and bottom up until a final list of objectives is agreed for the organisation (the 'Hoshin'), (2) the Nemawashi process (or Consensus Building), which spirals bottom up from the person promoting a project until the final approval of this project by the Top Management. Simulations of these models should enable us to better understand the Lean organisation and its success factors.

This paper is organised as follows: Sect. 2 presents the definition and properties of the Lean Organisation. Section 3 characterises the properties of complex systems which are of interest for our model, and presents relevant modelling approaches as well as the stakes of modelling organisations as complex systems. Section 4 introduces the concepts and models of the Hoshin Kanri and the Nemawashi processes. Section 5 specifies the performed simulations and presents their output. Section 6 discusses these outputs and their significance both for modelling Lean organisations as complex systems and for understanding them better. Section 7 concludes this work.

2 What Is Lean

In the 25 years since the publication of The Machine That Changed The World [21], based on the Massachusetts Institute of Technology's five-year study on the future of the automobile, the first book in the west revealing what then became known as Lean (the core idea of Lean is to maximise customer value while minimising waste),[1] the concepts and practices that originated at Toyota have been applied first in the manufacturing of automobiles, where they have been widely adopted, then in the manufacturing of goods in general, and from there it has been applied successfully to all organisation types. When supported from the top as in The Lean Turnaround [3] and applied consistently, it has delivered superior results. As examples of application to other sectors and organisation types, let's mention: Lean Start Ups [16], Lean IT [2], Lean Healthcare,[2] Lean Government.[3] The successful application of Lean to so many different environments suggests systemic properties of Lean that we believe are those displayed also by complex systems: many agents interacting with each other to produce results far superior to what each agent could achieve separately.

[1] Lean Enterprise Institute Website: http://www.lean.org/whatslean/.

[2] Marl Graban's blog: http://www.leanblog.org/.

[3] Lean Government Starter Kit: http://www.epa.gov/lean/government/.

3 Modelling Complex Systems

A complex system can be defined as any system consisting of a large number of interacting autonomous entities, creating several layers of collective organisations leading to emerging behaviours [5]. In this section, we highlight their core properties, introduce representative modelling approaches, and provide some hints related to the complex system modelling of organisations and enterprises.

3.1 Properties of Complex Systems

Several authors [8, 14] establish that Complex Adaptive Systems (CAS) are characterised by the following properties:

- *Emergence*: the whole is more than the sum of the parts. Agents produce together results that far exceed what they could do individually.
- *Co-evolution*: the agents evolve jointly.
- *Connectivity*: all entities are connected.
- *Distributed Control*: the control is distributed to the lowest possible level.
- *Far-from-equilibrium*: a system without external influences tends to equilibrium, but this is not the case when observing organisations that are constantly evolving based on external conditions, for example, creating new rules (a phenomenon called autopoiesis).
- *Non-linearity*: there is a strong dependency on initial conditions.
- *State of paradox*: different elements of the system are apparently opposed to each other.

These properties prove to be key to understand the complex behaviour of human-scale systems such as Lean organisations, as discussed in Sect. 6.

3.2 Complex Systems and Organisations

The modelling of organisations as complex systems poses the challenge of modelling discrete entities exhibiting characteristics of complex systems at the mesoscale, i.e. at the scale of visible events. Such modelling requires the analysis of three complementary domains: concepts, models, and empirical [18].

Concepts are typically expressed as ontologies [12]. Models can be built either as Complex Adaptive Systems (CAS) [14] or using stochastic approaches [10, 11]. CAS models provide efficient views for representing emerging behaviours from atomic interactions. In cases where emergent properties can be quantified, but the way they emerge is not well understood, stochastic approaches and probabilistic models such as fuzzy logic, probabilistic graphs, or Bayesian behaviours [8, 10, 11] enable to identify the relationship between entities. Vensim is a tool supporting this kind of approach [7]. Static relationship structures are best represented as networks [20].

The application of complexity models to the analysis of organisations is often limited to a conceptual level implying emerging structures [13], or key properties such as 'connectivity and interdependence', 'co-evolution', 'dissipative structures, far-from-equilibrium and history', 'exploration-of-the-space-of-possibilities', 'feedback', 'self-organisation and emergence', or 'chaos and complexity' [15]. Analysis of organisation interactions is a fertile domain for network models, in particular for understanding networks of collaborating enterprises [6] or corporate control mechanisms [20].

4 Modelling Lean

In the current models, Hoshin Kanri and Nemawashi are considered as isolated processes. The specific Nemawashi process is abstracted in the Hoshin Kanri, and only its output (item selected or not) is considered in the model. The first view for analysing complex systems, i.e. concepts [18], is given for both of them as ontologies.

4.1 Definitions

4.1.1 Hoshin Kanri

Hoshin Kanri (Hoshin means Compass and Kanri means Management in Japanese) is the process by which the objectives are set at various levels in the enterprise, at the function level (like Information Systems), at the Legal Entity Level (like 'French Legal Entity'), or at the Global Level (companywide). The Hoshin Kanri process is a very typical example of how the Lean organisation is working, because it involves interaction of agents at all levels of the organisation, spiralling up through the layers to enable good ideas from all employees to be adopted at a much higher level, and percolating top down to enable the organisation strategy to reach all employees who will have to play a part in realising it, as shown in Fig. 1. It shows the respect for people of the Lean Organisation, enabling all employees to express their ideas, giving strong value to the ideas related to their own area of expertise, that they are recognised to master more than anybody else.

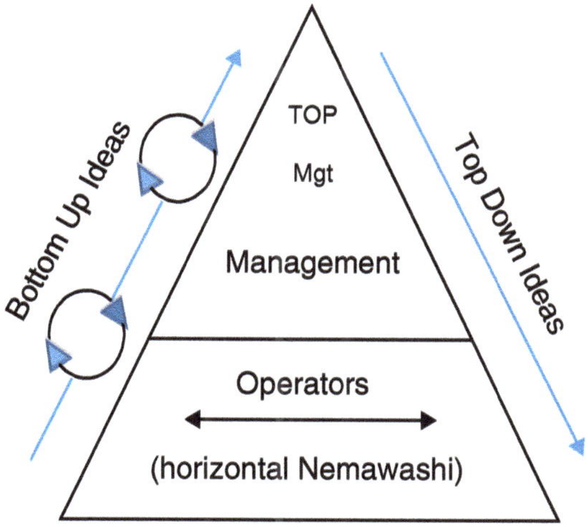

Fig. 1 The Hoshin Kanri process

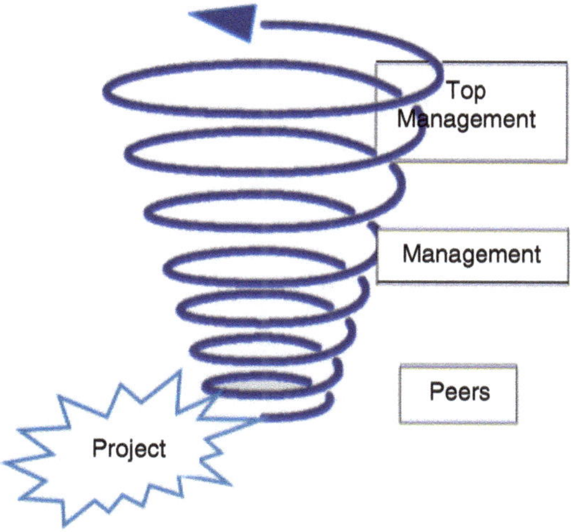

Fig. 2 The Nemawashi process

4.1.2 Nemawashi

Nemawashi is a Japanese word that conveys the meaning of a tree that is transplanted, taking enough earth around it to enable the tree to survive when planted elsewhere. When taking this analogy to an idea, it means to explain the idea (the tree) to all the stakeholders necessary for its implementation (the earth around the roots) so that it can be brought to implementation with the support (buy in) of all the stakeholders (the transplantation). It is not as such an idea unique to the Lean organisation, but it is used a lot as a technique in Lean because of the importance to value everybody's ideas and input in this organisation form. Figure 2 shows the principle of the Nemawashi process.

If we want to create a model of this Nemawashi process, we have to imagine a representation of the idea, typically an A3 document with visualisations and text. We can model this A3 by a set of items representing ideas: $(i_1, i_2, \ldots i_n)$.

The interaction with each stakeholder will result in changes (enrichment by the stakeholder's experience) that will be updated in the document. For example, if an item $j_p (1 \leq p \leq n)$ is deemed better than i_p, then it will replace it and the set of items will become: $(i_1, \ldots, j_p, \ldots i_n)$.

There will be several loops of this interaction, starting with the peers of the agent, then spiralling up the organisation as shown in Fig. 2, eventually converging to a better A3, where each stakeholder will recognise some of their own ideas, which will encourage them to sign the final document and support the project. If a stakeholder did not propose any improvements to the document, none of his points will be on the document, but it is fair to assume that he will approve it.

4.2 Concepts

Only the concepts required for the modelling of the processes considered are given here. A more complete ontology of the concepts required for modelling the Lean organisation and their relationship can be found in our previous work [12].

4.2.1 Hoshin Kanri

Hoshin Kanri is characterised by three entities: its participants (top management, middle management, and employees), the items, which are proposals that participants make for potential selection as strategic initiatives for the coming year, and the time, in particular Nemawashi days where the Hoshin Kanri is pushed further through consensus between the employees of the organisation at the different levels. Figure 3 shows the ontology for the Hoshin Kanri process.

4.2.2 Nemawashi

Nemawashi is characterised by three entities: its participants, identical to those of Hoshin Kanri, the items, which are also found at the Hoshin Kanri level, but also the specific project bound with these items and proposals, i.e. items being considered for adoption. Figure 4 shows the ontology for the Nemawashi process.

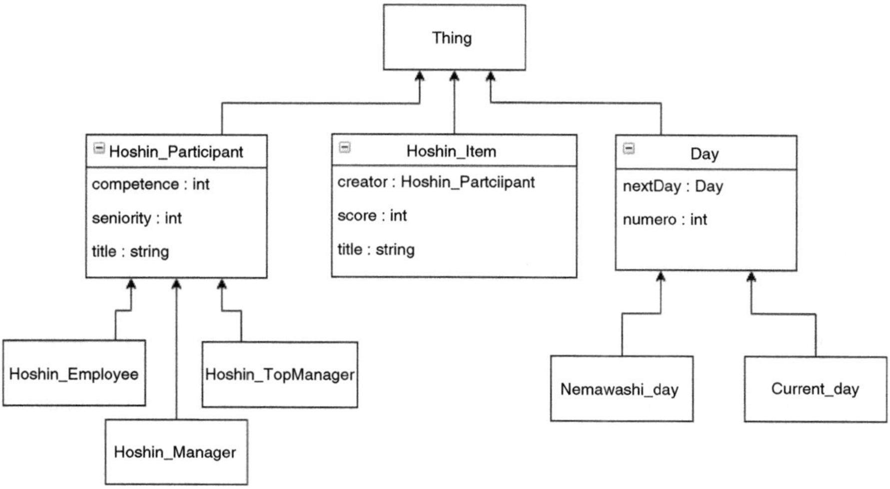

Fig. 3 Hoshin Kanri ontology

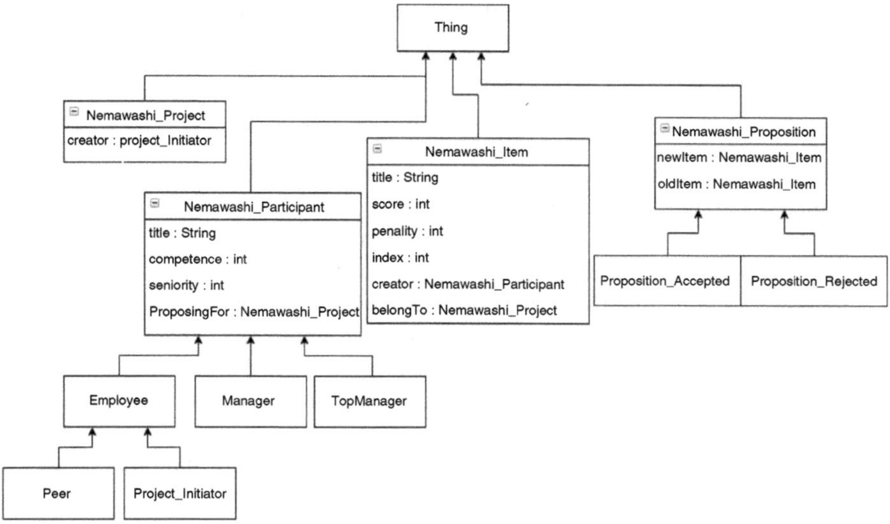

Fig. 4 Nemawashi ontology

5 Simulations

The models of both processes described in the previous section are implemented to demonstrate their behaviour in time and to challenge the hypothesis of the Lean organisation being a complex system.

The Hoshin Kanri is modelled as agents interacting in a stochastic manner. These agents are implemented first in an object oriented way in the Python[4] language, then abstracted as a set of probabilistic rules with Drools[5] embedding an identical behaviour.

The Nemawashi is implemented as a set of probabilistic rules only.

5.1 Hoshin Kanri

A Hoshin Kanri process is performed each year so as to determine the strategic initiatives to be taken and then enforced in the next business year. An initial set of proposals is generated, made public to the organisation, and all employees can propose their own improvements. Better proposals are kept, weaker ones are removed. The objective of the simulation is to evaluate the impact on the resulting decisions, based on the interactions between the organisation agents of:

1. the quality of initial proposals,
2. the emitter of initial proposals, either Top Management or all employees,
3. the seniority and skills of employees and managers,
4. the elapsed time.

Simulation parameters are driven from our experience:

- the Hoshin Kanri process is simulated on a period of 90 days, or 3 months, which is the typical time frame used for this.
- the agents are at three levels: Top Management, Management, and employees.
- two types of initialisations can be performed: either the Top Management proposes the initial items, or everybody in the organisation can do.
- the Hoshin items produced are scored based on a simple rating based on seniority and experience of each agent. This makes it more likely that Top Management or Management will have their proposed items retained; however, it does not make it a fatality.
- the frequency of items input accelerates towards the end of the process, which has been simulated here by a reverse Fibonacci sequence:

$$y_i = 90 - f_i, \text{ for } i = 1, 10, \text{ giving} : (89, 88, 87, 85, 82, 77, 69, 56, 35, 1) \quad (1)$$

- the model presented here is abstracted, because items are often not replaced by others as a whole, but interaction between the agents at various levels also uses the Nemawashi model to merge several items in a more valuable one.

---

[4]https://www.python.org/.

[5]http://www.drools.org/.

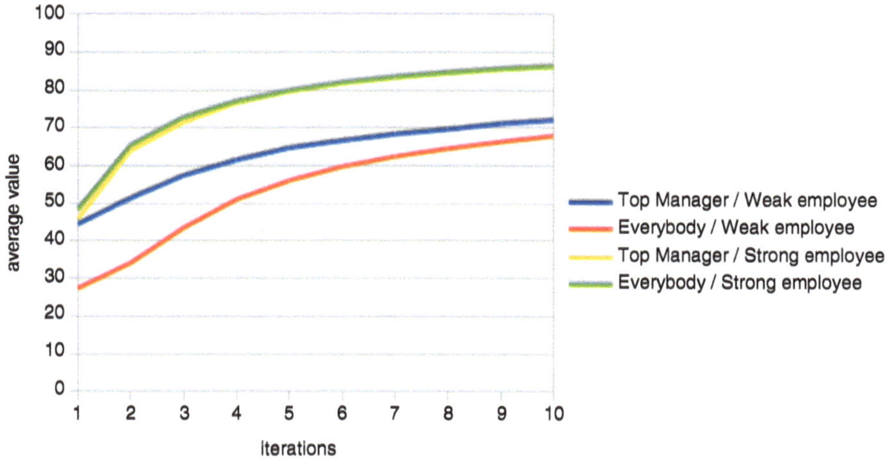

Fig. 5 Evolution in time of the average item value for Hoshin Kanri

The quality of proposals is represented on an arbitrary scale from 0 to 100, with 100 representing a higher quality and 0 a lower quality. This quantification enables to abstract the comparison process between two items: the better item is kept, the weaker is removed.

Figure 5 shows the evolution in time of the average item value for the Hoshin Kanri process, for varying initialisation processes (by Top Management/by everybody) and varying skill level of employees of the organisation (Weak Peers/ Strong Peers). When the Top Management issues the initial items, the resulting decision quality, as one can expect, depends on the feedback of the employees: if the employees have weaker skills, the resulting decisions will have a lower overall quality. When the employees have a high seniority and proficiency, an interesting phenomenon occurs: employee-driven Hoshin Kanri leads to results as good as Top Management-driven Hoshin Kanri. In this case, the presence of management seems to be useless. Actually, these results match interestingly enough a radical shift in the culture of Lean organisations: the role of leaders is to enable emergence, not to take (all) the decision themselves.

Figure 6 shows the evolution of the number of items remaining from the original proposal in the Hoshin Kanri process. This number converges for the different configurations, except when the Top Management initialises the process and the employees have a weaker seniority and skill level. In this case, as can be expected, fewer modifications are observed.

It is worth noting that employee-driven Hoshin Kanri with weaker employee seniority and skills lead to as many new proposals as when the employees are more experienced. However, if we compare these results with those from Fig. 5, we see that the same amount of energy will then be spent to achieve weaker results. Hence, when the employees have weaker experience, the Hoshin Kanri process should better be led by the Top Management.

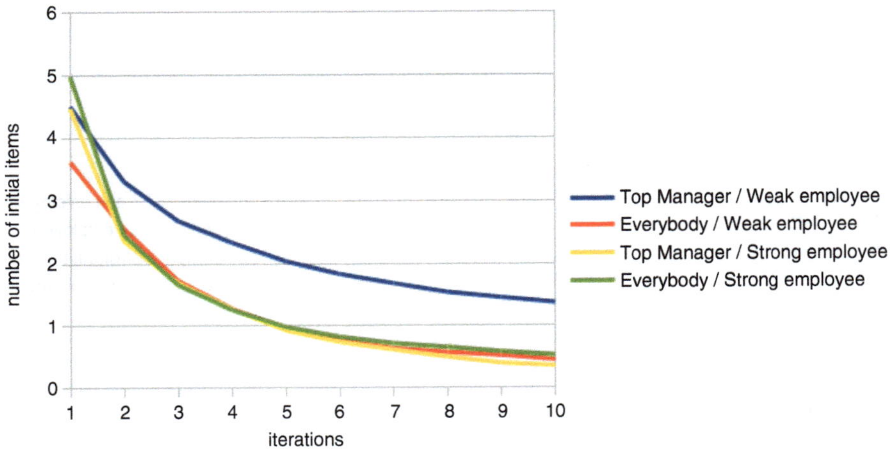

Fig. 6 Evolution of the number of items kept from the original Hoshin Kanri proposal

5.2 Nemawashi

The Nemawashi process is performed at each step of the Hoshin Kanri for building consensus. We focus here on the consensus building process itself, independently of its potential integration in the Hoshin Kanri.

The core simulation parameters are the seniority value (in years), as well as the competence value (a score of 1–10) of the Nemawashi_Participant.

The first step is the creation of a project by the initiator. He puts 20 items in it (for example, 20 items on an A3 document). Each item is generated with an id, a penalty value (initialised to 0), and a score.

The maximum score for the Nemawashi items is given by the following equation:

$$\text{max\_score} = \text{initiator's seniority} * \text{initiator's competence} \tag{2}$$

Then, the item score is randomly chosen between 0 and the maximum score. The penalty is incremented each time that a participant gives a proposition to replace it with a better item and when the initiator does not consider it. The higher the penalty, the higher the chance it will be replaced.

Once the project is initiated, the initiator will show it to his peers, then to the managers, and finally to the Top Manager to enhance it with the experience of all the participants. Each of them will see some items and if they have a more valuable idea, propose it to the initiator. The probability that the initiator accepts to swap the item is given by the following equation:

$$p = (\text{score of the new item} + (5 * \text{penalty of the project's item}))/100 \tag{3}$$

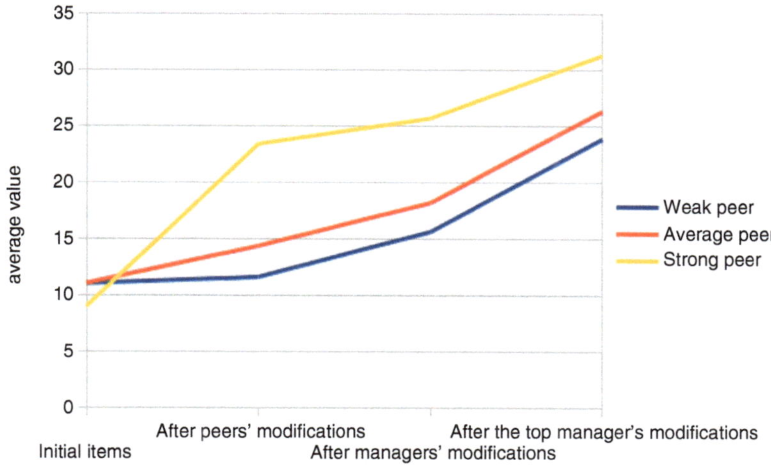

Fig. 7 Evolution in time of the average item value for Nemawashi

If the initiator rejects the new idea, the penalty of the project's item is incremented. Otherwise the items are exchanged. Moreover, if the initiator rejects all the items of a manager, there is 80 % chance that the manager won't sign (proposing new items again), and 10 % in the case he accepts at least one of the manager's ideas. If the manager is satisfied and doesn't challenge other items, he gives his agreement to the project and signs it. At the end, the top manager can reject the whole project with a probability of 5 %.

Figure 7 shows the evolution in time of the average item value for the Nemawashi process, according to the skill level of employees. Highly skilled employees achieve a good result even without management intervention. Employees with standard or below average skill levels achieve a less efficient Nemawashi, even with management intervention. Figure 8 shows the evolution of the number of items kept from the original proposal for the Nemawashi process. Initial items are rapidly withdrawn by skilled employees, whereas several iterations of management are required to improve the proposal set with less skilled employees.

6 Discussion

6.1 Modelling Lean Organisations as Complex Systems

Let's now consider again the properties of complex systems and see how the Lean organisation exhibits those properties, based on our simulations:

- *Emergence:* in the Nemawashi process, an individual can enrich his process gradually by spiralling up the layers of management and getting good advice

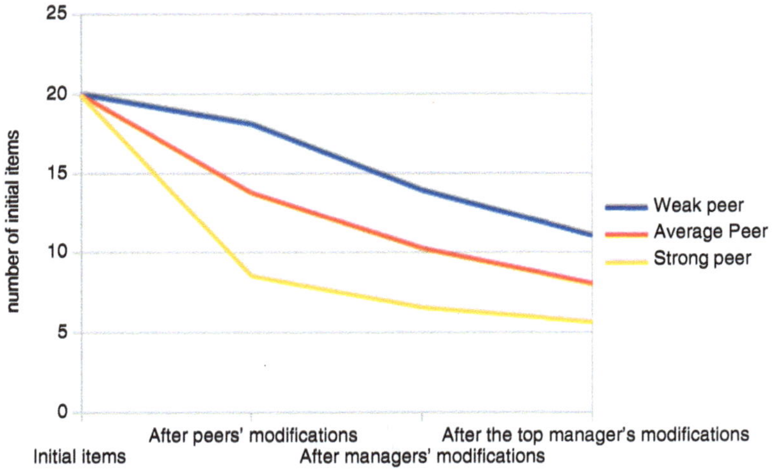

Fig. 8 Evolution of the number of items kept from the original Nemawashi proposal

integrated in his project. In Hoshin Kanri, even ideas sent top down by the Top Management are challenged and enriched by the whole organisation, creating a better set of objectives, with a better buy in from the whole organisation. The Lean organisation thus creates more emerging patterns than more traditional forms of organisation.

- *Co-evolution:* the modelled processes mandate systematic involvement of all levels of the organisation, with an impact that evolves based on their seniority and experience. This encourages the co-evolution of the agents.
- *Connectivity:* all entities are connected from the processes spiralling up and down and extending horizontally between peers.
- *Distributed Control:* the initiative is given to the agents (employees) to come up with ideas/projects, and defend them through the organisation. While comments are given at all levels to enrich the projects, the control is left with the initiator of the project, who is respected by the various layers of management.
- *Far-from-equilibrium:* the Lean organisation is never at equilibrium. The evolution of the world outside the organisation leads to Hoshin items proposed by Top Management and discussed within the organisation. Each project proposed can lead to major changes, which will be applied more effectively as different levels of management are involved in the Nemawashi process.
- *Non-linearity:* a big dependence on slightly different initial conditions is observed. When slightly different instructions are given at the beginning of the Hoshin process, the following top-down and bottom-up interactions may lead to a very different final Hoshin document, hence the importance to start the process with parameters that are carefully considered after deep reflection of the previous cycle (called Hansei—Reflection in Japanese).

- *State of paradox:* This issue expands beyond the current models. It is best illustrated by the paradox of Just in Time, mandating a continuous flow of logistics and production pulled by the customer and Jidoka ('automation with a human touch') which mandates to stop the same flow as soon as a defect occurs.

Building on this basis and on our work developing an ontology of Lean with Rules to operate on it, we shall further work on modelling several processes typical of the Lean Organisation, and gradually enrich the models with the experience derived from the practice of Lean in different contexts.

6.2 What We Learnt About Lean Organisations

Lean organisations aim at structuring an emergent system enabling the employees, or operators, to deeply impact the organisation strategy according to actual issues in the organisation. Figure 5 shows that, in optimal conditions, this is actually the case.

The model and simulation results isolate three critical success factors for emergent strategy definition through Hoshin Kanri and Nemawashi:

- the proficiency of employees, which enables them to make proposals as good as the management, thanks to a finer knowledge of the organisation, and reduces the communication overhead by getting quicker results; when the employees have weaker experience, the Hoshin Kanri process should better be led by the Top Management.
- the readiness of Management to accept emerging strategic proposals, to take advantage of this proficiency.
- the rigour in the execution of the emergent decision process, which can be realistic (the time pressure and increase of activity as a deadline is approaching is a natural tendency in all human structures) but needs to be successfully completed to fulfil the decision refinement process.

The model focuses on the quantified quality of the proposals. It does not take the alignment between employees and management into account, which is considered as a key success factor in many organisations [1, 9] and is thus an additional critical success factor.

7 Conclusions and Perspectives

A complexity-based model of a Lean organisation needs to entail three complementary views [18]: the conceptual view, the model view, and the empirical view. Based on the concepts defined in previous works [12], we introduce here a first model of two representative processes of the Lean enterprise: the Hoshin Kanri process, for collaborative choice of the management objectives, and the Nemawashi process, for consensus building. Qualitative behaviour of complex systems are observed in the Lean organisation: emergence, co-evolution, connectivity, distributed control, non-linearity, as well as state of paradox. Then Hoshin Kanri and Nemawashi are modelled using agents, on one hand, and probabilistic rules, on the other hand, to validate the field observations. Significant results are highlighted, in particular the emergent behaviour of the Hoshin Kanri process.

The proposed model enables to demonstrate a ground-breaking property of Lean organisations: through Nemawashi and Hoshin Kanri, emergent strategies can be defined effectively by proficient employees, with little to no added value brought by the management. Though companies like Toyota do not claim they can work without management, they confirm that managing a Lean organisation requires shifting from a traditional authoritative to a leader-servant governance style. A few smaller companies, however, like Favi[6] or Poult[7] for instance, in France, have already successfully implemented a 'freed' management style where radical emergence is leading to financial success for the company, confirming, in spite of the numerous practical obstacles, the credibility of the proposed model.

8 Scientific Validation

This paper has been unanimously validated in a collaborative review mode with the following reviewers:

- Paula Castaneda, National University of San Martin, Buenos Aires, Argentina.
- Ricardo Palma, University of Mendoza, Argentina.
- Imane Bouhaddou, Ecole Nationale Supérieure des Arts et Métiers (Meknés, Maroc).
- Ismaïla Diouf, Université Cheikh Anta Diop (Dakar, Sénégal).

References

1. Beer M, Voelpel SC, Leibold M, Tekie EB (2005) Strategic management as organizational learning: developing fit and alignment through a disciplined process. Long Range Plan 38(5):445–465
2. Bell SC (2012) Run grow transform: integrating business and lean IT. CRC Press, Boca Raton
3. Byrne A, Womack JP (2012) The lean turnaround: how business leaders use lean principles to create value and transform their company. BusinessPro collection. McGraw-Hill Education
4. Cassell C, Worley JM, Doolen TL (2006) The role of communication and management support in a lean manufacturing implementation. Manag Decis 44(2):228–245

[6]http://ayeba.fr/2014/01/linnovation-par-les-méthodes-de-management-le-cas-favi/.

[7]https://imatechnologies.wordpress.com/2013/06/15/poult-raconte-son-histoire-de-liberation/.

5. Collet P (2014) Mooc ose (optimisation stochastique evolutionnaire), Sept 2014
6. Csáji BC, Monostori L (2008) A complexity model for networks of collaborating enterprises. In: Proceedings of the 17th world congress of IFAC (International Federation of Automatic Control), COEX, Seoul
7. Größler A, Stotz M, Schieritz N (2003) A software interface between system dynamics and agent-based simulations: linking vensim® and repast®. In: Proceedings of the 21st system dynamics society international conference, Citeseer
8. Holland JH (2014) Complexity: a very short introduction. Very short introductions. Oxford University Press, Oxford
9. Joshi MP, Kathuria R, Porth SJ (2003) Alignment of strategic priorities and performance: an integration of operations and strategic management perspectives. J Oper Manag 21(3):353–369
10. Koller D, Friedman N (2009) Probabilistic graphical models: principles and techniques. MIT Press, Cambridge
11. Liu Z, Li H.-X. (2005) A probabilistic fuzzy logic system for modeling and control. IEEE Trans Fuzzy Syst 13(6):848–859
12. Masai P, Parrend P, Zanni-Merk C (2015) Towards a formal model of the lean enterprise. Procedia Comput Sci 60:226–235
13. Messaoudene Z, Gramdi J (2007) Proposition d'un cadre conceptuel et systémique des systèmes de production lean. In: 7è Congrès international de Génie Industriel, Trois Rivières, Québec
14. Miller JH, Page SE (2007) Complex adaptive systems: an introduction to computational models of social life. Princeton studies in complexity. Princeton University Press, Princeton
15. Mitleton-Kelly E (2003) Complex systems and evolutionary perspectives on organisations: the application of complexity theory to organisations. Elsevier Science Ltd, Amsterdam
16. Ries E (2011) The lean startup: how today's entrepreneurs use continuous innovation to create radically successful businesses. The lean startup: how today's entrepreneurs use continuous innovation to create radically successful businesses. Crown Business
17. Rother M (2010) Toyota Kata. MacGraw Hill, New York
18. Sanders L (2013) Problématiques géographiques, systémes complexes et modélisation. In: Rencontres nationales systémes complexes (RNSC)
19. Sugimori Y, Kusunoki K, Cho F, Uchikawa S (1977) Toyota production system and Kanban system materialization of just-in-time and respect-for-human system. Int J Prod Res 15(6):553–564
20. Vitali S, Glattfelder JB, Battiston S (2011) The network of global corporate control. PloS One 6(10):e25995
21. Womack JP, Jones DT, Roos D (1991) The machine that changed the World: the story of lean production. The MIT international motor vehicle program. HarperCollins, New York

An Artificial Immune Ecosystem Model for Hybrid Cloud Supervision

Fabio Guigou, Pierre Parrend, and Pierre Collet

1 Introduction

In the recent years, Artificial Immune Systems (AIS) have received some attention in applications related to data mining and clustering, such as document classification or intrusion detection. Their distributed and dynamic nature as well as their ability to build up experience makes them highly suitable for anomaly detection in large datasets. However, their computational cost still limits their industrial adoption.

While the AIS model is no newcomer in the field of bio-inspired computing, it has not yet received nearly as much attention as other models such as evolutionary algorithms, neural networks or swarm computing [5]. Its main drawback is the computational power required to perform classification and outlier detection [9, 24], typically by generating the antibodies and matching them with the antigens [7]. However, the increasing power of modern hardware has made them popular in some

F. Guigou (✉)
ICube Laboratory, Université de Strasbourg, Strasbourg, France

IPLine, Caluire-et-Cuire, France

Complex System Digital Campus (UNESCO Unitwin), Paris, France
e-mail: fguigou@ipline.fr

P. Parrend
ICube Laboratory, Université de Strasbourg, Strasbourg, France

ECAM Strasbourg-Europe, Schiltigheim, France

Complex System Digital Campus (UNESCO Unitwin), Paris, France
e-mail: pierre.parrend@ecam-strasbourg.eu

P. Collet
ICube Laboratory, Université de Strasbourg, Strasbourg, France

Complex System Digital Campus (UNESCO Unitwin), Paris, France
e-mail: pierre.collet@unistra.fr

P. Bourgine et al. (eds.), *First Complex Systems Digital Campus World E-Conference 2015*, Springer Proceedings in Complexity, DOI 10.1007/978-3-319-45901-1_6

niche applications such as intrusion detection [2] where their ability to adapt to a changing system, react more intensely to known patterns and work in distributed architectures was not found in other models.

In this paper, we present a new class of AIS-like bio-inspired frameworks, which we call Artificial Immune Ecosystems (AIE). We made up this term to refer to frameworks that retain the architectural and some conceptual elements of AIS, while mixing the strict biological model with other statistical tools. The works on software architecture related to computer immunology can be traced back to the late 1990s [7, 8]. We then propose Service Level Monitoring of complex IT networks as a possible field of application for such a framework and demonstrate that a monitoring engine based on AIE, by memorizing earlier failures, can raise an alarm much faster than the traditional suites used in cloud supervision when a key metric (disk usage, CPU load, free memory, packet loss, etc.) drifts away from its baseline behaviour. The investigation of possible immune responses to a detected anomaly is not in the scope of this paper.

The rest of the paper is organized as follows. In Sect. 2, the state of the art of monitoring in the context of hybrid cloud, as well as the application of AIS as a framework for engineering robust IT ecosystems, is discussed. Section 3 presents our AIE model. Section 4 gives the experimental setup and results. Section 5 evaluates and discusses these results. Section 6 concludes this work.

2 Background

2.1 Monitoring

In the recent years, virtual infrastructures provided by cloud operators have begun to replace the traditional physical servers and network devices. Their success is mainly due to their scalability, cost-efficiency and simplicity when compared to datacentre hosting. Infrastructure-as-a-Service (IaaS) is the convergence of cloud computing, cloud storage and cloud networking: the end user is provided with a complete virtualized ecosystem [17, 27]. Service providers no longer advertise network-only or VM-only platforms: they are now pushed towards an economic model in which they will have to operate core network infrastructure, system virtualization, network storage, WAN and Internet links and may even provide assistance in the LAN ecosystem. The latest model in IaaS is the hybrid cloud model: instead of keeping all his hardware on premises (private cloud) or fully externalizing it and accessing it through a public IP address (public cloud), the customer can mix those two paradigms, keep part of his computing local and reach delocalized resources through private or public addresses [23].

Service Level Monitoring (SLM) is a critical part of IT infrastructure management and a crucial point for modern cloud service providers [16, 20]. SLM is a prerequisite for enabling Service Level Agreements (SLA) [6], i.e. the contractual level of infrastructure quality: availability, throughput, CPU, memory, etc. SLA define the risk level a customer is ready to accept. Current SLM solutions (Nagios, ntop, Sensu, etc.) have two important shortcomings. On the one hand, they are limited by their essentially centralized nature: while the probes themselves are distributed across the network, they are still managed by a central server storing the data and raising alarms in case of anomaly. Since the data is centralized in a single point, the IO and CPU load are very high, which requires powerful dedicated servers. On the other hand, they are stateless and only react to values crossing a given threshold or custom scripts raising an alarm. Though these systems keep the measured data for later analysis, it is not used as a knowledge base to predict future failures. They have no ability to learn from earlier failures, which makes root cause analysis and recovery complex and lengthy in a time-critical environment.

To overcome these limitations, shorten recovery times and allow for SLM scalability, we use a new AIE model to propose a distributed, decentralized architecture that displays learning and knowledge sharing abilities. While other approaches [21] are very efficient at analyzing a signal or time series and detecting anomalies with respect to its history, they do not address the problem of learning and recognizing the failure modes. AIS have also been used in this context [4], but with the traditional antigen/antibody model which requires the input to be converted to a bit string and generates a lot of detectors, and therefore computational overhead.

2.2 *Artificial Immune Systems*

The field of AIS, also known as computer immunology [8], emerged in the end of the 1980s and in the beginning of the 1990s on the basis of the theory of idiotypic networks, which highlights the network-like connections between lymphocytes responsible for protecting the organisms against external threats [3]. First artificial models [7] draw on historical self vs. non-self selection in the natural immune system as conceptualized by Ehrlich [22]. The second generation of models [11] lowers the detection threshold by taking co-stimulation, i.e. the simultaneous presence of several signals, into account [14]. The third generation of models [1] still refines the analysis by taking the impact of aggression into account: this is the danger model [15].

The AIS domain has been a prolific field for theoretical endeavours [10], and knows a broad range of applications [2]. However, its main focus so far is the application for data analysis and classification [26] which proves to provide only limited performance [9, 24].

However, the mainstream AIS models are challenged by systemic models of natural immune systems, which characterizes core abilities of immunity such as memory, maturation or ageing [25], as emergent properties of the network built by lymphocytes. The immune system would be not only a closed auto-regulated system for detecting aggressions and degradations, but also a complete ecosystem built from the very interactions between its components and with its environments. This shift in the analysis of immune systems provides huge perspectives for exploiting the adaptability and distributed properties of immune systems for engineering ecosystems which are able to detect both known and de novo performance or security anomalies, and to perform suitable reactions in the context of evolving IT infrastructures.

3 Yet Another AIS Model

One of the main drawbacks of AIS is the computational cost associated with the generation and maturation of the antibodies and the number of comparisons that must be performed for each incoming data point. While modern hardware is capable of handling such loads in experimental setups, we do not expect the amount of data acquired by a typical monitoring server (hundreds of probes, each yielding one to a few dozen measures per minute) to be analysed in real time on typical low-end servers. To overcome this limitation, we propose an alternative model in which only the previously recorded anomalies (failure modes) are recorded and matched against. Therefore only a handful of patterns are matched against and no random generation and maturation has to happen.

We combine the possibility to analyse new points with respect to the history of the system, the known failure modes and a predefined threshold by defining three different analysis modes which will be detailed later: traditional mode, lightweight mode and full mode. We explicit the properties of these modes in Sect. 3.2.

3.1 Core Features

While taking some distance from the bio-inspired model of the immune system, in particular to avoid the aforementioned performance drawbacks, the immune ecosystem we propose retains most of the AIS features, such as distribution, dynamic adaptation to a changing self and increased reaction after a secondary exposure, and puts aside the antigen/antibody model. It is based on the notions of innate immunity (failure modes already recorded by other probes), acquired immunity (failure detection by an alarm threshold, then pushed to other probes) and natural history (the past behaviour of the supervised system). Since the data related to earlier failures of the monitored devices is recorded, it would even be possible to transfer them from one instance of the AIE to another, which would represent a very basic vaccination process. However, the concept of antibodies and antigens, usually modelled in AIS by binary strings, is replaced by statistical analysis: instead, new data is matched against recorded signatures and a baseline, and the matching algorithm itself can be adapted to the problem.

3.1.1 Distribution

Distribution is an often overlooked characteristic of immune systems [12] that takes all its importance in the context of infrastructure monitoring. In biological immune systems, it allows for an immediate local response on the infection site. It is materialized by the omnipresence of antibody cells in the lymphatic system. In AIS, it enables local data storage and processing, hereby saving bandwidth and gaining overall efficiency. At the same time, a synchronized knowledge base must be available to all nodes at any time. It lists all the potential threats to the system.

3.1.2 Input and Output

The input to our system consists of real-valued time series. The immune ecosystem is triggered every time a new data point is added to the series (e.g. every minute in the context of a supervision system) or when a given number of points have been received (which may help with performance issues). It outputs alarm signals and updates to the global knowledge base.

Figure 1 shows the data fed to the system over 6 weeks. This data shows the typical measures of the number of sessions on a firewall. The first 5 weeks are normal, with activity peaks on working days; week 6 shows an abnormally high number of sessions, which should be considered a failure.

The system also reacts to user input to classify the alarms and warnings in the knowledge base. This allows the operator to ignore mild transient threshold crossings as well as diagnose some abnormal behaviour as errors even though no threshold is crossed.

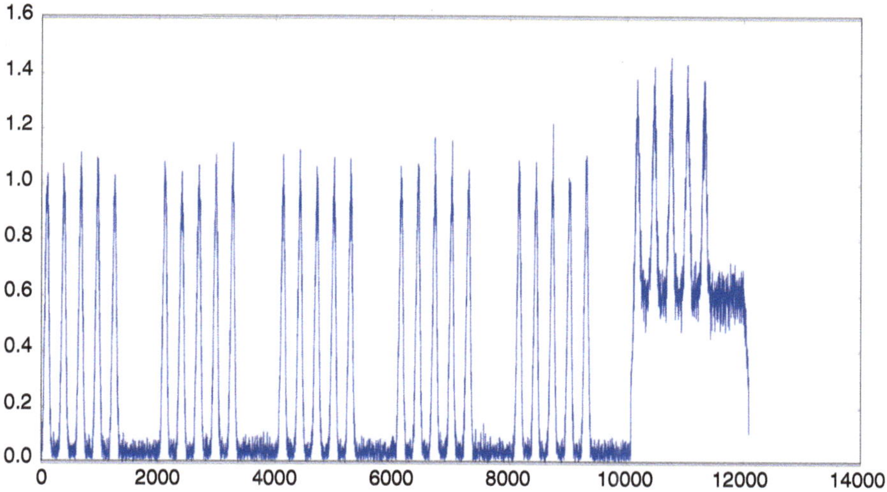

Fig. 1 Typical raw input data over 6 weeks

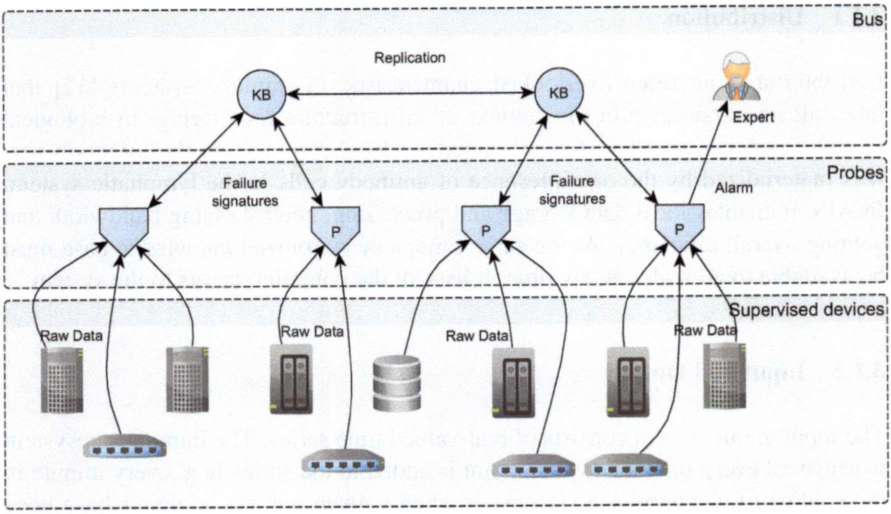

Fig. 2 Target architecture

3.2 Architecture

We propose a fully distributed architecture in which a messaging bus is used to communicate between nodes and interact with the human expert. The probes capture data from the supervised systems (servers, networking devices, links, etc.) and keep it locally for history and analysis. This creates a virtual database scattered across the nodes. At the same time, a shared knowledge base is cooperatively maintained by the probes. It keeps the signatures of all the failure events previously recorded. This knowledge base may be replicated for redundancy, up to the point where each node keeps a complete copy of it and can therefore operate even in case no other node is reachable. The messaging bus allows the probes to raise alarm signals and access and update the shared knowledge base. In the context of a distributed supervision system, it could also carry reconfiguration and deployment instructions for the probes.

Figure 2 shows the typical dataflow between the probes (P), devices, knowledge base (KB) and the expert. In our implementation, the probes, knowledge base and bus are run on the same servers, making the nodes homogeneous.

3.3 Operations

3.3.1 Overview

Our AIE has three operating modes, each providing a finer analysis: traditional mode, lightweight mode and full mode. The traditional mode is simply the usual operation of a regular SLM system, *á la* Nagios: each new data point is compared to a fixed threshold and, if that threshold is crossed, an alarm is raised. The lightweight mode adds a second check following the comparison: a window containing the k last data points is matched against the signatures of previous failures. Whenever a positive match is found, an alarm is raised, though with a smaller priority. It allows the system to detect the first signs of a potential failure before it actually happens. The full mode adds a third level of analysis by matching that same window against a long history (typically the last month). If no close enough match is found, then the window is considered an anomaly and a warning is raised. The human expert can choose to classify that event as a failure indication or ignore it. While this mode is by far the most computationally expensive, it is also the most powerful, since it may detect unknown failure modes before the incident takes place.

3.3.2 Details

We first need to introduce a set of parameters for our AIE. Let n be the baseline length of the probe, i.e. the number of points kept locally as history, k the analysis window size, with $k << n$, m the signature length, a the sensitivity factor of the full mode and b the sensitivity factor of the lightweight mode.

In a typical monitoring setup, with one incoming point per minute, we may want to keep a baseline history of 1 month, as patterns often last 1 week (e.g. peaks on working days and flatlines during the week-end). A sound window size for anomaly detection is about 2 h, which yields values of $n = 43{,}200, k = 120$. $m - k$ is the time during which the early symptoms of an incident can be detected before the threshold is crossed, for which 1 h is a reasonable choice, yielding $m = 180$. These values would be typical in a production but induce too many computations for experimental settings, therefore we worked with smaller values ($n = 8000, k = 24$ and $m = 36$).

In full mode, a history consisting of the last n data points is kept as the system baseline. Whenever a new point (or a given number of new points if computational cost is an issue) is received, a short window w of size k containing the last k samples is matched against the whole history w. A simple distance metric $d_i = \frac{1}{k}\sum_{j=0}^{k-1} |h_{i+j} - w_j|$ is computed at each position. We define $d_0 = \min(d)$ and $i_0 = \operatorname{argmin}(d)$, so that $d_{i_0} = d_0$. This allows the system to detect behaviour deviating from the baseline. If we define μ and σ as the mean and standard deviation of d_0 over the set of all newly inserted points, we can define a threshold $\tau = \mu + a\sigma$ so that when $d_0 > \tau$ we raise an alarm indicating an abnormal behaviour. In experimental setups, we found $1 < a < 2$ to yield acceptable results; the results we present were obtained with $a = 2$. Figure 3 shows the influence of the choice of a, assuming a normal distribution of the measured distances. This configuration defines the notions of self and non-self for the AIE by detecting the relative position of a measure in the probability distribution.

The lightweight mode avoids the computational cost of matching history. In this mode, the new data is matched against the known failure signatures (i.e. windows of m points that have been followed by a failure, with $m > k$) instead of the system

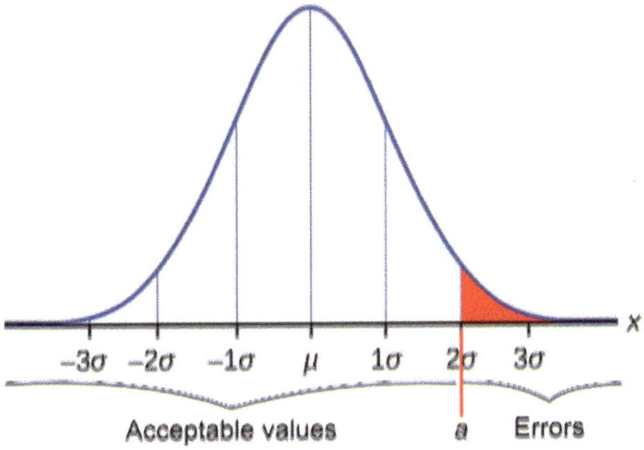

Fig. 3 Choice of a over a normal distribution

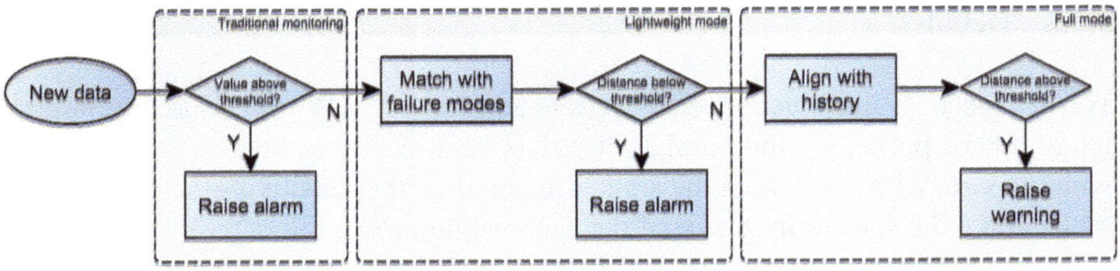

Fig. 4 Operating mode

baseline. If a positive match is found ($d_0 < b\delta_s$, where $\delta_s = \frac{\max(s)-\min(s)}{\max(s)}$ models the variability of the recorded signature), an alarm is raised. We used a sensitivity factor of $b = 0.1$ after a trial-and-error search.

Lastly, when an alarm threshold is crossed, an alarm is raised as in any traditional monitoring system and the m points preceding the failure are added to the knowledge base. Note that in full mode, whenever an anomaly is detected, its signature is recorded and the expert can decide to tag it as a failure, in which case it is added to the signatures matched in lightweight mode.

Figure 4 shows the workflow for the execution in full mode. Exiting after the first box is equivalent to traditional threshold-based alarming, after the second means running in lightweight mode.

4 First Experiments

4.1 Setup

We built an experimental setup using a single probe which was fed with artificial data inspired from real supervision time series. A Redis database was setup to

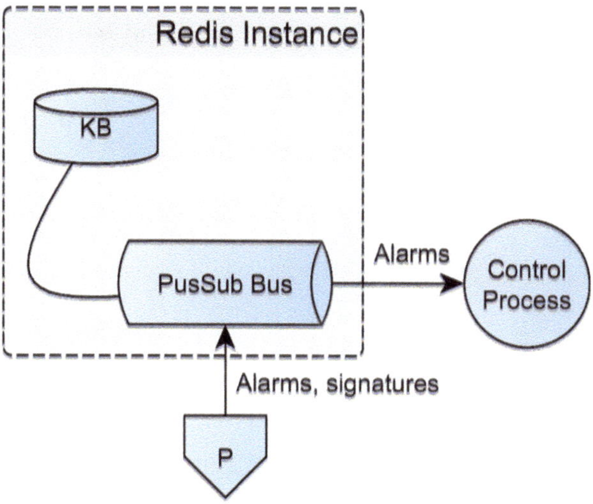

Fig. 5 Experimental setup

provide the communication channel and knowledge base. A control process listened to the PubSub bus to collect the alarm signals. We estimate the decoupling of these components sufficient to validate the architecture even with a single node, using multiple runs of one probe instead of parallel runs of a set of probes (the knowledge base being persisted between the runs). Figure 5 shows the data flow in the experimental setup. The data itself was generated using an alternation of bell curves and runs of zeros to which we added multiplicative and additive Gaussian noise, leading to the typical shape of CPU load plots.

4.2 Results

We used the data presented in Sect. 3. The system was run multiple times to train the lightweight mode. At the beginning of the failure event, Fig. 6 shows where the three stages send alarms. The failure signature is uploaded to the knowledge base, which allows the lightweight mode to react faster than the full mode after a primary exposure. The point corresponding to the lightweight mode was obtained after the first run, i.e. after the capture of a first incident window.

The experimental results show that the full mode, on an unknown failure, reacted 4 h and half before the threshold was crossed, and the lightweight mode 42 min before on a known failure. The full mode was able to evaluate up to five points per second on a single-core process running on an Intel i5 CPU; the lightweight mode processed hundreds of points per second per signature. We therefore conclude that the lightweight mode may run on monitoring servers, but that the full mode would require optimizations and could only be used on servers monitoring only a handful of devices.

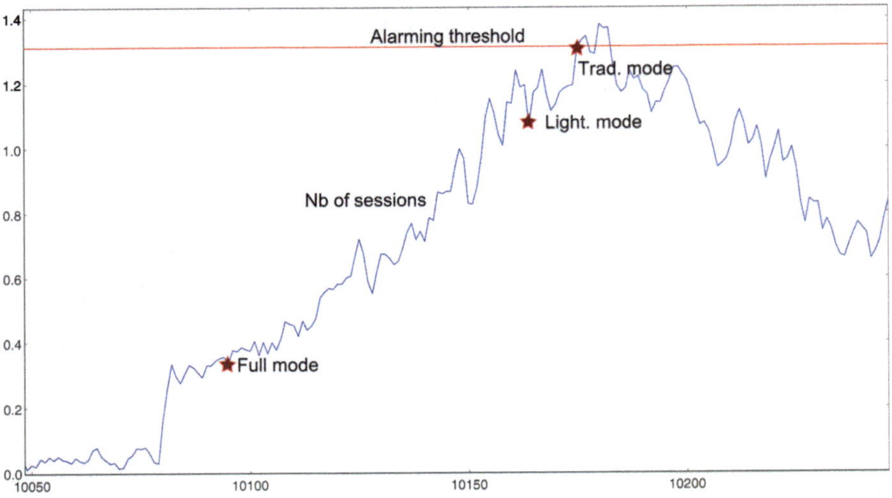

Fig. 6 Zoom on failure event, with the activation points and captured window

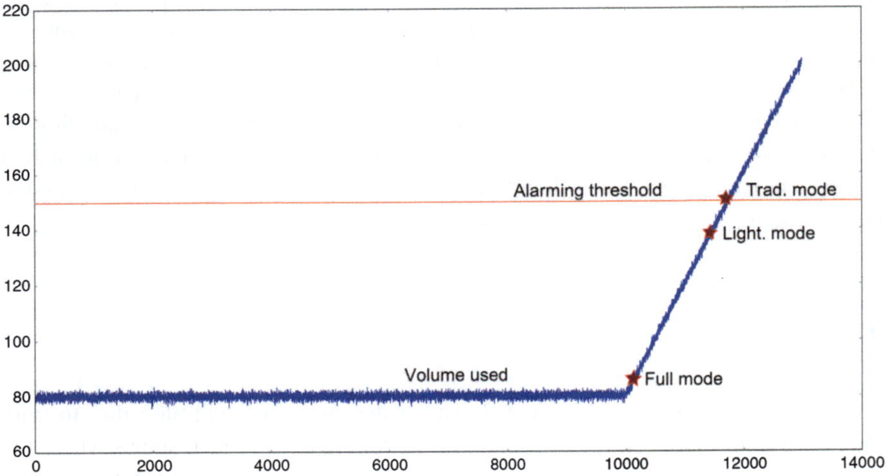

Fig. 7 Reaction to a complete change of behaviour

Figure 7 shows the reaction to a complete change of behaviour, which can be observed when monitoring the disk space used on a server. The space use begins to ramp up when a process begins to write a lot of logs. The full mode reacts at the very beginning of the incident, 3 days before the lightweight mode trained with the same failure.

5 Discussion

5.1 Core Properties of Artificial Immune Ecosystems

The AIE we propose enables identifying weak signals through the multiplication of lymphocytes (abstracted as an event counter) and through the detection of signal dynamics far below traditional warning thresholds. It embeds the core structural properties of AIS: distribution, input and output management, memory. Distribution is handled by the hierarchical structure of the AIE: local identification is performed through the lightweight mode, which performance overhead is compatible with the monitoring of typical servers; centralized detection is performed through the full mode which is more resource demanding and therefore limited to sensitive monitoring data and specific network nodes. Related to the input, only time series are supported so far, which already provides a great coverage for monitoring issues. Semantic data could be required for finer analysis in a second time. Related to the output, the ability of the current version of our AIE is to emit early warning to enable quick reaction of a human operator. No automated or semi-automated reaction is supported so far. Memory is implemented through storage of known failure modes in a shared knowledge base.

AIE abstract and formalize the structural features of immune systems. However, existing benchmarks [9, 24] highlight the fact that data classification and outlier detection algorithms based on AIS experience significant performance drawbacks with regard to statistical approaches like naive Bayesians. We are strongly convinced that a suitable abstraction of AIS, similarly to abstractions that exist in the domain of evolutionary algorithms, could provide a performant and adaptive solution for addressing evolving data as it appear in dynamic IT ecosystems.

5.2 Challenges

We have shown the ability of our specific AIE to detect unknown abnormal situations and known potential failures in scenarios where an error can be modelled by a threshold being crossed or an event not matching the past lifetime of the system. However, it is unable to raise alarms on more subtle events requiring correlation with other data sources or multiscale history matching. For instance, a high CPU load on a Sunday caused by an attack, if its shape is close enough to the normal load on a week day, would not give rise to any warning.

Moreover, we still need to experiment on our architecture in a real production environment where performance issues will arise and the actual data may display much more variability than our test data. We also need to study its behaviour on the long term, i.e. the growth rate of the knowledge base and the overall performance impact. Depending on these tests, we may introduce scaling functions in the signatures and a maximal lifetime for the failure modes. Our early tests have only validated the basic concepts; a mature software solution that could be deployed and used with any level of confidence by a cloud operator is yet to be designed and implemented.

Lastly, the only possible reaction of the system is to send a notification to a human expert, as with most SLM solutions. A more efficient way to react would be to activate more probes when a potential failure is detected [13, 18, 19], so as to collect more meaningful data, decide whether or not the alarm should be raised and help the expert with his analysis. Some control over supervised devices could also be handed to the AIE to allow some dynamic reconfiguration and prevent failures.

6 Conclusions and Perspectives

We have proposed a new computational and architectural framework inspired by the AIS and using statistical analysis where AIS tried to model antibodies and antigens: the AIE. To evaluate our proposal, we have implemented such an AIE for cloud infrastructure monitoring. By using multiple statistical tools, we made it possible to trade anticipation for performance and run only parts of the system where a large number of probes would have to be run in traditional AIS.

We still need to deploy this prototype in a real production environment to assess its actual efficiency and precision when confronted to real-world events. It will be up to the end users to determine whether the configuration (sensitivity factors) is simple enough to be used on an industrial scale and if the errors (false alarms) in the first deployments are acceptable.

To conclude, we expect this new framework to open new perspectives to the field of artificial immunity by allowing experts to tune the comparison operator to the problem instead of tuning the problem to the AIS antibody/antigen model. We have already proved its interest in terms of architecture and modularity by building a distributed monitoring system with learning ability and acceptable performance. We plan to apply this framework and architectural implementation to other monitoring problems involving the distributed analysis of time series.

7 Scientific Validation

This paper has been unanimously validated in a collaborative review mode with the following reviewers:

- Masanori Sugisaka (Alife Robotics: Japan)
- Juan-Julio Merelo (University Granada: Spain)
- Claudia Eckert (Open University: UK).

Acknowledgements The work presented here has been funded by IPLine SAS, by the French ANRT in the frame of CIFRE contract 2015/0079 and by the French Banque Publique d'Investissement (BPI) under program FUI-AAP-19 in the frame of the HuMa project.

References

1. Aickelin U, Cayzer S (2008) The danger theory and its application to artificial immune systems. arXiv preprint arXiv:0801.3549
2. Aickelin U, Dasgupta D, Gu F (2014) Artificial immune systems. In: Search methodologies. Springer, Berlin, pp 187–211
3. Calenbuhr V, Bersini H, Stewart J, Varela FJ (1995) Natural tolerance in a simple immune network. J Theor Biol 177(3):199–213
4. Dasgupta D, Forrest S (1995) Novelty detection in time series data using ideas from immunology. In: Proceedings of the international conference on intelligent systems
5. Dasgupta D, Majumdar N, Nino F (2007) Artificial immune systems: a bibliography. Computer Science Division, University of Memphis, Technical report
6. Dillon T, Wu C, Chang E (2010) Cloud computing: issues and challenges. In: 2010 24th IEEE international conference on advanced information networking and applications (AINA). IEEE, New York, pp 27–33
7. Forrest S, Perelson AS, Allen L, Cherukuri R (1994) Self-nonself discrimination in a computer. In: IEEE symposium on security and privacy. Oakland, pp 202–212
8. Forrest S, Hofmeyr SA, Somayaji A (1997) Computer immunology. Commun ACM 40(10): 88–96
9. Haidar AA (2011) An adaptive document classifier inspired by t-cell cross-regulation in the immune system. Ph.D. thesis, Citeseer
10. Haidar AA, Six A, Ganascia J-G, Thomas-Vaslin V (2013) The artificial immune systems domain: identifying progress and main contributors using publication and co-authorship analyses. Adv Artif Life, ECAL 12:1206–1217
11. Hofmeyr SA, Forrest S (1999) An immunological model of distributed detection and its application to computer security. The University of New Mexico
12. Hosseinpour F, Amoli PV, Farahnakian F, Plosila J, Hämäläinen T (2014) Artificial immune system based intrusion detection: innate immunity using an unsupervised learning approach. Int J Digit Content Technol Appl 8(5):1
13. Jeswani D, Korde N, Patil D, Natu M, Augustine J (2010) Probe station selection algorithms for fault management in computer networks. In: 2010 second international conference on communication systems and networks (COMSNETS), pp 1–9, Jan 2010
14. Lafferty KJ, AJ Cunningham (1975) A new analysis of allogeneic interactions. Immunol Cell Biol 53(1):27–42
15. Matzinger P (1994) Tolerance, danger, and the extended family. Annu Rev Immunol 12(1): 991–1045
16. Molina-Jimenez C, Shrivastava S, Crowcroft J, Gevros P (2004) On the monitoring of contractual service level agreements. In: First IEEE international workshop on electronic contracting, 2004. Proceedings, July 2004, pp 1–8
17. Moreno-Vozmediano R, Montero RS, Llorente IM (2012) IaaS cloud architecture: from virtualized datacenters to federated cloud infrastructures. Computer 45(12):65–72
18. Natu M, Sethi AS (2006) Active probing approach for fault localization in computer networks. In: 2006 4th IEEE/IFIP Workshop on end-to-end monitoring techniques and services, April 2006, pp 25–33
19. Natu M, Sethi AS (2008) Application of adaptive probing for fault diagnosis in computer networks. In: Network operations and management symposium, April 2008. NOMS 2008. IEEE, New York, pp 1055–1060

20. Sahai A, Machiraju V, Sayal M, van Moorsel A, Casati F (2002) Automated SLA monitoring for web services. In: Feridun M, Kropf P, Babin G (eds) Management technologies for e-commerce and e-business applications. Lecture notes in computer science, vol 2506. Springer Berlin, pp 28–41
21. Salvador S, Chan P (2005) Learning states and rules for detecting anomalies in time series. Appl Intell 23(3):241–255
22. Silverstein AM (2005) Paul Ehrlich, archives and the history of immunology. Nat Immunol 6(7):639–639
23. Sotomayor B, Montero RS, Llorente IM, Foster I (2009) Virtual infrastructure management in private and hybrid clouds. IEEE Internet Comput 13(5):14–22
24. Stibor T, Mohr P, Timmis J, Eckert C (2005) Is negative selection appropriate for anomaly detection? In: Proceedings of the 7th annual conference on Genetic and evolutionary computation. ACM, New York, pp 321–328
25. Thomas-Vaslin V (2014) A complex immunological idiotypic network for maintenance of tolerance. Front Immunol 5:369
26. Timmis J, Neal M, Hunt J (2000) An artificial immune system for data analysis. Biosystems 55(1):143–150
27. Vaquero LM, Rodero-Merino L, Morán D (2011) Locking the sky: a survey on IaaS cloud security. Computing 91(1):93–118

Analysis of a Planetary Scale Scientific Collaboration Dataset Reveals Novel Patterns

Soumya Banerjee

1 Introduction

Scientific collaboration networks are an important component of scientific output and contribute significantly to expanding our knowledge and contribute to gross domestic product. Collaboration networks have been modelled before [1, 2]. Here we examine a novel dataset for the Mendeley collaboration network [3]: We find novel clusters of countries with different characteristics of collaboration. We propose a dynamical model that explains these characteristics. We also find interesting patterns in the behaviour of the countries that may reflect foreign policies and contemporary geopolitics.

2 Collaboration Data

We use a planetary scale collaboration data freely available for download from Mendeley Labs [3]. Analysis of this data with k-means clustering reveals the following patterns:

a. We plot the percentage of external connections that each country has vs. the distinct number of countries each country is connected with. The plot shown below (Fig. 1) has a distinctive shape. Each point on the plot is a distinct country.

S. Banerjee (✉)
Harvard University, Boston, MA, USA

Ronin Institute, Montclair, NJ, USA

Complex Biological Systems Alliance, North Andover, MA, USA

University of Oxford, Mathematical Institute, Oxford, UK
e-mail: neel.soumya@gmail.com

P. Bourgine et al. (eds.), *First Complex Systems Digital Campus World E-Conference 2015*, Springer Proceedings in Complexity, DOI 10.1007/978-3-319-45901-1_7

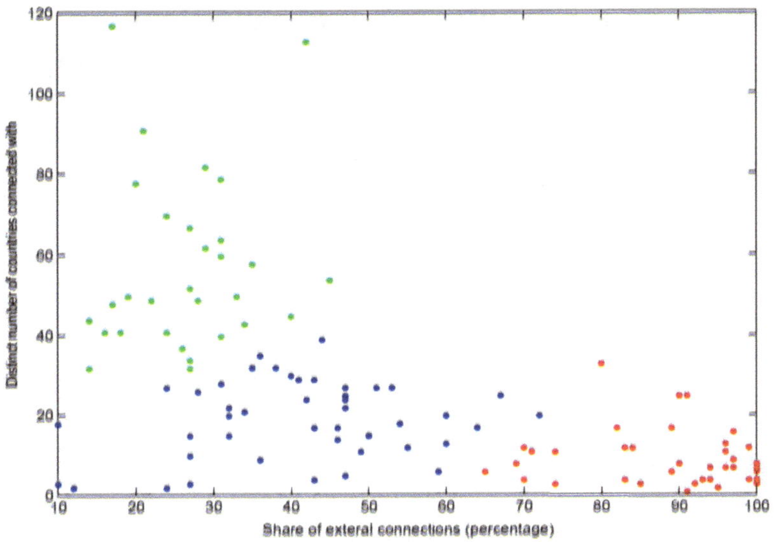

Fig. 1 Plot of the percentage of external connections that each country has vs. the distinct number of countries each country is connected with. Clustering is done with k-means and shows three distinct clusters

b. There are three clusters corresponding to three quadrants; it is interesting to note that there is no country in the fourth quadrant (top right: the UK is close to the fourth quadrant).
c. There are no countries with high share and high distinct number of countries connected with. Countries with a larger share of foreign connections and less distinct countries connected with are less developed nations.
d. The country with the highest number of distinct number of countries connected with and largest share of external connections is the UK (top rightmost in the plot).
e. The country with the lowest number of distinct number of countries connected with and lowest share of external connections is Algeria (bottom leftmost in the plot).

In order to explain these patterns, we propose a dynamical model for how the links are actually formed; for example, poor countries might preferentially seek out collaborations with richer countries or richer countries might help disadvantaged nations (e.g., Liberia has 100 % external connections). We simulate the time evolution of connections. Our dynamical model correctly predicts the broad observations of scientific connections of countries.

3 Mathematical Model

Our dynamical model has two compartments: developing countries (x) and developed countries (y). Developing countries grow their scientific expertise and self-connections at a rate proportional to $\alpha * x$ (self-growth) and also by interacting with developed nations at a rate proportional to $\beta * x * y$. Developed nations (y) are assumed to have reached equilibrium of growth. The model, represented as differential equations is shown below:

$$\frac{dx}{dt} = \alpha x + \beta xy$$
$$\frac{dy}{dt} = 0$$

Shown below (Fig. 2) is a simulation of how scientific connections of developing countries evolve over time for a chosen set of parameters.

For the same set of parameters we also simulated how the percentage of foreign connections should vary (over time) with the number of self-connections (Fig. 3). This plot accurately recapitulates the patterns in the empirical data (Fig. 1). It shows there are only three quadrants. It is consistent with and effectively recapitulates empirical data.

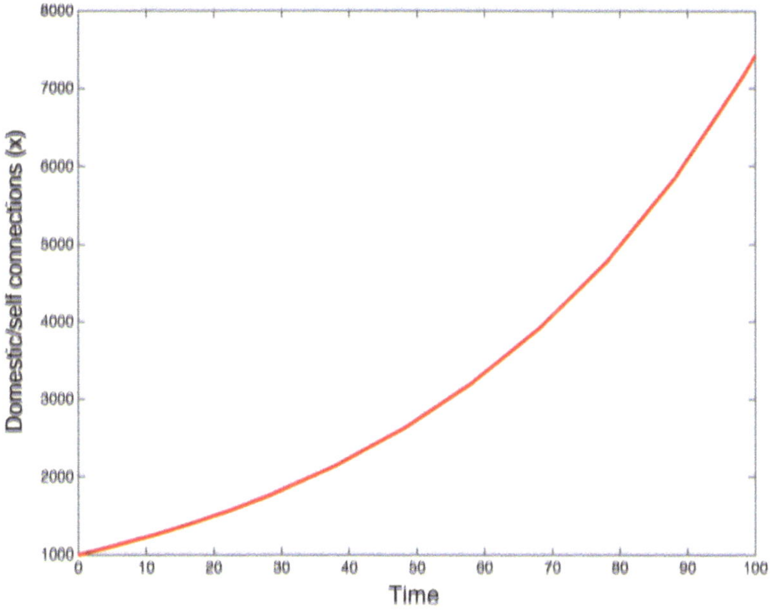

Fig. 2 Simulated plot of how scientific connections of developing countries evolve over time

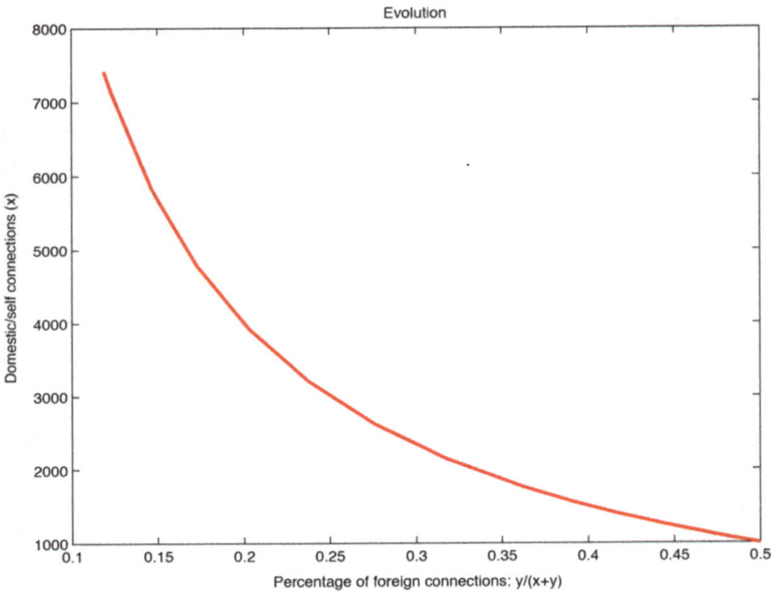

Fig. 3 Simulated plot of self-connections vs. percentage of foreign connections

4 Data and Statistical Methodology

All data is freely available from Mendeley Labs [3]. The table of the complete collaboration network is available for download [3]. The data was curated by Mendeley and a connection is defined as co-membership of two researchers in a Mendeley group. Further details of data pre-processing are available online [3].

All statistical analysis was conducted in MATLAB.

5 Discussion

We find novel patterns in the planetary scale collaboration dataset (Fig. 1). The table of the complete collaboration network is available for download [3]. We note some interesting patterns in the data:

1. The UK has a very high percentage of foreign connections; this maybe because of its colonial past.
2. Less developed or poorer countries usually have a very high percentage of foreign connections.
3. Iran has a large number of foreign connections; it is interesting that this is so despite foreign sanctions imposed against it.

4. The largest or richest countries have more distinct foreign connections. This may suggest that the rest of the world wants to collaborate and form connections with others; this could also be driven by interest in issues relevant to poorer countries like tropical diseases, socio-economic research into poverty, and archeological research in countries in Africa.
5. India has more foreign connections than China.
6. The USA has a lower percentage of foreign connections but in absolute numbers it has the highest number of connections.
7. The highest percentage of foreign connections is usually occupied by very poor countries (presumably they are trying to build capability in science and technology by collaboration), e.g., Liberia has 100 % foreign connections.
8. Among the developed nations, the UK has the largest percentage of foreign connections (probably reflecting its colonial heritage).
9. Some countries like El Salvador have a low percentage of foreign connections (this could be as a result of the fact that the country was embroiled in civil war for a long time). The policy implications are that it shows a need for targeted relief at starting active science and research programs in these nations.
10. South Korea, Japan, Taiwan, Japan, and Germany have a low percentage of foreign connections (they have a large number of absolute connections and it is possible that they have invested very heavily in their own science programs). This is interesting since these are highly developed countries.
11. Liberia has 100 % external connections: this suggests that more efforts need to be taken to develop its own scientific infrastructure.
12. Cuba has very few connections (this is again possibly due to sanctions imposed against it).

In summary we propose a dynamical model that explains current trends of clustering in scientific collaboration network. Taken together our data and models show signatures of past and contemporary foreign policies and geopolitics on the scientific collaboration landscape.

Our analysis also points to the power and generality of dynamical systems in modelling diverse sociological systems [4, 5].

Our model and analysis gives insights and guidelines into how scientific development of developing countries can be guided. This is intimately related to fostering economic development of impoverished nations and creating a richer and more prosperous society.

6 Scientific Validation

This paper has been unanimously validated in a collaborative review mode with the following reviewers:

1. Maria Jose Jorente, Universidade Estadual Paulista
2. Mariana Cantisani, Federal University of Paraíba.

References

1. Newman MEJ (2001) The structure of scientific collaboration networks. Proc Natl Acad Sci 98:404–409
2. Pan RK, Kaski K, Fortunato S (2012) World citation and collaboration networks: uncovering the role of geography in science. Scientific reports
3. Mendeley Labs Collaboration Map. http://labs.mendeley.com/collab-map/. Accessed June 2015
4. Banerjee S, Hentenryck PV, Cebrian M (2015) Competitive dynamics between criminals and law enforcement explains the super-linear scaling of crime in cities. Palgrave Commun 1
5. Banerjee S (2016) A biologically inspired model of distributed online communication supporting efficient search and diffusion of innovation. Interdisciplinary Descr Complex Syst 14:10–22

A Formal Model to Compute Uncertain Continuous Data

Jérôme Dantan, Yann Pollet, and Salima Taibi

1 Introduction

Current researches in the domain of ICT describe and extend existing formalisms to design systems that have to manage more and more data, because of the increasing number of sensors and means of storage. Many actors of scientific domains have to cope with data that may be uncertain, imprecise, or incomplete, to assist humans in their decisions. For this, they have to merge data from many data sources. In the last 40 years many well-known mathematical approaches to model imperfect data have been applied, such as probability based calculus. But increasing amounts of data have to be processed so that there is not enough time for data cleaning step. Decisions of experts from various fields are based on aggregations of data. We have therefore to take into account their imperfection by a rigorous approach. We propose algebraic structures for uncertain data modeling covered either by probability or possibility theories.

J. Dantan (✉)
Esitpa, Agri'terr, Mont-Saint-Aignan, France

CNAM, CEDRIC, Paris, France
e-mail: jdantan@esitpa.fr

Y. Pollet
CNAM, CEDRIC, Paris, France
e-mail: yann.pollet@cnam.fr

S. Taibi
Esitpa, Agri'terr, Mont-Saint-Aignan, France
e-mail: staibi@esitpa.fr

P. Bourgine et al. (eds.), *First Complex Systems Digital Campus World E-Conference 2015*, Springer Proceedings in Complexity, DOI 10.1007/978-3-319-45901-1_8

2 State of the Art

For a long time, uncertainty modeling remained addressed by the probability theory, which is the mathematical study of phenomena characterized by randomness and uncertainty. However, this approach is little suitable for total ignorance representation, and objective interpretations of probabilities, assigned to such events remain difficult when handled knowledge are no longer linked to random and repeatable phenomena [1]. As against, it is possible to model uncertainty, thanks to the possibility theory.

The possibility theory [1, 2] removes the strong probability additive constraint and associates the events of Ω to a possibility measure denoted Π and a necessity measure denoted N, that are both applications from Ω to $[0; 1]$, respectively, satisfying: $\Pi\,(A \cup B) = \max\,(\Pi(A), \Pi(B))$ and $N\,(A \cap B) = \min\,(N(A), N(B))$. The relationship between the possibility of an event and its opposite is given by $\Pi(A) = 1 - N\,(\neg A)$ and total ignorance is then given by $\Pi(A) = \Pi\,(\neg A) = 1$, which implies: $N(A) = N\,(\neg A) = 0$. This approach also allows representing imprecision using notions of fuzzy sets and distributions of possibilities. Thus, a fuzzy set [3] F of a set E is defined by a membership function μ_F from E to $[0; 1]$, which associates each element x of E its membership degree $\mu_F(x)$, to the subset F (i.e., x belongs "more or less" to F). When this membership function is normalized (i.e., an x value from E such as $\mu_F(x) = 1$ exists), $\mu_F(x)$ can then interpreted as the chance that F takes the value x ($\mu_F(x)$ is then a possibility distribution).

In [4], we provided a formalism for both representing and manipulating quantities which may have a finite number of possible or probable values. Such quantities are values of the $\mathbb{R}$ set. We provided an algebraic structure to operate the chained computations on such quantities with properties similar to $\mathbb{R}$, that does not allow the classical approaches based on fuzzy sets seen in the literature. In this paper, we provide an extension of this approach to continuous quantities that are, on the one hand, probabilistic and, on the other, possibilistic. The continuous mixed quantities are not in the scope of this article. In addition, we restrict ourselves to quantities belonging to the $\mathbb{R}$ set.

3 Approach

The provided approach provides a formalism for both representing and manipulating rigorously quantities which may have a finite number of possible or probable values with their interdependencies. Then, we defined an algebraic structure to operate chained computations on such quantities with properties similar to $\mathbb{R}$.

Let Ω be a universe, with both a possibility measure Π and a probability measure P, each having values belonging to $\mathbb{R}$. The considered values are, respectively, denoted: $a_1, a_2, \ldots, a_{n1}$, with possibilities $\alpha_1 = \Pi(a_1), \ldots, \alpha_{n1} = \Pi(a_{n1})$ and b_1,

$b_2, \ldots, b_{n2}$, with probabilities $\beta_1 = P(b_1), \ldots, \beta_{n2} = P(b_{n2})$. In the following of this article, we denote D1-Distributions on $\mathbb{R}$: D1D($\mathbb{R}$) with possibilistic bases $B_I = \{X_{I,i}; i = 1, \ldots\}$ (I fixed) and probabilistic bases $B^J = \{X^{J,j}; j = 1, \ldots\}$ (J fixed). The types of structures here considered are aggregation (Cartesian product) of elementary types taken among $\mathbb{R}$ space. We define two types of bases, generating the D1D vector space of D1D on the infinite space but countable $\mathbb{R}$ space.

We have defined an internal product on vectors: $Z = X \cdot Y$ that checks the following properties:

- $X \cdot X = X$ (idempotency);
- $X \cdot 0 = 0_v$, where 0_v is the null vector (absorbing element);
- $i_1 \neq i_2$ implies $X_{I,i1} \cdot X_{I,i2} = 0_v$;
- $j_1 \neq j_2$ implies $X^{I,i1} \cdot X^{I,i2} = 0_v$;
- $X_{I1,i1} \cdot X_{I2,i2} \neq 0_v$;
- $X_{I1,i1} \cdot X^{I2,i2} \neq 0_v$;
- $X^{I1,i1} \cdot X^{I2,i2} \neq 0_v$.

With such two types of vectors bases, we are able to distinguish the sources of uncertainty during combinations of values and then make rigorous computations.

4 Background: Uncertainty on Discrete Quantities

We define a purely possibilistic D1D as a value "a" which may have a finite number n_1 of possible values: $(a_1, \ldots, a_{n1})$ of K^{n_1} with their respective associated possibilities $\alpha_1 = \Pi(a_1), \ldots, \alpha_{n1} = \Pi(a_{n1})$ of $[0, 1]^{n_1}$.

The canonical form of a purely possibilistic D1D "a" is the following expression: $= \sum_{i=1}^{n} a_i/\alpha_i \cdot X_{I,i}$, with a_i are the possible values of a, α_i are the possibilities, associated with each value a_i (one of which at least is equal to 1) and $X_{I,i}$ (I fixed) correspond to the partition of the universe Ω corresponding to values of quantity a.

We define a purely probabilistic D1D as a value b which may have a finite number n_2 of probable values: $(b_1, b_2, \ldots, b_{n2})$ of K^{n_2} with their respective associated probabilities $\beta_1 = P(b_1), \ldots, \beta_{n2} = P(b_{n2})$ of $[0, 1]^{n_2}$. The n_2 values $(b_1, b_2, \ldots, b_{n2})$ completely define the probability distribution b/β_i on $\mathbb{R}$, associated with the probabilistic variable $b \cdot \beta_i$ is equal to zero except on values $b_1, b_2, \ldots, b_{n2.}$

The canonical form of a purely probabilistic D1D "b" is the following expression: $b = \sum_{i=1}^{n} b_i/\beta_i \cdot X^{J,i}$, with b_j are the probable values of b, β_j are the probabilities, associated with each value b_j (the sum of β_j is equal to 1) and $X^{J,j}$ (J fixed) correspond to the partition of the universe Ω corresponding to values of quantity b.

4.1 Internal Composition Law (+)

The following expression: $a_1 + a_2 = \sum_{i=1}^{n_1} a_{1,i}/\alpha_{1,i} \cdot X_{1,i} + \sum_{j=1}^{n_2} a_{2,j}/\alpha_{2,j} \cdot X_{2,j}$ is evaluated with a composed sum of $(a_{1,i} + a_{2,j})/(\alpha_{1,i} \min \alpha_{2,j}) \cdot X_{1,i} \cdot X_{2,j}$, i.e., $a_1 + a_2 = \sum_{i=1}^{n_1} \sum_{j=1}^{n_2} (a_{1,i} + a_{2,j}) / (\alpha_{1,i} \min \alpha_{2,j}) \cdot X_{1,i} \cdot X_{2,j}$.

The following expression: $b_1 + b_2 = \sum_{i=1}^{n_1} b_{1,i}/\beta_{1,i} \cdot X^{1,i} + \sum_{j=1}^{n_2} b_{2,j}/\beta_{2,j} \cdot X^{2,j}$ is likewise evaluated with a composed product of $(b_{1,i} * b_{2,j}) / (b_{1,i} \cdot b_{2,j}) \cdot X^{1,i} \cdot X^{2,j}$, $b_1 + b_2 = \sum_{i=1}^{n_1} \sum_{j=1}^{n_2} (b_{1,i} + b_{2,j}) / (\beta_{1,i} * \beta_{2,j}) \cdot X^{1,i} \cdot X^{2,j}$.

As a special case, if both operands are expressed in the same basis (i.e., they are dependent, or in other words linked), then the formula is simplified. We check that it can take into account rigorously dependencies to not artificially explode the number of possible values.

4.2 Internal Composition Law (*)

Similarly to +, the product internal composition law is evaluated with composed products of $(a_{1,i} * a_{2,j})/(\alpha_{1,i} \min \alpha_{2,j}) \cdot X_{1,i} \cdot X_{2,j}$ (possibility) and $(b_{1,i} + b_{2,j}) / (\beta_{1,i} * \beta\,(2,j))$ (probability). Finally, we showed that D1D ($\mathbb{R}$) has a vector space structure on $\mathbb{R}$.

5 Continuous Quantities

5.1 Combinations of Continuous Possibilistic Quantities (Trapezoids)

In this case, we restrict ourselves to possibilistic coefficients and to trapezoids. So this is purely continuous possibilistic D1D. We then get the following canonical expression: $a = \sum_{i=1}^{n} a_i \cdot T(\lambda_{1i} \cdot a_i, \lambda_{2i} \cdot a_i) / \alpha_i \cdot X_{I,i}$, where $T(\lambda_{1i} \cdot a_i, \lambda_{2i} \cdot a_i)$ is a trapezoid centered on a_i, with kernel $a_i \pm \lambda_{1i} \cdot a_i$ and support $a_i \pm \lambda_{2i} \cdot a_i$.

The resulting computations are degraded trapezoids. Considering that the symmetric trapezoid centered on a_i with kernel $a_i \pm \lambda_{1i} \cdot a_i$ and support $a_i \pm \lambda_{2i} \cdot a_i$,

$\hat{a}_i$ (i.e., $a_i \cdot T(\lambda_{1i} \cdot a_i, \lambda_{2i} \cdot a_i)$) is stable for the four main algebraic operations +, −, *, /. Here are the values for the resulting cores and supports four main algebraic operations:

- Addition: $a + b = \sum_i \begin{bmatrix} \alpha i \\ 0 \end{bmatrix} / a_i \cdot T(\lambda_{1i} \cdot a_i, \lambda_{2i} \cdot a_i) \cdot X_{1,i} + \sum_j \begin{bmatrix} \eta j \\ 0 \end{bmatrix} / b_j \cdot T(\kappa_{1j} \cdot b_j, \kappa_{2j} \cdot b_j) \cdot X_{2,j}$. We then get the following expression:

$$a+b=\sum_{i,j} \begin{bmatrix} \alpha i \min \eta j \\ 0 \end{bmatrix} / (a_i+b_j) \cdot T(\lambda_{1i} \cdot a_i+\kappa_{1j} \cdot b_j, \lambda_{2i} \cdot a_i+\kappa_{2j} \cdot b_j) \cdot X_{1,i} \cdot X_{2,j}.$$

- Subtraction: $a - b = \sum_i \begin{bmatrix} \alpha i \\ 0 \end{bmatrix} / a_i \cdot T(\lambda_{1i} \cdot a_i, \lambda_{2i} \cdot a_i) \cdot X_{1,i} - \sum_j \begin{bmatrix} \eta j \\ 0 \end{bmatrix} / b_j \cdot T(\kappa_{1j} \cdot b_j, \kappa_{2j} \cdot b_j) \cdot X_{2,j}$. We then get the following expression:

$$a-b=\sum_{i,j} \begin{bmatrix} \alpha i \min \eta j \\ 0 \end{bmatrix} / (a_i-b_j) \cdot T(\lambda_{1i} \cdot a_i+\kappa_{1j} \cdot b_j, \lambda_{2i} \cdot a_i+\kappa_{2j} \cdot b_j) \cdot X_{1,i} \cdot X_{2,j}.$$

- Product: $a * b = \sum_i \begin{bmatrix} \alpha i \\ 0 \end{bmatrix} / a_i \cdot T(\lambda_{1i} \cdot a_i, \lambda_{2i} \cdot a_i) \cdot X_{1,i} * \sum_j \begin{bmatrix} \eta j \\ 0 \end{bmatrix} / b_j \cdot T(\kappa_{1j} \cdot b_j, \kappa_{2j} \cdot b_j) \cdot X_{2,j}$. We then get the following expression:

$$a*b=\sum_{i,j} \begin{bmatrix} \alpha i \min \eta j \\ 0 \end{bmatrix} / (a_i * b_j) \cdot T(a_i \cdot \kappa_{1j}+b_j \cdot \lambda_{1i}, a_i \cdot \lambda_{2i}+b_j \cdot \kappa_{2j}) \cdot X_{1,i} \cdot X_{2,j}.$$

- Division: $a/b=\sum_i \begin{bmatrix} \alpha i \\ 0 \end{bmatrix} / a_i \cdot T(\lambda_{1i} \cdot a_i, \lambda_{2i} \cdot a_i) \cdot X_{1,i} / \sum_j \begin{bmatrix} \eta j \\ 0 \end{bmatrix} / b_j \cdot T(\kappa_{1j} \cdot b_j, \kappa_{2j} \cdot b_j) \cdot X_{2,j}$. We then get the following expression:

$$a/b=\sum_{i,j} \begin{bmatrix} \alpha i \min \eta j \\ 0 \end{bmatrix} / (a_i/b_j) \cdot T\left((a_i \cdot \kappa_{1j}+b_j \cdot \lambda_{1i}) / b_j^2, (a_i \cdot \lambda_{2i}+b_j \cdot \kappa_{2j}) / b_j^2\right).$$

5.2 *Combinations of Continuous Probabilistic Quantities (Gaussian Distribution)*

In this case, we restrict ourselves to probabilistic coefficients and to Gaussian distributions. We then get the following expression: $b = \sum_{i=1}^{n} b_i \cdot N(\lambda_i \cdot b_i) / \beta_i \cdot X^{J,i}$. The probability density on $\mathbb{R}$ is then the weighted sum of n Gaussian densities. In the internal composition law + (sum) case, to combine, e.g., $(a+b)$, with

$b = \sum_{i=1}^{n} b_i \cdot N(\lambda_i \cdot b_i) / \beta_i \cdot X^{J,i}$, we have to make a convolution of two density functions of the following formula: $\sum_i p_i \cdot N(\sigma_i(x))$. We reduce such computations to Gaussian densities $N(\sigma_1) * N(\sigma_2)$ convolutions, that result to Gaussian densities of the form $N\left(\sqrt{\sigma_1^2 + \sigma_2^2}\right)$.

In the internal composition law * (product) case, a residue term appears when calculating the product of two Gaussian distributions. This term might be ignored considering that random variations are small compared to the main values. However, in other cases, this residue does not give a Gaussian distribution but a Bessel function, which means that the law continuous probabilistic D1D are not stable by the law *.

6 Conclusion

The provided approach provides a formalism for both representing and manipulating rigorously quantities which may have possible or probable values with their interdependencies. Then, we define an algebraic structure to operate chained computations on such quantities with properties similar to $\mathbb{R}$.

We have extended our formalism on continuous quantities. In special cases such as trapezoids for possibilities and normal distributions for probabilities, some algebraic properties of D1D have been maintained. However, combinations of continuous quantities including probabilistic ones require additional assumptions that make the computations not mathematically rigorous.

The next steps are to compute mixed continuous quantities, by considering either the trapezoidal possibility distributions as intervals of cumulative distribution functions [5] or probability–possibility transformations [6].

7 Scientific Validation

This paper has been unanimously validated in a collaborative review mode with the following reviewers:

- Prof Pierre Collet, PhD, HDR, Head of the Department of Computer Science, Faculty of Mathematics and Computer Science, Strasbourg University Unistra (Strasbourg, France)
- Prof Henri Prade, PhD, HDR, CNRS Research Advisor, IRIT (Toulouse Institute of Computer Science Research), Paul Sabatier University (Toulouse, France).

References

1. Dubois D, Prade H (1988) Théorie des possibilités, Application à la représentation des connaissances en informatique. Masson, Paris (in French)
2. Zadeh LA (1978) Fuzzy sets as a basis for a theory of possibility. Fuzzy Set Syst 1:3–28
3. Zadeh LA (1965) Fuzzy sets. In: Information and control, vol 8. Academic Press, New York
4. Dantan J, Pollet Y, Taïbi S (2015) Combination of imperfect data in fuzzy and probabilistic extension classes. J Environ Account Manag 123–150. doi:10.5890/JEAM.2015.06.004
5. Destercke S, Dubois D, (2009) The role of generalised p-boxes in imprecise probability models. In: Proceedings of 6th international symposium on imprecise probability. Presented at ISIPTA '09, GBR, Durham, pp 179–188
6. Dubois D, Foulloy L, Mauris G, Prade H (2004) Probability-possibility transformations, triangular fuzzy sets, and probabilistic inequalities. Reliab Comput 10(4):273–297

Bayesian Causalities, Mappings, and Phylogenies: A Social Science Gateway for Modeling Ethnographic, Archaeological, Historical Ecological, and Biological Variables

CS-DC'15 Panel on Synthesis of Ecological, Biological, and Ethnographic Data 9–13

Douglas White, Paul Rodriguez, Eric Blau, Stuart Martin, Lukasz Lacinski, Thomas Uram, Feng Ren, Wesley Roberts, and Tolga Oztan

1 Introduction

We use DEf Wy procedures to begin to integrate analysis of ethnographic, archaeological, ecological, historical empires, and biological variables into complex system analyses of human evolutionary (deep structure and processes) and causal analysis of networks of variables at a given set or series of time periods (surface structure). At this time our datasets of societal samples and variables relate to specific places and times, and the focus of these variables has been largely sociocultural but it is imperative—given the role of biology in evolutionary processes—that our data are collated to variables (and models) from bioecological sciences pinpointed or encompassing where possible the spatiotemporal sites of our datasets. This is an ambitious undertaking but one successful example of this is a result of Botero

D. White (✉)
University of California, Irvine, Irvine, CA, USA
e-mail: douglas.white@uci.edu

P. Rodriguez
San Diego Supercomputer Center, University of California, San Diego, San Diego, CA, USA

E. Blau • S. Martin • L. Lacinski • T. Uram
Argonne National Laboratory, Lemont, IL, USA

F. Ren
Xiamen University, Xiamen, Fujian, China

W. Roberts
Carnegie Library of Pittsburgh, Pittsburgh, PA, USA

T. Oztan

P. Bourgine et al. (eds.), *First Complex Systems Digital Campus World E-Conference 2015*, Springer Proceedings in Complexity, DOI 10.1007/978-3-319-45901-1_9

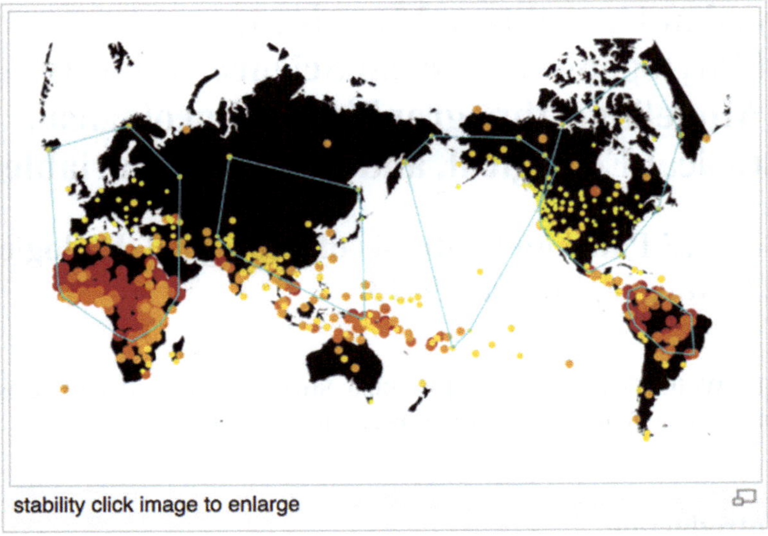

Fig. 1 Ethno atlas stability biovariable

et al. [1] who predict the High Moral God variable from multiplicative relationships between bioecological stability and abundance variables (Figs. 1 and 2) when added to the Ethnographic Atlas dataset of 1265 societies [2]. The idea that multiplying features of two subcontinents can get predictive results is dubious.[1]

Specifically, when the lower-values (small yellow datapoints) of the bioecological variables are multiplied the resultant variable predicts "Moral High God" religions such as Christianity (Fig. 1) and Islam (Fig. 2) which together constitute the majority of "Moral High God" religions, as shown by White [2]. The deficiency of the Botero et al. model is that their results are invalidated by endogeny due to failure to apply appropriate tests of whether their predictors of the deep evolutionary background effects of language phylogeny and spatial diffusion alongside predictors of specific independent variables are correlated with the error terms of their regression.

[1]This was a feature of a 2014 publication of the PNAS predicting moral god religion using the data of the Ethnographic Atlas [3]. The direction of this article is to explore the use of Bayesian learning graphs, in an early stage of analysis, using the more extensive dataset of the Standard Cross-Cultural Sample [4]. Our variables SuperjhWriting and AnimXbwealth substitute for the PNAS variables; in doing so they do not multiply but rather distinguish a dominant ecological region of pastoralism, associated with Islam as a major Moral God religion from the large states and empires present in both the Islamic and the Christian regions of West Eurasia and North Africa.

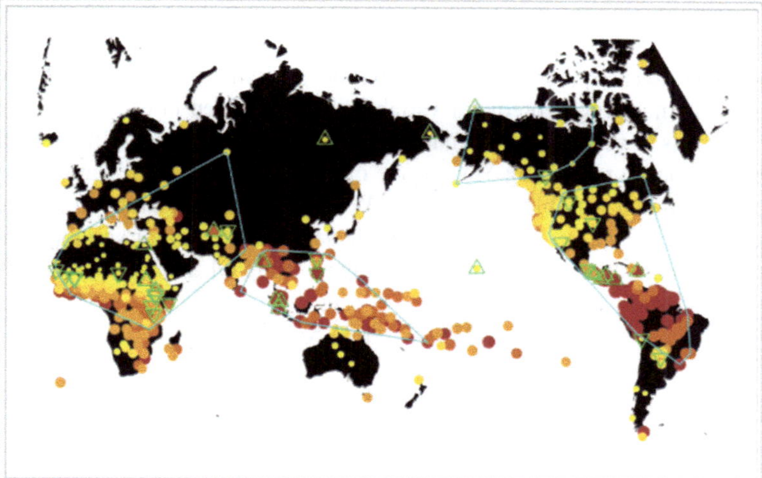

Fig. 2 Ethno atlas abundance biovariable

2 Methods

When we use DEf Wy R software implementation with appropriate diagnostics (Dow, personal communication), we are able to solve diagnostics for exogeneity by use of two-stage OLS to separate the evolutionary network effects from independent variable effects. By producing Wy variables explicitly, as well as imputing missing values, their method enables further analysis and comparisons, such as using a Bayesian Network to examine the network of variables. Dow and Eff [5] provide a succinct description of their method:

Dow [6] recently proposed that Galton's Problem be formulated as a network autocorrelation effects regression model, where the autocorrelated dependent variable is included as an endogenous predictor variable. It is straightforward to extend the usual regression model to include an additional variable that incorporates the effects of trait transmission on the distribution of percentage of married females in monogamous marriages, the dependent variable (y) in our proposed model:

$$y = \alpha + \lambda \mathbf{W}y + \mathbf{X}\beta + \varepsilon \tag{1}$$

where ε is an $n \times 1$ vector of normally distributed error terms with zero mean, $\mathbf{X}$ is an $n \times k$ matrix of exogenous variables, β is a $k \times 1$ vector of regression coefficients, α is the intercept, and λ is the scalar network autocorrelation effect coefficient. The first independent variable on the right of the equals sign is the product of the square $n \times n$ trait transmission matrix $\mathbf{W}$ and the $n \times 1$ vector of scores on the dependent y variable.

OLS regression requires that all of the independent variables be uncorrelated with the errors, ε, otherwise all of the estimated regression coefficients will be biased. Two-stage least squares (2SLS) estimation procedures are commonly used to deal with endogenous predictor variables. The first step in the 2SLS estimation of Eq. (1) is to generate an estimate of $\mathbf{W}y$ that is independent of the ε term. This can be done by regressing $\mathbf{W}y$ on one or more "instrumental" variables, which are independent variables that predict $\mathbf{W}y$ but are uncorrelated with ε. Kelejian and Prucha [7] suggest $\mathbf{WX}$ as a suitable set of instrumental variables,[2] so that the following equation is estimated at stage 1 using OLS:

$$\mathbf{W}y = a + \mathbf{WX}c + e \tag{2}$$

Which allows computation of $\hat{y}$, an exogenous predictor for $\mathbf{W}y$:

$$\widehat{y} = \widehat{a} + \mathbf{WX}\widehat{c} \tag{3}$$

The vector of predicted scores is then entered into Eq. (1) and the following stage 2 equation is estimated, again using OLS regression:

$$y = \alpha + \lambda\widehat{y} + \mathbf{X}\beta + \varepsilon \tag{4}$$

The estimates from this second stage model permit valid inferences about the effect of trait diffusion net of the functional associations (assessed by the estimated λ and its associated significance level), and the functional associations net of diffusion (the estimated β coefficients and their significance levels).

It is straightforward to extend the 2SLS approach to handle multiple $\mathbf{W}$ matrices simultaneously. However, network matrices representing commonly posited diffusion processes can be highly correlated with each other in any particular study, thus the use of multiple matrices may result in problems of multicollinearity and disentangling their separate contributions may be difficult. Dow and Eff [8] suggest one approach to handling this problem: combine multiple matrices into a single $\mathbf{W}$ matrix.

3 Results

Table 1 gives a result of DEf Wy applied to a "Stages of Religious Evolution" (SRE) variable v2013 [9], using the 186 society Standard Cross-Cultural Sample [4] where the number of variables contributed by many other researchers has grown to 2109.

[2]The Wy variable is thus intended to be endogenous by definition.

Table 1 CoSSci DEf01 Wy result for an SCCS model with a Box-Cox power coefficient for the Dependent Variable v2013, Evolution of God [9], where lambda (λ) is the Box-Cox coefficient of predictability

R model	Coef	Std coef	VIF	Relimp	P val	Hcpval	*Bootpval*	Star	DepVar = "v2013bc": Box-Cox stages of religious evolution
(Intercept)	−0.0462	NA	NA	NA	0.8375	0.8183	0.8275100000		Box-Cox transformation; lambda = 0.65673706
AnimXbwealth	0.1086	0.2635	1.5554	0.0718	0.0000	0.0000	0.0000100000	*****	AnimXbwealth v206*v208.d1 AnimalHusb*Bridewealth
bio.16Sq	−0.0944	−0.1514	1.0391	0.0265	0.0034	0.0042	0.0099200000	***	BIOCLIM: precipitation of wettest quarter Sq
mnnppXv208	0.0491	0.2152	1.2185	0.0315	0.0001	0.0011	0.0013100000	***	mnnppXv208 mean net primary production w/in 50 km
SuperjhWriting	0.0839	0.2803	1.8066	0.1239	0.0000	0.0000	0.0000000000	*****	SuperjhWriting v237*v149 JurisHierarchyXWriting
v208	0.1401	0.3714	6.0379	0.0136	0.0029	0.0025	0.0026300000	***	Mode of marriage (Atlas 6)
v208Xv232	−0.0242	−0.2980	8.4717	0.0226	0.0435	0.0254	0.0302900000	**	v208 = Bridewealth of Dowry * v232 = Cultivation
v2137	−0.3021	−0.1567	2.3074	0.0210	0.0418	0.0644	0.0564000000	*	Food Production: Planting (task present==1, absent==0)
v232	0.2190	0.4617	4.4483	0.0846	0.0000	0.0000	0.0000100000	*****	Intensity of cultivation
Wy	0.4828	0.2844	1.5758	0.1519	0.0000	0.0000	0.0000000000	*****	Network lag term

(continued)

Table 1 (continued)

R model	Coef	Std coef	VIF	Relimp	P val	Hcpval	*Bootpval*	Star	DepVar = "v2013bc": Box-Cox stages of religious evolution			
Totry	mnnppXv208:v208Xv		SuperjhWriting:mnnppXv208						**v2013BoxCox.xlsx**			
Diagnostics	Fstat	df	P value	Star	Diagnostics							
RESET test. H0	0.0081	328,111,838	0.9282		RESET test. H0: model has correct functional form							
Wald test. H0	0.8911	651	0.3455		Wald test. H0: appropriate variables dropped							
Breusch–Pagan test. H0	1.8299	132,818	0.1761		Breusch–Pagan test. H0: residuals homoskedastic							
Shapiro–Wilk test. H0	0.9791	1,330,112	0.3224		Shapiro–Wilk test. H0: residuals normal							
LM test. H0	0.0033	270,315	0.9543		LM test. H0: autocorrelated error (geographic) not present							
LM test. H0	1.5456	67,539,214	0.2138		LM test. H0: autocorrelated error (language) not present							
LM test. H0	0.6841	21,444,398	0.4082		LM test. H0: autocorrelated error (ecology) not present							
Hausman test. H0	0.6780	118,163	0.4103		Hausman test. H0: Wy is exogenous							
Sargan test. H0	0.1975	34,020	0.6567		Sargan test. H0: residuals uncorrelated with instruments							
Other Stats	Distance	Language	*X*	Weak.	R2.final	R2.UR.model		Nimp	Nobs	BClambda		
X	1.0000	0.0000	NA	41.63	0.5478	0.5684		5	186	0.65673706		

More recent analyses lead to inclusion of more independent variables with exogeneity in diagnostics and higher $R^2 = 0.72$. The analysis employs a Box-Cox dependent variable with coefficient $y^{(\lambda)} = (y^{\lambda} - 1)/\lambda = 9.656$ that calculates an optimal power threshold for the model ($R^2 = 0.5478$, modified to $R^2 = 0.72$ with the additional variables

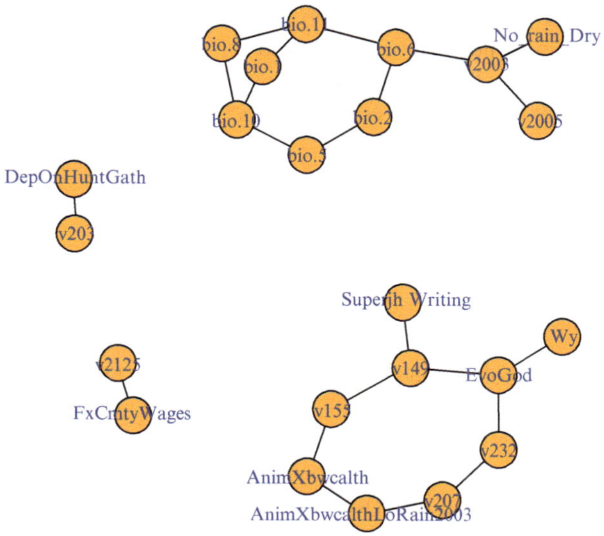

Fig. 3 Edge selection over bootstrapped samples of Bayes network shows separate graph clusters after thresholding. Variables are: v2013 EvoGod (the dependent variable of the DEf01 Wy model) | Superjurisdictional HierarchyXWriting = SuperjhWriting | v149 Writing | v155 Money | Wy = Evolutionary effects of DEf01 Wy diffusion and language phylogeny | v232 Intensity of Cultivation | v207 Dependence on Agriculture | AnimXbwealth | AnimXbwealthLoRain2003. Some of the latter variables are connected to the EvoGod DEf01 Table 1 Wy variable but not intrinsically. This graph includes dependent and independent variables in Table 1, and additional set of "unrestricted variables" (to be explored). Additional clusters show a set of // climate variables bio.1 ...bio.11 | v2005 Water | 2003 Rain | No_rain_Dry // and two pairs of variables: // DepOnHuntGath | v203 Dependence on Gathering // 2125 Wage labor | FxCmntyWages //. Wy is correlated with the EvoGod variable as shown in the figure, which is also a feature of Table 1

Group significance tests may be used to select among these candidate variables those that pass the Holm-Bonferroni (H-B) test using the OLS derived *p*-values. We then further analyzed the network of variables using Bayesian networks learning procedures with bootstrapping to produce the analysis shown in Fig. 3, which includes the SRE v2013 dependent and independent variables and a set of candidate model variables. In this step, all variables are analyzed for potential inter-relations using Bayesian networks with structure learning methods as implemented in the R library (bnlearn). Structure learning finds likely edges by combining a search for likely conditional dependencies and global network scores [10]. Choosing edges are made more robust by taking random bootstrap samples of the data, learning a network for each sample, and then taking statistics over the samples of networks. High Performance Computing (HPC) resources at San Diego Supercomputer Center were used to enable a very large number bootstrap samples.

It is worth noting that the Wy variables from two-stage OLS provides a kind of latent variable for the Bayesian network structure learning. This helps alleviate a potential concern with hidden causes among variables in Bayesian networks. One of the deficiencies of comparing DEf01 Wy results with Bayesian structure learning is that some variables need to be discretized for Bayesian learning of conditional probability tables. The Bnlearn package contains discretization functions to best transform variables to maintain correlation between independent and dependent variables.

4 Discussion: Deep (DEf01 Wy) and Surface (Bnlearn) Causality

DEf01 Wy regression often renders, in two stages, a deep model of causality where past history is shown to have affected, as in Table 1: diffusion, and linguistic ancestry, characteristic of *deep evolutionary effects* on human societies contrasted with surface causalities as in Fig. 3. This approach is potentially applicable to all analyses where indicators of historical or evolutionary processes are significant.

4.1 Endogeneity and Exogeneity: Generic and Specific

Endogeneity "includes omitted variables, omitted selection, simultaneity, common-method variance, and measurement errors—renders estimates causally uninterpretable" [11, p. 1086]. This includes regression results in which the independent variables are correlated with residual errors. DEf01 Wy results are potentially replete with measures of significant endogeneity that destroy attempts to obtain results that represent potential causality, as noted in Antonakis et al. [11] for, e.g., the Ramsey RESET test (p. 18), Hausman test (p. 19), Breusch–Pagan Heteroskedasticity (p. 28), shown as Diagnostics in Table 1 but that also include the Shapiro–Wilk and Breusch–Godfrey autocorrelation LM tests for distance, language, and ecology.

Exogeneity, where dependent and independent variables are uncorrelated with regression errors, is intrinsic to validity in Bayesian network learning, although omitted variables, omitted selection, simultaneity, common-method variance, and measurement errors may still produce endogeneity. If properly specified on all these dimensions, library(bnlearn) results, recently enhanced in numerous publications for implementation [12–14], provide determinant results that we call *surface causality*, as above: causality that omits autocorrelation. These are distinct from the effects of common histories of the units in a sample, or what has been called *Galton's problem* in the social sciences.

The only other alternative approaches to causal estimates in observational studies that provide results comparable to experiments are those of Cook et al. [15], Shadish and Cook [16], and Thistlethwaite and Campbell [17], but these approaches (see [11]), like controlled experimental studies that don't measure deep histories as does DEf01 Wy, measure surface rather than deep causalities.

5 Scientific Validation

This paper has been unanimously validated in a collaborative review mode with the following reviewers:

- Karla de Jesus, from Centre of Research, Education, Innovation and Intervention in Sport, Faculty of Sport, University of Porto, Porto, Portugal
- Joao Ranhel, from Universidade Federal de Pernambuco
- Klaus Jaffe, from Universidad Simon Bolivar, Caracas.

6 Summary

The ongoing goal of this project has been articulated as possibilities of collaboration between bioecological scientists such as Botero et al. [1], and exemplified by Brown's [18] innovation in developing an early framework in macroecology that has been continued by many others, including Lomolino et al. [19], and Harcourt [20], a member of our panel, and other members of our panel.

Acknowledgments Douglas R. White thanks the Santa Fe Institute for hosting multiple 1–2 week Causality Working Groups engaging Tolga Oztan, Peter Turchin, Amber Johnson, and many others on this topic in 2010–2014 and to Jürgen Jost and the MPI for Mathematics in the Sciences for hosting of our working group in June 2011. We thank Anthon Eff for his immense work in building the R code prior to and as used in CoSSci.

References

1. Botero C, Gardner B, Kirby K, Bulbulia J, Gavin M, Gray RD (2014) The ecology of religious beliefs. Proc Natl Acad Sci U S A 111(47):16784–89
2. White DR (2015) Oscillatory complexity in human history: earth's asymmetric biogeography and ethnographic data. Conference paper CS-DC'15, session on synthesis of ecological, biological and ethnographic data, ASU, Arizona
3. Murdock GP (1967) Ethnographic atlas. Pittsburgh University Press, Pittsburgh
4. Murdock GP, White DR (1969) Standard cross-cultural sample. Ethnology 8(4):329–369
5. Dow M, Eff A (2013) Determinants of monogamy. J Soc Evol Cult Psychol 7(3):211–238
6. Dow MM (2007) Galton's Problem as multiple network autocorrelation effects. Cross-Cult Res 41:336–363
7. Kelejian HH, Prucha IR (1998) A generalized spatial two-stage least squares procedure for estimating a spatial autoregressive model with autoregressive disturbances. J Real Estate Financ Econ 17(1):99–121
8. Dow MM, Eff EA (2009) Cultural trait transmission and missing data as sources of bias in cross-cultural survey research: explanations of polygyny re-examined. Cross-Cult Res 43(2):134–151
9. Sanderson SK, Roberts WW (2008) The evolutionary forms of religious life: a cross-cultural, quantitative analysis. Am Anthropol 110(4):454–466

10. Scutari M, Denis J-B (2014) Bayesian networks, with examples in R, Texts in statistical science. Chapman & Hall/CRC, London
11. Antonakis J, Bendahan S, Jacquart P, Lalive R (2010) On making causal claims: a review and recommendations. Leadersh Q 21(6):1086–1120
12. Højsgaard S, Edwards D, Lauritzen S (2012) Graphical models with R. Springer, New York
13. Nagarajan R, Scutari M, Lèbre S (2013) Bayesian networks in R with applications in systems biology. Use R!, vol 48. Springer, New York
14. Scutari M, Nagarajan R (2013) On identifying significant edges in graphical models of molecular networks. Artif Intell Med 57(3):207–217, http://www.aiimjournal.com/article/S0933-3657(12)00154-6/abstract?cc=y=
15. Cook TD, Shadish WR, Wong VC (2008) Three conditions under which experiments and observational studies produce comparable causal estimates: new findings from within-study comparisons. J Policy Anal Manage 27(4):724–750
16. Shadish WR, Cook TD (2009) The renaissance of field experimentation in evaluating interventions. Annu Rev Psychol 60:607–629
17. Thistlethwaite DL, Campbell DT (1960) Regression-discontinuity analysis: an alternative to the ex post facto experiment. J Educ Psychol 51(6):309–317
18. Brown JH (1995) Macroecology. University of Chicago Press, Chicago
19. Lomolino MV, Brown JH, Whittaker R, Riddle BR (2010) Biogeography, 4th edn. Sinauer Associates, Sunderland
20. Harcourt A (2012) Human biogeography. University of Chicago Press, Chicago
21. Antonakis J, Bendahan S, Jacquart P, Lalive R (2014) Causality and endogeneity: problems and solutions. In: Day DV (ed) The Oxford handbook of leadership and organizations. Oxford University Press, New York, pp 93–117
22. Dow MM, Anthon Eff E (2009) Multiple imputation of missing data in cross-cultural samples. Cross-Cult Res 43(3):206–229
23. Holm S (1979) A simple sequentially rejective multiple test procedure. Scand J Stat 6(2):65–70

A General Approach to the Linear Stability Analysis of Miscible Viscous Fingering in Porous Media

Tapan Kumar Hota, Satyajit Pramanik, and Manoranjan Mishra

1 Introduction

Flow of miscible fluids in a porous medium is found in a wide range of natural processes as well as industrial applications [5]. These displacement flows feature an interfacial instability that occurs when a less viscous fluid invades into the more viscous one. Finger like patterns are resulted at the interface, hence this is called viscous fingering (VF) instability. This hydrodynamic instability which is also known as Saffman–Taylor instability [14] has been a subject of many theoretical and experimental studies [3, 8, 10, 12] for many years. Stability analysis of VF has been performed extensively over the years [1, 5, 7, 11]. Traditionally, the modal analysis along with quasi-steady-state approximation (QSSA) has been used in different coordinate systems [7, 11]. Although, at the early times QSSA in a self-similar coordinate system successfully predicts the unconditional stability of perturbations, it fails to describe the early period dynamics of the disturbances [6]. Another approach which is more practical is the amplification theory (AT), in which the linear equations are solved as an initial value problem. However, this approach has a drawback of choosing the representative initial condition. In this article, using non-modal stability theory, we unify the frozen time method and AT to explore two important aspects of VF instability, namely the effect of non-normality of the governing non-autonomous linear operators and the optimal amplifications of the disturbances. The paper is organized as follows. In Sect. 2, the mathematical formulation of the physical model is presented. Section 3 describes the non-modal stability analysis and the numerical method to calculate the energy growth function $G(t)$. In Sect. 4 we present the results and discussions. Finally, the conclusions are given in Sect. 5.

T.K. Hota (✉) • S. Pramanik • M. Mishra
Department of Mathematics, Indian Institute of Technology Ropar, Rupnagar, Punjab, India
e-mail: tapan.hota@iitrpr.ac.in; satyajitp@iitrpr.ac.in; manoranjan@iitrpr.ac.in

P. Bourgine et al. (eds.), *First Complex Systems Digital Campus World E-Conference 2015*, Springer Proceedings in Complexity, DOI 10.1007/978-3-319-45901-1_10

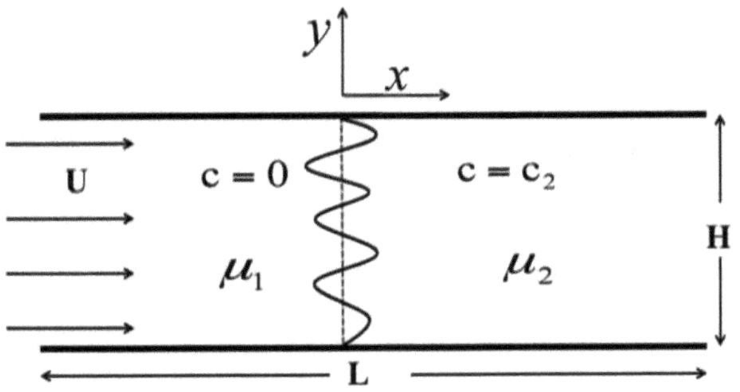

Fig. 1 Schematic of the flow configuration with coordinate system. Initially the interface is flat (*dotted line*) and then a wave like infinitesimal perturbation is applied

2 Mathematical Formulations

Consider a uniform rectilinear displacement of miscible fluids in a two-dimensional, homogeneous, porous medium with constant permeability [see Fig. 1]. The fluids are assumed to be incompressible, miscible, non-reactive, and neutrally buoyant and the dispersion is isotropic. The non-dimensional governing equations for the prescribed two-dimensional flow in a reference frame moving with velocity U are given by the following coupled nonlinear partial differential equations (PDEs) [5, 16]:

$$\nabla \cdot \boldsymbol{u} = 0, \ \nabla p = -\mu(c)(\boldsymbol{u} + \boldsymbol{i}), \ \frac{\partial c}{\partial t} + \boldsymbol{u} \cdot \nabla c = \nabla^2 c, \tag{1}$$

where p is the dynamic pressure, $\boldsymbol{u} = (u, v)$ is the Darcy's velocity, c is the concentration of the solvent, $\boldsymbol{i}$ is the unit vector along the x (downstream) direction, and $\mu(c)$ is the fluid viscosity which depends exponentially on the solute concentration, i.e., $\mu(c) = \exp(Rc)$ [16]. Here $R = \ln(\mu_2/\mu_1)$ is the log-mobility ratio, where μ_1 and μ_2 correspond to the viscosity of the less and more viscous fluid, respectively. The coupled PDEs (1) are provided with the following initial and boundary conditions in the moving frame of reference [5, 11]:

Initial conditions:

$$\mathbf{u} = (0,0), \quad c(x, y, t = 0) = \begin{cases} 0, & x \le 0 \\ 1, & x > 0, \end{cases} \quad \forall y. \tag{2}$$

Boundary conditions:

$$\mathbf{u} = (0,0), \qquad \frac{\partial c}{\partial x} = 0, \quad |x| \to \infty, \qquad \text{streamwise direction} \tag{3}$$

$$u \text{ is arbitrary}, \quad \frac{\partial c}{\partial y} = 0 = \frac{\partial v}{\partial y}, \quad \forall x, \qquad \text{spanwise direction} \tag{4}$$

The base state of the flow is assumed to be a pure diffusion of the concentration along the axial direction that can be written as

$$\boldsymbol{u}_b = (u_b, v_b) = \mathbf{0},\ c_b(\xi) = \frac{1}{2}\left[1 + \operatorname{erf}\left(\frac{\xi}{2}\right)\right],\ \mu_b = \mu_b(\xi), \tag{5}$$

where $\operatorname{erf}(z) = \frac{2}{\sqrt{\pi}} \int_0^z e^{-\eta^2} \mathrm{d}\eta$ is the error function and $\xi := x/\sqrt{t}$ is the similarity variable transformation. Introducing the infinitesimal perturbations in terms of the Fourier modes of the form $(u', c')(\xi, y, t) = (u', c')(\xi, t)e^{iky}$ and using standard procedure of linear stability analysis along with central difference scheme to discretize the spatial variable, ξ, the nonlinear coupled PDEs (1) can be written as [6, 11]

$$\frac{\mathrm{d}c'}{\mathrm{d}t} = \mathcal{A}(k, R, t)c', \tag{6}$$

where k is the non-dimensional wave number and $\mathcal{A} = \mathcal{A}(k, R, t)$ is the stability matrix.

For such non-autonomous system [Eq. (6)], there are two distinct regions in the evolution of disturbances [15]. The first is the transient region, and the second is the asymptotic long time region. In mathematical sense, instability behaviors in the second region are determined by the eigenvalues of the stability matrix $\mathcal{A}$. In the first region, the transient growth may be quite substantial, such that the nonlinear region may be reached before the growth of eigenvalue mode. Thus, the eigenvalue approach is insufficient to study the stability analysis. For nonnormal and time independent matrices, $\mathcal{A}$, the stability analysis have explored by many researchers (see [15] and the references within). Recently, few works have discussed non-modal stability analysis (NMA) with unsteady base state flow [2, 4, 6, 13]. But in the case of VF, many aspects of stability matrix $\mathcal{A}(k, t)$ remained unexplored compared to its autonomous counterpart.

3 Transient Behaviors and Non-modal Analysis

In the framework of NMA, there are two major and broader aspects which are of ample interest, the responses to external excitations and the transient energy growth of initial conditions. Mathematically, the former can be studied from the structure of "ϵ-pseudospectra" [17] which is given by $\Lambda_\epsilon(\mathcal{A}) = \{z \in \mathbb{C} \,:\, \sigma_{\min}(z - \mathcal{A}) \leq \epsilon\}$, where $\sigma_{\min}(\mathcal{A})$ denotes the smallest singular value of $\mathcal{A}$, and $0 < \epsilon \ll 1$. And the latter can be analysed from the energy growth function $G(t)$ that identifies the optimal growth of energy at time t. In addition, it is often useful in hydrodynamic stability problem to know the initial growth rate of the energy growth function. This can be obtained from numerical abscissa of the stability matrix $\mathcal{A}$ [17]. Following [6, 13] $G(t)$ can be described as

$$G(t) = G(t, k, R) := \max_{c'_0} \|\Phi(t_0; t)c'_0\|_2 = \|\Phi(t_0; t)\| = \sup_j s_j(t), \tag{7}$$

where $\Phi(t_0; t)$ is the propagator matrix or matrizant satisfying the matrix differential equation

$$\frac{\mathrm{d}}{\mathrm{d}t}\Phi(t_0; t) = \mathcal{A}(k, R, t)\Phi(t_0; t), \tag{8}$$

and it is related to the concentration perturbation field by $c'(t) = \Phi(t_0; t)c'_0$ with $c'(t_0) = c'_0$ being an arbitrary function. Here, s_j's are the singular values of $\Phi(t_0; t)$ and $\|\cdot\|_2$ is the standard Euclidean norm. It is important to note that the pseudospectra and the growth function are dependent on the definition of norm [17]. From mathematical and physical considerations, an appropriate measure of the disturbance is indispensable. In this article the standard Euclidean norm $\|\cdot\|_2$ and associated inner product have been used. Further, as the velocity perturbation is slaved to the concentration perturbation, the optimal amplification $G(t)$ has been calculated only for the concentration perturbation c' [4]. For stability analysis we consider the following quantities: the spectral abscissa and numerical abscissa are given by $\alpha(\mathcal{A}) \equiv \max\{\Re(\lambda(\mathcal{A}))\}$, $\eta(\mathcal{A}) \equiv \max\{\lambda(\mathcal{A} + \mathcal{A}^{\mathcal{T}})/2\}$, respectively, where $\lambda(\mathcal{A})$ denotes the eigenspectrum of $\mathcal{A}$, $\Re$ denotes the real part, and $\mathcal{A}^{\mathcal{T}}$ denotes the transpose of the matrix $\mathcal{A}$. In the present paper, the computational domain has been chosen to be $[-50, 50]$ with step size 0.2. The initial value problem (8) is solved by Runge–Kutta fourth order method, and a matlab GUI EigTool [17] has been used to draw the pseudospectra of the stability matrix $\mathcal{A}$. The detail of the numerical procedure can be found in [6].

4 Results and Discussion

In stability analysis, the eigenspectrum $\lambda(\mathcal{A})$ is the principal aid to give an insight into how a system behaves. If the stability matrix $\mathcal{A}$ is non-normal (i.e., $\mathcal{A}\mathcal{A}^{\mathcal{T}} \neq \mathcal{A}^{\mathcal{T}}\mathcal{A}$), then the pseudospectra $\Lambda_\epsilon(\mathcal{A})$ are likely to explain the system behavior better than the eigenvalues, $\lambda(\mathcal{A})$. Thus, $\Lambda_\epsilon(\mathcal{A})$ will help us to understand the response to the external excitations in the parameter space k and R. Figure 2 shows the parabolic profile (the dashed line) of the numerical range, $W(\mathcal{A}) = \{x^{\mathcal{T}}\mathcal{A}x : \|x\| = 1$, of the stability matrix $\mathcal{A}$. It is clearly visible that the boundary of $W(\mathcal{A})$ strictly contains the spectrum of $\mathcal{A}$. This reflects that $\mathcal{A}$ cannot be unitarily diagonalizable; or in other words, their eigenfunctions are not orthogonal. In the inset figures (Fig. 2), it is illustrated that the pseudospectra protrude strongly into the right half-plane, which imply that the evolution process will be susceptible to large transient effects. This is due to the non-normality of the stability matrix, and it signifies that the system is unstable, which is not captured by analysing the spectrum alone. This implies that at the initial time the energy of the disturbance is

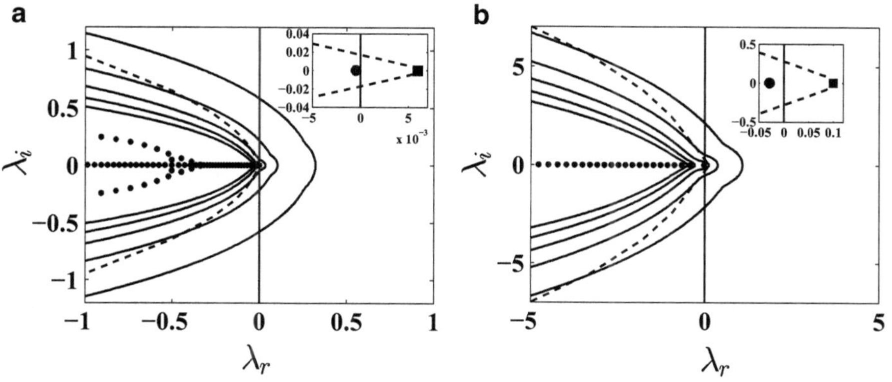

Fig. 2 ϵ-pseudospectra of $\mathcal{A}$ for (**a**) $R = 1, k = 0.08, t = 31$, (**b**) $R = 4, k = 0.3, t = 2.5$. *Black dots* (•): eigenvalues; *black square* (■): numerical abscissa; *dashed line*: boundary of the numerical range; *solid lines*: contours from innermost to outermost representing levels from $\epsilon = 10^{-2.5}$ to $10^{-0.5}$ with increment $10^{-0.5}$. λ_i, λ_r are the imaginary and real part of eigenvalues, respectively. *Inset figure* shows the numerical abscissa, $\eta(\mathcal{A})$ (■), and spectral abscissa, $\alpha(\mathcal{A})$(•)

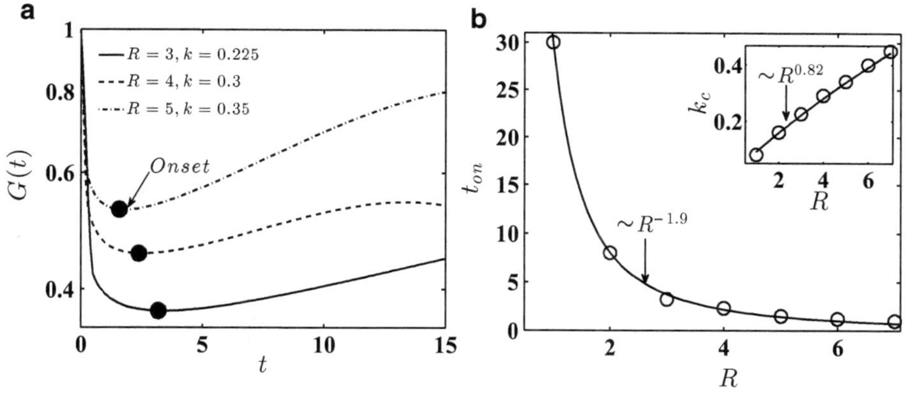

Fig. 3 (**a**) Optimal amplification $G(t)$ for different R and the corresponding critical wave numbers. The *black dot* denotes the onset t_{on}. (**b**) The variation of the onset of instability t_{on} and the critical wave number k_c (*inset*), for different values of R determined from NMA. The *circles* (○) are simulation data and continuous lines are fitted curves

growing faster than what is anticipated from the spectral abscissa $\alpha(\mathcal{A})$. It can be conjectured that with increasing R, the boundary of numerical range becomes wider and the eigenvalues are more sensitive to the external forcing, as confirmed from the contours of pseudospectra.

Next, we present the optimal amplification of the perturbations to understand the transient effects. For this purpose we plot the energy growth function $G(t)$ for $(R, k) \in \{(3, 0.225), (4, 0.3), (5, 0.35)\}$ in Fig. 3a. This figure shows that after a initial diffusion dominated period and each curve experiences a substantial energy growth. The onset of instability is the time when $G(t)$ starts increasing (the first local minimum) and it is shown as black dot (•). Thus, NMA clearly distinguishes the domain where initial perturbations are damped or have no time to grow significantly due to diffusion, and a domain exhibiting strong convection. Further, to understand how the VF mechanism responses to external forcing, the spatial structures of the optimal input can be obtained from the singular value decomposition of the propagator matrix $\Phi(t_0; t)$ [6, 13, 15]. Using matlab curve fitting tool (cftool), with 95 % confidence bounds, we found empirical relationships between the onset of instability, t_{on}, and R, as well as the critical wave number, k_c (the minimum wave number at which the instability sets in), and R. The obtained results are shown in Fig. 3b. It is shown that $t_{\mathrm{on}} \sim R^{-1.9}$, i.e., nearly a inverse square of R, whereas $k_c \sim R^{0.82}$. This empirical relationships are in good agreement with Tan and Homsy [16], who measured $t_{\mathrm{on}} \sim R^{-2}$, and $k_c \sim R$, for a step-like initial concentration profile. The non-normality of the stability matrix $\mathcal{A}$ could be the source of observed difference between the two cases.

5 Conclusion

A unified approach of linear stability analysis for the VF is presented. It is shown that at early times diffusion dominates, which causes the energy decay before it starts amplifying due to strong convection. For this purpose we have studied different components of the spectrum, such as the spectral abscissa, boundary of numerical range, and ϵ-pseudospectra. To understand the transient growth of energy, we studied the optimal amplification of disturbances for various flow parameters. It can be concluded that such approach not only presents a comprehensive stability analysis algorithm, but also explains the physical mechanism appropriately. It will be very interesting to apply this procedure for other unsteady base state problems such as VF in liquid chromatographic condition [12], analysing the effect of precipitation reactions in CO_2 sequestration techniques [10], or in understanding the effect of external forces, e.g., magnetic field [9] to VF to name a few.

6 Scientific Validation

This paper has been unanimously validated in a collaborative review mode with the following reviewers:

- Agota Toth, from University of Szeged
- Denis Grebenkov
- Soumya Banerjee, from Broad Institute of MIT and Harvard, Ronin Institute.

Acknowledgements The authors would like to thank the three reviewers for their helpful critics and suggestions. This helped us in improving the standard of the manuscript. S.P. gratefully acknowledges the National Board for Higher Mathematics, Department of Atomic Energy, Government of India for the financial support through a Ph.D. fellowship.

References

1. Ben Y, Demekhin EA, Chang HC (2002) A spectral theory for small amplitude miscible fingering. Phys Fluids 14:999
2. Daniel D, Tilton N, Riaz A (2013) Optimal perturbations of gravitationally unstable, transient boundary layers in porous media. J Fluid Mech 727:456–487
3. De Wit A, Bertho Y, Martin M (2005) Viscous fingering of miscible slices. Phys Fluids 17:054114
4. Doumenc F, Boeck T, Guerrier B, Rossi M (2010) Transient Rayleigh-Bénard-Marangoni convection due to evaporation: a linear non-normal stability analysis. J Fluid Mech 648:521–539
5. Homsy GM (1987) Viscous fingering in porous media. Annu Rev Fluid Mech 19:271–311
6. Hota TK, Pramanik S, Mishra M (2015) Nonmodal linear stability analysis of miscible viscous fingering in porous media. Phys Rev E 92:053007. doi:10.1103/PhysRevE.92.053007
7. Kim MC (2012) Linear stability analysis on the onset of the viscous fingering of a miscible slice in a porous media. Adv Water Resour 35:1–9
8. Mishra M, Martin M, De Wit A (2008) Differences in miscible viscous fingering of finite width slices with positive or negative log mobility ratio. Phys Rev E 78:066306
9. Mishra M, Thess A, De Wit A (2012) Influence of a simple magnetic bar on buoyancy-driven fingering of traveling autocatalytic reaction fronts. Phys Fluids 24:124101
10. Nagatsu Y, Ishii Y, Tada Y, De Wit A (2014) Hydrodynamic fingering instability induced by a precipitation reaction. Phys Rev Lett 113:024502
11. Pramanik S, Mishra M (2013) Linear stability analysis of Korteweg stresses effect on the miscible viscous fingering in porous media. Phys Fluids 25:074104
12. Rana C, De Wit A, Martin M, Mishra M (2014) Combined influences of viscous fingering and solvent effect on the distribution of adsorbed solutes in porous media. RSC Adv 4:34369
13. Rapaka S, Chen S, Pawar RJ, Stauffer PH, Zhang D (2008) Non-modal growth of perturbations in density-driven convection in porous media. J Fluid Mech 609:285–303
14. Saffman PG, Taylor G (1958) The penetration of a fluid into a medium or Hele-Shaw cell containing a more viscous liquid. Proc R Soc Lond Ser A 245:312–329
15. Schmid PJ, Henningson DS (2001) Stability and transition in shear flows. Applied mathematics science, vol 142. Springer, New York
16. Tan CT, Homsy GM (1986) Stability of miscible displacements in porous media: rectilinear flow. Phys Fluids 29:3549
17. Trefethen LN, Embree M (2005) Spectra and pseudospectra: the behavior of nonnormal matrices and operators. Princeton University Press, Princeton

References

Viscosity Scaling in Hydrodynamic Instabilities in Porous Media

Satyajit Pramanik

1 Introduction

Viscous fingering (VF) in various fluid flow problems of industrial and environmental interest, such as oil recovery [1], pollution remediation [2], CO_2 sequestration [3], and chromatography separation [4–7], have drawn attention of many researchers over many decades. This hydrodynamic instability is featured when a less viscous fluid displaces a more viscous and hence less mobile one in porous media [1]. Based on the geometry of the interface featuring the instability in the form of finger-like structure, VF problems can be broadly classified into two categories. The first one being the classical Saffman–Taylor instability in two semi-infinite fluids separated by a flat interface [1], while the other is the displacement of a finite sample, typically in chromatography column [4]. In the latter case, the sample is confined in a rectangular region that can feature VF either at the frontal or rear interface depending on the viscosity contrast in the downstream direction. The influence of positive and negative log-mobility ratio was investigated in the context of solute adsorption on porous matrix [8]. In the absence of adsorption and for constant diffusivity of the solute concentration, Mishra et al. showed that the onset of instability and the subsequent dynamics of the fingers are identical for more and less viscous slices [9]. Pramanik and Mishra [10] showed that in the presence of the Korteweg stress [11, 12] the fingering dynamics of a miscible slice of more viscosity are identical to that of a less viscosity. This was achieved with an appropriate scaling. In this paper, we are interested to explore whether such dynamics are also possible with concentration dependent diffusivity in absence of the Korteweg stress

S. Pramanik (✉)
Department of Mathematics, Indian Institute of Technology Ropar, Rupnagar 140001, Punjab, India
e-mail: satyajitp@iitrpr.ac.in

P. Bourgine et al. (eds.), *First Complex Systems Digital Campus World E-Conference 2015*, Springer Proceedings in Complexity, DOI 10.1007/978-3-319-45901-1_11

effect. In this context, we propose a modified viscosity scaling that helps to answer this question when the diffusivity is inversely proportional to the viscosity of the fluid. We structure this paper as follows. Mathematical formulation and numerical solution of the present problem are discussed in Sect. 2, followed by results and discussion and conclusion in Sects. 3 and 4, respectively.

2 Mathematical Formulation and Numerical Solution of the Problem

Consider a rectilinear displacement of a finite sample of viscosity μ_2 by another fluid of viscosity μ_1 at a uniform velocity Ue_x, where e_x is the unit vector in the x-increasing direction. The displacing fluid consists of the same sample solvent having no solute concentration in it, i.e., $c = 0$, while the finite sample has solute concentration $c = c_2$. We assume that the variation of the diffusion coefficient with concentration is governed by the generalized Stokes–Einstein relation [13, 14],

$$D(c) \cdot \mu(c) = \text{constant}, \tag{1}$$

so that $D(c) = D_0/\mu(c)$, where D_0 denotes the diffusion coefficient of an infinitesimally small amount of the displaced fluid in the displacing fluid. Fluids are assumed to be incompressible, miscible, and neutrally buoyant. For nondimensionalization we choose D_0/U, D_0/U^2, U, $\mu_{\text{ch}}D_0/\kappa$, μ_l, c_2 as the characteristic length, time, velocity, pressure, viscosity, and concentration, respectively [10]. Here $\mu_l = \min\{\mu_1, \mu_2\}$, κ is the permeability of the porous media, and D_0 represents a reference diffusivity. The related dimensionless equations in a frame of reference moving with the velocity Ue_x can be written as [1],

$$\nabla \cdot \boldsymbol{u} = 0, \quad \nabla p = -\mu(c)(\boldsymbol{u} + e_x), \quad \frac{\partial c}{\partial t} + \boldsymbol{u} \cdot \nabla c = \nabla \cdot (D(c)\nabla c), \tag{2}$$

where $D(c)$ represents concentration dependent dispersion. The viscosity of the fluids depends exponentially on the solute concentration, i.e., $\mu(c) = e^{Rf(c)}$, where $R = \ln(\mu_m/\mu_l)$ is the log-mobility ratio and $f(c)$ is c (more viscous slice) or $1 - c$ (less viscous slice), and $\mu_m = \max\{\mu_1, \mu_2\}$. Therefore, the dimensionless form of the concentration-dependent diffusivity becomes $D(c) = 1/\mu(c)$. For two-dimensional incompressible flow we define the stream-function $\psi(x, y, t)$, such that $u = \psi_y, v = -\psi_x$ are the longitudinal and transverse component of the Darcy velocity vector $\boldsymbol{u}$. Therefore, we can write (2) as,

$$\frac{\partial c}{\partial t} + J(x, y, t) = 0, \quad \nabla^2 \psi = -Rf'(c)N(x, y, t). \tag{3}$$

The nonlinear terms $J(x, y, t)$ and $N(x, y, t)$ in (3) are defined as,

$$J(x,y,t) = \psi_y c_x - \psi_x c_y - e^{-Rf(c)}\left(\nabla^2 c - Rf'(c)|\nabla c|^2\right), \tag{4}$$

$$N(x,y,t) = \left(\nabla\psi + e_y\right)\cdot\nabla c, \tag{5}$$

where e_y is the unit vector in y-direction.

Equations (3) are solved using a Fourier pseudo-spectral method [4, 15] to analyze the influence of variable diffusivity on VF of more and less viscous miscible slices. We apply discrete Fourier transform (DFT) to $c(x, y, t), \psi(x, y, t), J(x, y, t)$, and $N(x, y, t)$ that converts (3) into differential algebraic equations,

$$\frac{\mathrm{d}\hat{c}_{m,n}(t)}{\mathrm{d}t} + \hat{J}_{m,n}(t) = 0, \quad \hat{\psi}_{m,n}(t) = \hat{N}_{m,n}(t)/(k_m^2 + k_n^2), \tag{6}$$

where $k_m = 2\pi m/L_x$, $k_n = 2\pi n/L_y$, $m, n = 0, 1, 2\ldots$ Here, L_x and L_y are the dimensionless length and width of the computational domain, respectively. Equation (6) is solved using Adams–Bashforth predictor and trapezoidal corrector method in the Fourier space to obtain the Fourier modes at next time step $t + \Delta t$. Inverse DFT has been performed for $\hat{c}_{m,n}$ and $\hat{\psi}_{m,n}$ to update the concentration and stream function at time $t+\Delta t$, $c(x, y, t+\Delta t)$, $\psi(x, y, t+\Delta t)$. Details of the numerical method used can be found in the paper of Tan and Homsy [15].

3 Results and Discussion

First we investigate the influence of variable diffusivity on miscible VF in a finite slice. The spatial structure of a more viscous slice of dimensionless width $l = 256$ is presented for variable diffusivity (Fig. 1a) and constant diffusivity (Fig. 1b) with $R = 2$. Figure 1 depicts that the fingering dynamics become more complex for variable diffusivity as compared to constant diffusivity case. In the former case the wavelength of the fingers becomes shorter and the fingers propagate faster, which is readily evident from an early interaction of the fingers with the frontal interface.

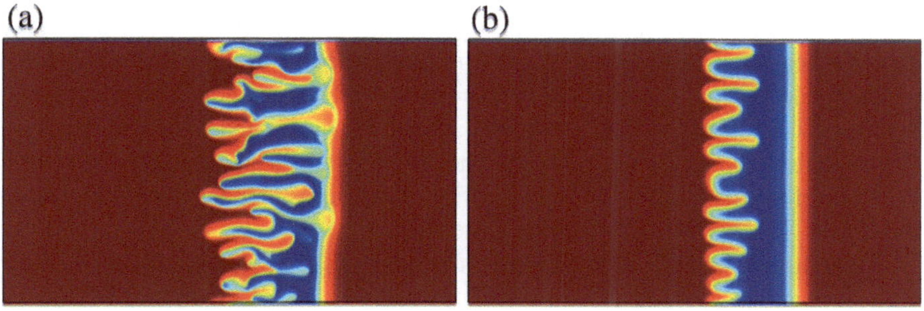

Fig. 1 Spatial structure of the solute concentration at $t = 800$ in the moving frame of reference for $R = 2, l = 256$: (**a**) variable diffusivity and (**b**) constant diffusivity

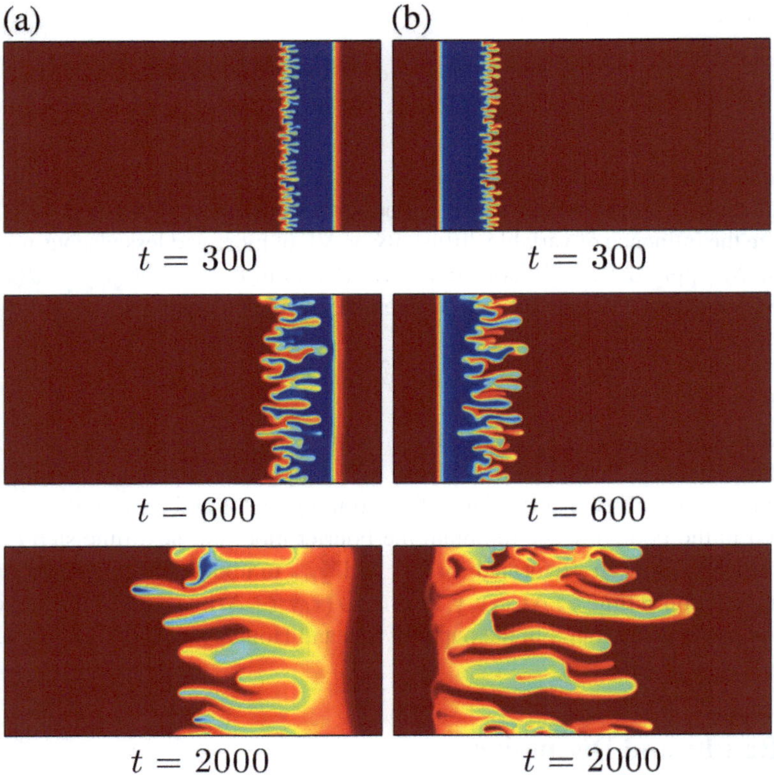

Fig. 2 Spatial structure of the solute concentration at successive times in the moving frame of reference for $l = 256$: (**a**) $\mu(c) = e^{2c}$ and (**b**) $\mu(c) = e^{2(1-c)}$

Figure 2 shows the spatio-temporal evolution of the solute concentration of more and less viscous slice for $R = 2,\ l = 256$ with concentration dependent diffusivity. This figure depicts that, before interacting with the stable interface, the fingering dynamics in more and less viscous slices are identical. We also perform simulations for a less viscous slice by considering $\mu_{\text{ch}} = \mu_1$ [9], such that $\mu = e^{-2c}$. It is observed that the finger patterns differ significantly in this case (for brevity not shown) as compared to that shown in Fig. 2b. In particular, the onset of instability delays when choosing $\mu = e^{-2c}$, which is attributed to the fast diffusion of the solute concentration as compared to $\mu = e^{2(1-c)}$. Thus we conclude, in order to compare between two similar flows one should be careful about the choice of the dimensionless values in such a way that they represent the same dimensional values. With the prescribed scaling of the viscosity of the fluids this has been achieved in the case of variable diffusivity VF in finite slice.

The temporal evolution of degree of mixing $\chi(t)$ [16] has been investigated for both more and less viscous finite sample of width $l = 256$. We mentioned in Sect. 2 that the displacement of a more or less viscous finite sample is represented by

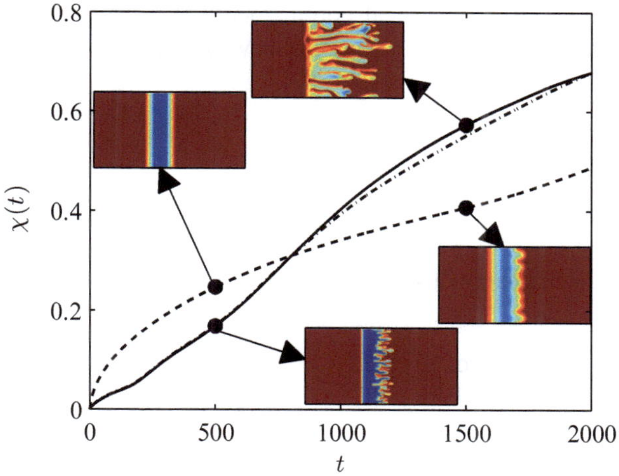

Fig. 3 Temporal evolution of the degree of mixing for $l = 256$ with $\mu(c) = e^{2c}$ (*dash-dotted line*), $\mu(c) = e^{2(1-c)}$ (*continuous line*), and $\mu(c) = e^{-2c}$ (*dashed line*). *Inset* images show the concentration distribution for the last two cases at $t = 500$ and 1500

$\mu(c) = e^{Rc}$ or $\mu(c) = e^{R(1-c)}$, respectively, where the log-mobility ratio $R > 0$. Figure 3 depicts that the temporal evolution of the degree of mixing, $\chi(t)$, for these two cases is almost identical, even after the interaction of the fingers with the respective stable interface. The displacement of a less viscous finite sample can alternatively be represented by $\mu(c) = e^{Rc}$, $R < 0$ [9]. It is shown that at early times, say at $t = 500$ in Fig. 3, mixing due to diffusion for $R < 0$ is more than the mixing due to VF for $R > 0$. However, at later times, for example, at $t = 1500$, VF enhances the fluid–fluid interface and hence the degree of mixing becomes larger for $R > 0$ than its $R < 0$ counterpart. This is due to the concentration-dependent diffusivity. For $R = 2$, the dimensionless diffusion coefficient is $e^{-2(1-c)}$, while for $R = -2$, the dimensionless diffusion coefficient is e^{2c}. Therefore, when the displacement of a less viscous finite sample is represented by $\mu(c) = e^{-2c}$, the finite slice diffuses faster than that when the displacement is represented by $\mu(c) = e^{R(1-c)}$. This leads to a faster diffusive mixing for $\mu(c) = e^{-2c}$ than that for $\mu(c) = e^{2c}$.

4 Conclusion

The influence of viscosity scaling on miscible VF in a finite slice with viscosity dependent diffusivity is investigated through numerical simulations. Choosing an appropriate characteristic viscosity we show that the fingering dynamics of a more viscous slice is identical to that of a less viscous one. We believe, the present scaling analysis shall be important to investigate the stability of a variable viscosity buoyantly unstable layer in vertical porous media [17] or miscible VF with non-monotonic viscosity profiles [18], and pave new way to understand fluid mixing in porous media.

5 Scientific Validation

This paper has been unanimously validated in a collaborative review mode with the following reviewers:

- Aǵota Tóth, University of Szeged,
- Sarah Klein, Université Paris-Sud.

Acknowledgements The author is grateful to Manoranjan Mishra for fruitful discussion and careful reading of the manuscript. The author gratefully acknowledges the financial support from the National Board for Higher Mathematics, Department of Atomic Energy, Government of India through a Ph.D. fellowship.

References

1. Homsy GM (1987) Viscous fingering in porous media. Annu Rev Fluid Mech 19:271–311
2. Welty C, Kane III AC, Kauffman LJ (2003) Stochastic analysis of transverse dispersion in density-coupled transport in aquifers. Water Resour Res 39:1150
3. Huppert HE, Neufeld JA (2014) The fluid mechanics of carbon dioxide sequestration. Annu Rev Fluid Mech 46:255–272
4. De Wit A, Bertho Y, Martin M (2005) Viscous fingering of miscible slices. Phys Fluids 17:054114
5. Mishra M, Martin M, De Wit A (2009) Influence of miscible viscous fingering of finite slices on an adsorbed solute dynamics. Phys Fluids 21:083101
6. Rana C, Martin M, De Wit A, Mishra M (2014) Combined influences of viscous fingering and solvent effect on the distribution of adsorbed solutes in porous media. RSC Adv 4:34369–34381
7. Rana C, Mishra M (2014) Fingering dynamics on the adsorbed solute with influence of less viscous and strong sample solvent. J Chem Phys 141:214701
8. Mishra M, Martin M, De Wit A (2010) Influence of miscible viscous fingering with negative log-mobility ratio on spreading of adsorbed analytes. Chem Eng Sci 65:2392–2398
9. Mishra M, Martin M, De Wit A (2008) Differences in miscible viscous fingering of finite width slices with positive or negative log mobility ratio. Phys Rev E 78:066306
10. Pramanik S, Mishra M (2015) Viscosity scaling of fingering instability in finite slices with Korteweg stress. Europhys Lett 109:64001
11. Pojman PA, Whitmore C, Turco Liveri ML, Lombardo R, Marszalek J, Parker R, Zoltowski B (2006) Evidence of an effective interfacial tension between miscible fluids: isobutyric acid-water and 1-butanol-water in a spinning drop tensiometer. Langmuir 22:2569–2577
12. Chen C-Y, Wang L, Meiburg E (2001) Miscible droplets in a porous medium and the effects of Korteweg stresses. Phys Fluids 13:2447–2456
13. Probstein RF (2003) Physicochemical hydrodynamics. Wiley, New Jersey

14. Sahu KC (2013) Double diffusive effects on pressure-driven miscible channel flow: influence of variable diffusivity. Int J Multiphase Flow 55:24–31
15. Tan CT, Homsy GM (1988) Simulation of non-linear viscous fingering in miscible displacement. Phys Fluids 31:1330
16. Jha B, Cueto-Felgueroso L, Juanes R (2011) Fluid mixing from viscous fingering. Phys Rev Lett 106:194502
17. Manickam O, Homsy GM (1995) Fingering instabilities in vertical miscible displacement flow in porous media. J Fluid Mech 288:75–102
18. Manickam O, Homsy GM (1993) Stability of miscible displacements in porous media with nonmonotonic viscosity profiles. Phys Fluids 5(6):1356–1367

MOOC as a Complex System

Natália Nakano, Mariana Cantisani Padua, and Maria José Vicentini Jorente

1 Introduction

During the second decade of the twenty-first century, Information Society is being shaped by a series of new factors. Social media, new mobile technology, and the possibility of access to Internet at low costs have impacted access, production, dissemination, and sharing of information, and consequently constructing knowledge in a way humanity has never witnessed before.

Transformations in the way we live and relate to each other were levered by the information explosion after the advent of Internet, especially with the use of social media and mobile technology, which, at first, provoked an unbalance. When a system is out of balance, different factors influence it "in order to establish a new, sustainable equilibrium" [1]. In our study, we believe that understanding the emerging phenomenon of MOOCs from a complex and interdisciplinary point of view, we can direct participants, tutors (curators), and educational institutions in achieving balance.

Considering that the linear transmission of information held in our schools is inappropriate for our current reality, researchers worldwide are studying and searching for nonlinear forms of teaching that fit the nonlinear characteristics of the human mind and social relationships. In addition, researchers also have to consider the technological possibilities and usage for teaching and learning purposes. The Massive Open Online Course (MOOC) may be a possibility for purposes of knowledge construction in the Digital Era we are currently experiencing.

MOOCs are open online courses offered on the Internet, often at no costs for the participant. They are offered by the most prestigious universities in the world, taught by the most popular professors from these universities on the internet,

N. Nakano • M.C. Padua (✉) • M.J.V. Jorente
UNESP, Av. Hygino Muzzi Filho, 737, Bairro: Mirante, 17.525-000 Marília, SP, Brazil
e-mail: natinakano@gmail.com; mariana.cantisani@gmail.com; mjjorente@yahoo.com.br

P. Bourgine et al. (eds.), *First Complex Systems Digital Campus World E-Conference 2015*, Springer Proceedings in Complexity, DOI 10.1007/978-3-319-45901-1_12

for whoever—with an internet connection—wishes to take it. MOOC institutions around the world include Coursera, EdX, Udacity in the USA, MiriadaX in Spain, and Veduca in Brazil.

We treat MOOCs as a complex system from the perspective of Design, and therefore the understanding of principles of new media development (or technological artifacts) must include complex aspects to the development of these systems.

Murray [2] claims "we recognize objects in the world by perceptual patterns associated with them (how they look and sound) and by the behavioral patterns they exhibit (the bird flies away as we draw near). [. . .] We can abstract general patterns from multiple specific experiences (kinds of animals, categories of food, the platonic idea of a chair), or apply patterns from one kind of experience to another [. . .]."

Developing systems comply with the idea of driving a car, for example: whoever has driven a car in his country has driven a car everywhere. The ignition, the gears, and the engineering all follow more or less the same principle. We claim that, although there is similarity, the understanding patterns are part of a distributed cognition in which culture is spread and shared. If a complex vision of these platforms is employed, design may help systems developers to go beyond and include cultural and contextual characteristics of different groups and thus facilitate interaction [2].

In this study we assume that the knowledge is constructed when interaction and conversation occur preceded by stimuli and reasoning on a phenomenon or object [3]. According to Wagensberg [4], "knowledge is a mental representation (necessarily finite) of a complexity (necessarily infinite) capable of crossing reality to reach another mind. Then, by definition, by this definition, there is no knowledge that cannot be transmitted. When a complexity cannot travel from one mind to another, perhaps it is then an idea, a intuition, an experience or a sight . . . , but it is not knowledge yet. Knowledge is the form that an idea acquires to survive the journey between two minds. Creating knowledge is treating ideas for the journey."

2 MOOCs from Design Perspective

In this complex scenario of technological innovations, re-shaped connections and relationships, discussions on new educational and interdisciplinary formats, the objective of this paper is to discuss a new form of knowledge dissemination and educational format (as currently presented by literature), as a complex digital environment, beyond what education and computer science have studied. We aim to discuss the emergence of MOOCs as a complex phenomenon from interdisciplinary perspectives.

MOOCs present a pedagogic format that embraces complexity made possible by emerging technologies and may be the key to a new order in information and knowledge dissemination. MOOCs allow language convergence, which allows different learning styles to take advantage from formats that foster comprehension and learning and hence knowledge construction [1].

MOOCs are self-organized [1], which means they are open to information flow, the participants are free to interact to each other and the tutor (or curator), free to bring information to the forums, free to relate, and connect to each other, and as a result a new complex phenomenon emerges. In other words, people are free to make their own decisions on the system. The participants are influenced by the relationships and the digital environment, and the converging languages, which altogether represent a new form of culture that re-shapes the individual, changing their sense of reality and vision of the world. MOOCs were primarily intended for knowledge dissemination; however, the creators of the platform could not foresee, at that time, the consequences of the connections and relationships could transform the way people interact and connect within the possibilities of the system.

Thus, the interaction occurs through the interface. Therefore the importance of design to worry on the organization and structuring of the information to provide ways for the participant to make the best decisions on the system and feel comfort and be able to construct their own path to have the best experience possible. Murray [2] presents three foundational design principles: "1. All things made with electronic *bits* and computer *code* belong to a single new *medium*, the *digital medium*, with its own unique *affordances*. 2. Designing any single artifact within this new medium is part of the broader collective effort of making meaning through the invention and refinement of digital media *conventions*. 3. When we expand the meaning-making conventions that make up human culture, we expand our ability to understand the world and to connect with one another."

Wagensberg applies a complex approach to his work when organizing the structure of a physical environment—a science museum—which can analogically be used in all kinds of physical and digital environments that aim to construct and disseminate knowledge. In our research, we have studied his successful principles [3] brought into practice to an environment that displays real objects and phenomenon in order to teach about reality and natural phenomenon.

The experiences provoked by Wagensberg's theories in his ideas for science museums can be projected to what a person experiences in an MOOC to be able to construct knowledge if the digital environment is well designed and the complex principles and theories of Design are applied.

As Wagensberg [3] argues, "the priority is to create a difference between before and after a visit [to a science museum] that will change the attitude in relation to all these activities and others related to science, such as: traveling, strolling around a bookstore, asking in the classroom, selecting TV channels, etc. A museum should offer more questions than answers."

The environment should offer stimuli for the process of constructing knowledge. According to the author, stimuli is reached with three classes of interactivity:

(1) Manual interactivity or emotion teaser (which he names Hands on)
(2) Mental interactivity or intelligible emotion (named Minds on)
(3) Cultural interactivity or cultural emotion (named Love on)

The author claims that touching real objects is experimenting nature and conversing with it, while thinking is conversing to yourself. Mental activity means practicing the intelligibility of science, relating the experiment in an artificial environment, in this case the science museum to real world experiences and daily life experiences. Understanding the world triggers emotions, but not any kind of emotion, intelligible emotions. As for cultural interactivity, a good environment shall provide cultural emotions, i.e., emotions triggered by the individual's identity prioritized in the museum.

Similarly, Design, especially the field of Design entitled User Experience, aims to provoke good sensations and optimal digital experiences while interacting with a digital system no matter for shopping, finding your location, or learning and producing knowledge.

An MOOC does not have a specific audience to satisfy, there is no previous knowledge required, it is intended to take knowledge to anyone—young or elderly—with an internet connection, no background level of education or culture is required. Therefore, MOOCs are claimed by some authors to democratize education, and empower people as these courses take education to people with no access to good, face-to-face, brick-and-mortar schools and consequently, promote huge educational and social changes [5]. The features of MOOCs do not substitute the experiences of books, films, videos, and lectures, on the contrary, a good experience with MOOCs make the participant longer for books, films, videos, and lectures.

Considering that even though learning and understanding are individual activities but take place in some kind of conversation, knowledge is achieved individually, i.e., depends on the individual's cognition, only if stimuli to conversation is provided [6]. Wagensberg claims he does not remember talking/conversing during the 20 years he was inside a brick-and-mortar classroom. Unlike the traditional mainstream classroom, the participants to an MOOC are invited to express their opinions, doubts, ideas, and share information in social media environments such as blogs and Facebook.

As for the development of digital environments for distance/distributed learning purposes, Rhode [7] proposes a hierarchy Instructional Design should follow to help people learn better: theories, models, and application. His ideas are summarized in Fig. 1. According to the author, an Instructional Design to a digital environment should be built upon conceptual and procedural models, which, in turn, are built upon theoretical foundations.

Analyzing the basis of the pyramid, within the learning theories in practice, we would like to highlight the importance of considering the innovative Connectivism Theory proposed by Siemens in 2004 [8]. Connectivism advocates that learning happens when connection takes place in the nodes of digital environments. The theory is in line with the theories of chaos and complexity especially as connectivism is believed to increase learners abilities to perceive connections between different fields, and allows participants to choose what they want to learn and how. Siemens [8] highlighted that instructional design and learning should be designed for adaptability, for "patterning, wayfinding and sensemaking" and focus on "content, context, and connections."

In the Hierarchy of Instructional Design, interaction would be placed under the conceptual and procedural models. On conceptual and procedural models, one must

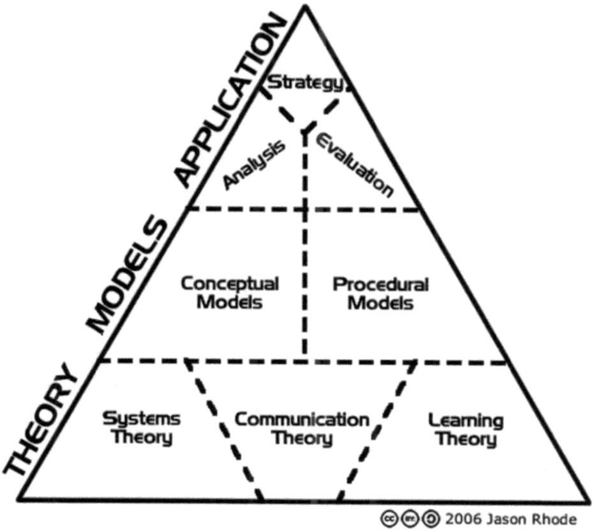

Fig. 1 Hierarchy of Instructional Design (Source: Jason Rhode, https://www.flickr.com/photos/jrhode/408667599/sizes/o/)

reflect on Morin's [9] studies to consider knowledge influenced by the brain and the cognition processes, culture and society, and spirit (emotions, conscience, and psyche), which altogether lead to strategies and action.

Action and strategy is represented in the pyramid proposed by Rhode [7]; however, the broader context or influencing environment in which the Instructional Design is inserted is not represented in his model. Concerning the development of a new kind of framework to facilitate the design and the process of building an MOOC environment, we highlight that an interdisciplinary approach is desired, if not demanded. Ideally, educators, media professionals, designers, computer scientists, and information scientists work together to create effective information environments based on current sound theories that mirror complexity theory.

3 Considerations

Models, schemes, and protocols are widely accepted in Cartesian and positivist view and even regarded efficient as they facilitate and organize the steps to be taken based on the assumption that these protocols promote patterns and annihilate flaws. However, they reduce some phenomena to simplistic paths that are impossible to be regarded as simple and reduced.

Based on the theoretical investigation of complexity and MOOCs we concluded that the results aimed for the courses and knowledge acquisition will be different in different cultures, and that perhaps these platforms should be reflected upon including, in addition to the language translations—which has already been done by some MOOC facilitators—but also include cognition, cultural/societal, and spiritual

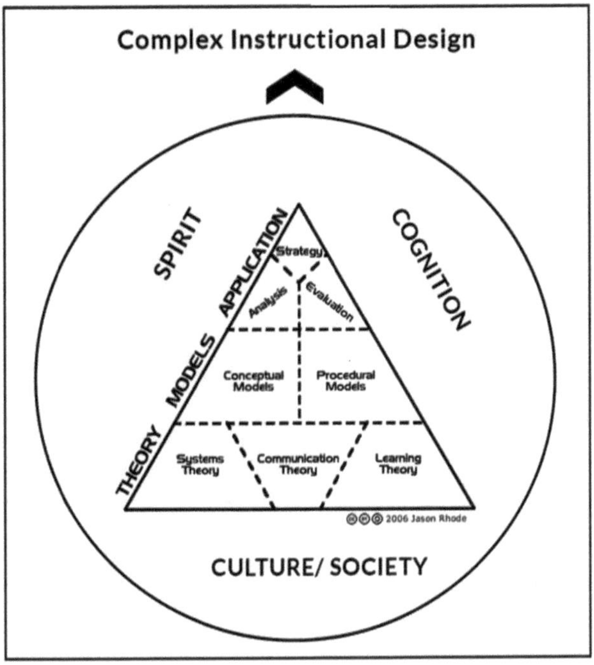

Fig. 2 Adapted from Rhode *Hierarchy of Instructional Design*

aspects, to which the Design may contribute. We propose an adaptation to Rhode's model, presented in Fig. 2, which includes the context in which the Instructional Design must be considered.

Even though MOOCs are a global phenomenon and the possibilities Internet has brought to humanity break through the barriers of time and space in knowledge construction, one must consider the human, cultural, environmental and spiritual characteristics of the individual, and groups of individuals when conceiving a system to disseminate information and promote knowledge construction.

If the MOOC Instructional Design is the same for different groups, these groups may not succeed depending on the group's context, which, in turn, may lead to the false idea that MOOCs are inappropriate for that complex context. Therefore, different Instructional Design frameworks must be designed for different groups, respecting the environment, culture/society, and spirit of that group, in addition to respecting the complex process of knowledge construction.

4 Scientific Validation

This paper has been unanimously validated in a collaborative review mode with the following reviewers:

- Soumya Banerjee/Harvard University/Boston, MA
- Kelly de Jesus/Universidade do Porto.

References

1. de Waard I, Abajian S, Gallagher MS, Hogue R, Keskin N, Koutropoulos A, Rodriguez OC (2011) Using mLearning and MOOCs to understand chaos, emergence, and complexity in education. Int Rev Res Open Distrib Learn 12:94–115
2. Murray JH (2011) Inventing the medium: principles of interaction design as a cultural practice
3. Wagensberg J. Principios fundamentales de la museología científica moderna. http://www.redined.mec.es/oai/indexg.php?registro=005200130021
4. País EE. Las palabras del conocimiento. http://elpais.com/diario/2003/05/14/futuro/1052863208_850215.html
5. McCuen@aacu.org: MOOCs and democratic education. https://www.aacu.org/liberaleducation/2013/fall/carver-harrison
6. País EE. Conversar, conversar. http://elpais.com/diario/2000/10/04/futuro/970610410_850215.html
7. Rhode J, Ph.D. Hierarchy of Instructional Design. http://www.jasonrhode.com/hierarchy-of-instructional-design
8. Siemens G. Connectivism: a learning theory for the digital age. http://er.dut.ac.za/handle/123456789/69
9. Morin E (2008) O método 3. O conhecimento do conhecimento. Tradução de Juermir Machado da Silva

Emerging Dance Movements Under Ecological Constraints in Contact Improvisation Dancers with Different Background

C. Torrents, J. Coterón, A. Ric, and R. Hristovski

1 Introduction

Contact improvisation is a form of dance improvisation which involves two bodies in contact. The improvisational characteristics of CI are such that the generation of movements is not based on fixed and standardized movements or techniques, since this dance form requires a body that responds to the physical exchange of weight and contact [1]. Theoretical approaches most used to explain creative behavior in dance are insufficient to explain how dancers create new configurations of movements during improvisation. Dance improvisation can be thought as a bank of emergent human movement forms and an expression of continuous exploration and discovery of idiosyncratic gestures, postures, and actions supported by immediate affordances, i.e., opportunities for action.

Hristovski et al. [2], in an empirical study based on the principles of coordination dynamics and using the framework of complex systems theories [3], showed how the manipulation of task and personal constraints may enable the emergence of innovative and functional tactical behaviors in athletes during sport performance. Under a specific mix of ecological constraints, performers in boxing generated certain behavioral states. It was observed that changing the set of task constraints led to spontaneous and novel behaviors, in this case, specific striking techniques

C. Torrents (✉) • A. Ric
INEFC–University of Lleida, Lleida, Spain
e-mail: carlotat@yahoo.com; angelric10@outlook.com

J. Coterón
Universidad Politécnica de Madrid, Madrid, Spain
e-mail: j.coteron@upm.es

R. Hristovski
Sts. Cyril and Methodius University, Skopje, Republic of Macedonia
e-mail: robert_hristovski@yahoo.com

P. Bourgine et al. (eds.), *First Complex Systems Digital Campus World E-Conference 2015*, Springer Proceedings in Complexity, DOI 10.1007/978-3-319-45901-1_13

that had not been practiced or taught before. These observations of the emergence of human tactical innovative behaviors are important from the perspective of Boden's [4] definition of "transformational creativity," since they explain how, when manipulating constraints, a transformation within a neurobiological action system may be enabled without reference to abstract rule-governed behaviors. Specifically in dance, Torrents et al. [5] studied couples of dancers improvising under different instructional constraints and observed that some of them had a significant effect on the type of configurations performed by the dancers, as well as on their creative behavior. Participants were all expert dancers, so the influence of the background in this phenomena remains unknown.

The aim of this study was to examine the effects that the manipulation of instructional task constraints has on the emergence of exploratory behavior in expert dancers and novices while improvising, and also to analyze the spontaneous emergence of specific movements in relation to the task constraints imposed.

2 Methods

One couple of contemporary dancers (one woman of 24 years old, 47 kg and 160 cm and a man of 20 years old, 168 cm and 64 kg) and one couple of Physical Activities and Sports Science students without dance background (two men: 20 ± 1 years old; 71 ± 7 kg and 175 ± 7 cm) participated voluntarily in the study. Participants warmed up individually. They danced together in duets lasting 480 s, and in a limited space of 12×12 m. They danced under two different instructional constraints:

- Instructional constraint (IC) 1: When dancing, try to keep your pelvis as close as possible to your partner
- IC 2: When dancing, try to keep your pelvis as far away as possible from your partner.

Four duets were video-recorded and video images were analyzed by two independent observers. The observers took into account the dance movements performed by each of the two dancers. Sequences of actions/postures were analyzed to determine their structural and dynamical characteristics. Action or posture and movements were defined on a coarse-grained scale. A total of 49 categories, corresponding to elementary movement/posture skill components (see Table 1), were taken into account. A value of 1 was ascribed to the active category and a value of 0 to the inactive one. This enabled the formation of a temporal 49×480 binary matrix with a time resolution of 1 s. Each 1-second window was defined as a 49-component binary vector (column) representing the full action configuration during the same time interval. It is important to note that each full action configuration ($L_{max} = 49$) was composed of combinations of skill and movement of shorter length ($1 < L < L_{max}$), which in turn comprised elementary skill/movement categories, each with a length $L_{min} = 1$.

A principal component analysis (PCA) [6] was performed on the 49 (category) × 480 (time point) binary matrices defining movement/posture configurations. This was done with the aim of detecting the structure of the dancer's action landscape. The number of significant principal components was determined by identifying those that accounted for ≥80 % of the explained variance.

The average dynamic overlap, $\langle q_d(t) \rangle$, was calculated as an average cosine autosimilarity of the overlaps between L_{max} configurations with increasing time lag in order to determine the dynamic properties of a dancer's complex movement patterns [7]. This measure captures the average similarity of configurations at ever increasing time distances (i.e., time lags) from each other. Hence, it is capable of detecting the rate and breadth of exploratory behavior on different time scales. The

Table 1 Observational instrument defined on a coarse-grained scale based on action or posture skills and movements [5]

1–6: **Support stability skills involving the floor**: Actions of stability or balance using the floor that can be supported by...	Upper limbs (one or two)
	Lower limbs (one or two)
	Head
	Pelvis
	Torso
	Back
7–12: **Support stability over a partner**: Actions of stability or balance supported over the... of the partner	Upper limbs (one or two)
	Lower limbs (one or two)
	Head
	Pelvis
	Torso
	Back
13–16: **Axial stability skills**: Turns around the...	Longitudinal axis of the body
	Horizontal transverse axis of the body
	Horizontal anteroposterior axis of the body
	A combination of axes
17: Jumping	
18: Being lifted or sustained by the partner	
19, 20: Level changing from...	Middle to down or falling
	Down to middle or rising from the floor
21–24: Locomotor skills involving...	Walking
	Movement on all fours
	Rolling
	Sliding
25: Receiving the partner	
26: Colliding with the partner	
27: Leading the partner	
28: Lifting or sustaining the partner	
29: Avoiding or escaping from the partner	

(continued)

Table 1 (continued)

30–47: Positions or actions of the different parts of the body	One leg is bent
	Both legs are bent
	One leg is moving
	Both legs are moving
	One arm is bent
	Both arms are bent
	One arm is moving
	Both arms are moving
	One arm is relaxed, following gravity line
	Both arms are relaxed
	Body is almost aligned
	Body is bent forwards
	Body is bent backwards
	Body is bent to the right
	Body is bent to the left
	Body is moving
	Body is inverted (more than 45°)
	Head is moving
48: There is a change in the direction of the movement	
49: There is a change in the global position of the body	

average dynamic overlap was fitted by the following equation, which is derived for systems with an intricate hierarchical structure:

$$\langle q_{\mathrm{d}}(t)\rangle = (1 - q_{\mathrm{stat}})\, t - \alpha + q_{\mathrm{stat}} \tag{1}$$

where q_{stat} is the asymptotic (i.e., stationary) value of the dynamic overlap, t is the time lag, and α is the dynamic exponent. q_{stat} detects the long-term exploratory breadth of the dancer, and α the rate of exploration.

In order to examine the coupling dynamics of the duets we conducted a cross-correlation analysis of the Hamming distance time series. Hamming distance time series provide information about the size of individual reconfigurations in each time point, and hence, cross-correlating the time series tells us about the time profile and the strength of duet couplings.

3 Results

In general, the results from the component scores analysis show that certain movements attract the system and appear with a high frequency, while others are statistically rare and short-lived reconfigurations that constitute fluctuations (see Fig. 1). These skills varied depending on the constraint and more clearly on the background of the dancers. First level PCA revealed 11 and 8 primary PCs for expert contemporary dancers in IC1, and 13 and 11 for IC2; Novice dancers obtained five or six primary PCs in both conditions, all results accounting for more than 80 % of the total variance.

The highest level PC contained the dominant long-term persistent actions, such as: leading the partner using the foot-floor surface as support with bipedestrian locomotion and support of the partner with the upper limbs. These results are similar to that obtained in a previous study with expert dancers [5]. Nevertheless, when analyzing the rest of the first level PCs, we were able to see that contemporary dancers were more influenced by the constraints imposed in the execution of specific skills. For instance, under pelvis-close constraint lifts were more time-persistent, although novices explored less coordination patterns. The PCs of the novices with pelvis-far constraint revealed that they dominantly used the support with the partner using upper limbs, while experts performed lots of combinations in all cases.

Figure 2 displays the average values obtained for all dancers and constraints of the stationary part of the dynamic overlap. The dynamic overlaps of pelvis-close constraints were always significantly higher than pelvis-far constraints (in both couples), while the values obtained by expert dancers were always significantly lower than amateurs (in both constraints). Hence, combined skill level and instructional constraints together revealed differences in the exploratory breadth of performers. It is interesting to notice that the observational instrument coupled to the measure of dynamic overlap is able to detect the differences in the exploratory breadth *within* expert and novice couples as well as *between* them. For example, N2 showed lower exploratory breadth than N1 performer in both, the pelvis-close and pelvis-far conditions. Similarly, E2 showed lower exploratory behavior than E1 in both conditions too.

The cross-correlation of Hamming distances time series showed consistently weak but significant lag 0 (synchronous reconfigurations) until lag 3 (leading–

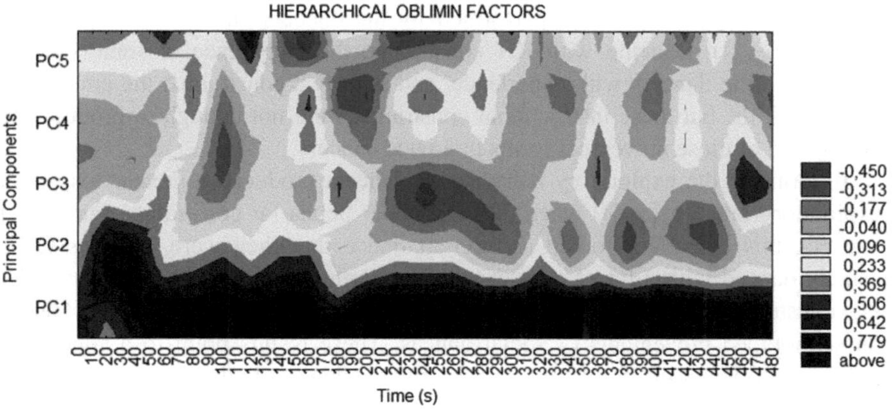

Fig. 1 Typical potential landscape of action configurations of one of the novice dancers. The first principal component appears with higher frequency compared with the rest

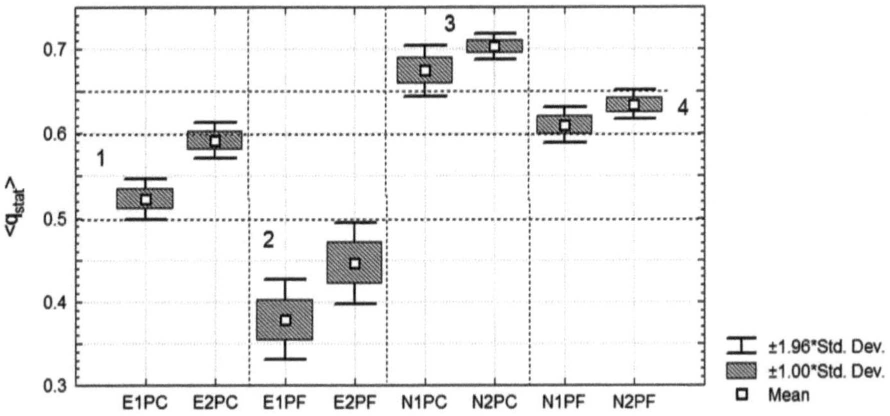

Fig. 2 The profiles of the stationary values of the dynamic overlap $\langle q_{\text{stat}} \rangle$ for expert E1 and E2 dancers and novice (N1 and N2) dancers, under 2 instructional constraints, pelvis-close (PC) and pelvis-far (PF)

following reconfigurations) correlations in expert dancers under both ICs but just with pelvis far in novels. Changes in reconfiguration sequences illustrated the versatility of the strength and length of the temporal couplings between actions in expert dancers. However, it is suggested that ICs influenced coupling dynamics of novel dancers, as dancing with pelvis-close disrupted the couple to be adhered to synchronous and leading–following reconfigurations.

4 Discussion

Instructional constraints, as well as the level of the dancers, had a significant effect on the type of configurations performed by the dancers and on the exploratory breadth ($\langle q_{\text{stat}} \rangle$). Dancing with IC1 constrained the emergence of some specific CI skills in the dancers with experience, and with IC2 constrained the emergence of other skills, but these instructional constraints were not enough to produce the emergence of those movement forms in novice dancers.

In terms of the exploratory behavior all dancers produced a greater long-term exploratory breadth $\langle q_{\text{stat}} \rangle$) when dancing under pelvis-far constraints. For expert dancers, this constraint forced a different dance that they were used, as CI is usually danced with a lot of contact with the partner. For novices, dancing under PC constraint was really difficult. It created a very repeating behavior based on stereotyped movements of social dances. These results point to a significant reduction of degrees of freedom when forcing the dancers to dance so close.

5 Scientific Validation

This paper has been unanimously validated in a collaborative review mode with the following reviewers:

- Reviewer 1: Christophe Schnitzler, University of Strasbourg
- Reviewer 2: Ludovic Seifert, University of Rouen.

References

1. Albright AC (2003) Contact improvisation at twenty-five. In: Albright BC, Gere D (eds) Taken by surprise. A dance improvisation reader. Wesleyan University Press, Middletown
2. Hristovski R, Davids K, Araújo D, Button C (2006) How boxers decide to punch a target: emergent behavior in nonlinear dynamical movement systems. J Sports Sci Med CSSI:60–73
3. Kelso JAS (1995) Dynamic patterns – the self-organisation of brain and behavior. MIT Press, Cambridge
4. Boden MA (1996) What is creativity. In: Boden MA (ed) Dimensions of creativity. MIT Press, Cambridge
5. Torrents C, Ric A, Hristovski R (2015) Creativity and emergence of specific dance movements using instructional constraints. Psychol Aesthetics Creativity Arts 9:65–74
6. Westerhuis JA, Kourti T, MacGregor JF (1998) Analysis of multiblock and hierarchical PCA and PLS models. J Chemom 12:301–321
7. Hristovski R, Davids K, Araújo D, Passos P, Torrents C, Aceski A, Tufekcievski A (2013) Creativity in sport and dance: ecological dynamics on a hierarchically soft-assembled perception-action landscape. In: Davids K, Hristovski R, Araújo D, Balagué N, Button C, Passos P (eds) Complex systems in sport. Routledge, London, pp 261–274

Adaptability in Swimming Pattern: How Do Swimmers Modify Propulsive Action as a Function of Speed?

Christophe Schnitzler, Ludovic Seifert, Chris Button, Gavin Kennedy, and Huub Toussaint

List of Acronyms

I^+	Force impulse over 5 s
$I^+{}_{/\text{arm}}$	Force impulse over 1 cycle in the right hand
SF	Stroke frequency
Fpull	Force (in N) developed during pull phase
Fpush	Force (in N) developed during the push phase
PrPs	Absolute time of propulsive phase duration (in s)
PrP%	Time of the relative duration of the propulsive phase

1 Introduction

Complex behaviour arises from the inter-relationship, interaction, and interconnectivity of elements within a system and between a system and its environment [1]. The emergence of behaviour in sport has been characterized as an adaptive complex

C. Schnitzler (✉)
L.I.S.E.C Laboratory UPRES EA 2310, University of Strasbourg, Strasbourg, France
e-mail: cschnitzler@unistra.fr

L. Seifert
Faculty of Sports Sciences, C.E.T.A.P.S. Laboratory UPRES EA 3832, University of Rouen, Mont-Saint-Aignan, France

C. Button • G. Kennedy
University of Otago, Dunedin, New Zealand

H. Toussaint
Academy of Physical Education, University of Applied Sciences Amsterdam, Amsterdam, The Netherlands

P. Bourgine et al. (eds.), *First Complex Systems Digital Campus World E-Conference 2015*, Springer Proceedings in Complexity, DOI 10.1007/978-3-319-45901-1_14

system which depends upon organismic, task, and environmental constraints [2]. Put up simply, moving forward represents the task constraint in swimming. To achieve this goal, propelling actions to overcome drag are produced. Environmental constraints are mostly linked to water physical properties (density and fluidity), which supports propulsion but also generates drag; last, organismic constraints consist in individual biometric characteristics, but also level of expertise. To move forward, swimmers thus generate force impulses with several part of their body. If skill level influences drag–speed relationship [3], it is unclear from the literature how it affects force impulse generation.

Mathematically, an impulse is defined as the time integral of the resultant force acting on a body [4]. Over a fixed period of time, the total impulse, I^+, expressed in Newton.second (N.s), is the integral over time of the total force production (Eq. 1)

$$I^+ = \int_{t2}^{t1} F(t)dt \, (\text{N.s}) \quad (1)$$

In swimming, speed is produced by a series of propulsive impulses generated by both arms and legs. If we consider swim speed being relatively constant from cycle to cycle, the total impulse I^+ can be considered as being the sum of this discrete impulse (Eq. 2):

$$I^+ = n \times \left(I^{+/}{}_{\text{right arm}} + I^{+/}{}_{\text{left arm}} + I^{+/}{}_{\text{right leg}} + I^{+/}{}_{\text{left leg}}\right) \quad (2)$$

n: number of cycle during the period considered

$I^{+/}{}_{\text{right arm}} + I^{+/}{}_{\text{left arm}} + I^{+/}{}_{\text{right leg}} + I^{+/}{}_{\text{left leg}}$: discrete impulses from arms (right and left) and legs (right and left) during a swim cycle.

If a swimmer wants to increase his/her average swim speed, he/she will have to find ways to increase his total propulsive force impulse (I^+), or decrease his hydrodynamic drag. This goal can be achieved by increasing the frequency of the impulses, or increasing the magnitude of within cycle impulses, or by simultaneously increasing both of these factors. Within cycle, the different impulses (e.g. $I^{+/}{}_{\text{right arm}}$) can be increased either (a) by increasing the amplitude of the force, (b) by increasing the duration of the force, and (c) by increasing both amplitude and duration.

Past studies showed that when a swimmer increases his/her speed, an increase in stroke frequency (SF) and a modification in relative duration of propulsive time (in percentage) occurs. But no study so far examined how the magnitude of the within cycle impulses was affected, and if these adaptations were skill-dependent.

In the present study, we aim to provide an insight of swimmers' adaptability as a function of skill level by for the first time recording simultaneously kinematical and kinetical parameters in ecological conditions.

Table 1 Main characteristics of the participants

Expertise level	Training/week (hours)	Age (years)	Hand surface area (cm^2)	Maximal speed ($m.s^{-1}$)	% of world record speed (100 m)
Low ($n=6$)	0.5	32.5 ± 4.0	165 ± 25	1.24 ± 0.05	45.4 ± 3.7
Medium ($n=6$)	4	27.0 ± 7.5	172 ± 16	1.54 ± 0.1	69.3 ± 4.9
High ($n=8$)	14	18.7 ± 2.9	159 ± 14	1.82 ± 0.05	82.5 ± 2.6

2 Methods

Twenty male swimmers sub-divided into three distinct categories: low-, medium-, and high-level of expertise participated in the present study (see Table 1).

The swim trials took place in a motorized aquatic flume. Participants had to swim at four individual-specific speeds relative to their maximal speed: v_1 (60 %), v_2 (70 %), v_3 (85 %), and v_4 (100 %). The kinematic parameters taken into account were: stroke frequency (SF), stroke length (SL), stroke index (SI), propulsive phase duration (PrP%), and coordination parameters (IdC). They were determined using the method described by Seifert et al. [5].

To determine force parameter four pairs of mono axial pressure sensors (Kyowa, Tokyo, and Japan) were glued to the surface of a glove. Due to technical limitations, only the results from one hand (the dominant one) could be recorded. Total resultant force was then calculated by applying Takagi and Wilson's method [6].

To calculate force impulse, video and force curve were superimposed on a single graphical user interface. The force signal was cropped and reconstructed so that only the force developed during propulsive time determined previously (pull and push phases) was taken into consideration. Force impulse over 5 s (I^+) was calculated by numerical integration. The discrete impulse for 6 hand cycle was determined and then averaged to obtain an estimate of the average impulse per hand cycle ($I^+{}_{/\text{stroke}}$). Also, from the force curve, the two main peak forces were determined during pull (Fpull) and push (Fpush) phase, and taken as indicators of the magnitude of the force within the cycle. The absolute duration (PrPs, in seconds) of the propulsive phase of the arm was also measured.

To analyse the data, a series of two-ways ANOVAs for repeated measures [repeated factor: pace] were used to compare the mean values for each variable as a function of pace and expertise level. To detect significant differences, Turkey's post hoc tests were used. The threshold for significance was set at the 0.05 level of confidence. For the statistical analysis, Minitab 15.1 was used.

3 Results

When increasing swim speed, all swimmers increased SF, I^+ and IdC ($p < 0.05$). Swim speed, SL, and SI discriminated among skill levels. The IdC was higher in low- and high-skilled level than in medium-level group ($p < 0.05$).

Figure 1 examines the determinant of I^+ as a function of skill and pace. In all populations, I^+ increases over pace is obtained by a subsequent increase in stroke frequency. However force impulse per cycle ($I^+_{/\mathrm{stroke}}$) only increased in medium (63 ± 6.7 vs 83 ± 7.3 N.s, $p < 0.05$)—and high (63 ± 7.2 vs 78 ± 7.2 N.S, $p < 0.05$)-expertise level from pace 1 (slow) to pace 4 (sprint). Absolute propulsive time per stroke (PrPs) decrease of 24 % when swim speed increases in all populations. Only medium- and high-expertise levels were capable of increasing both peak pull and peak push forces when increasing swim speed (high level: 58 vs 74 N for pull, 69 vs 82 N for push phase; medium level: 55 vs 82 N for the pull phase; medium level: 55 vs 82 N for the pull phase, 65 vs 97 N or the push phase). This suggests that, for medium and high level swimmers, $I^+_{/\mathrm{stroke}}$ is increased by an increase in the magnitude of the force developed rather than by the propulsive time duration.

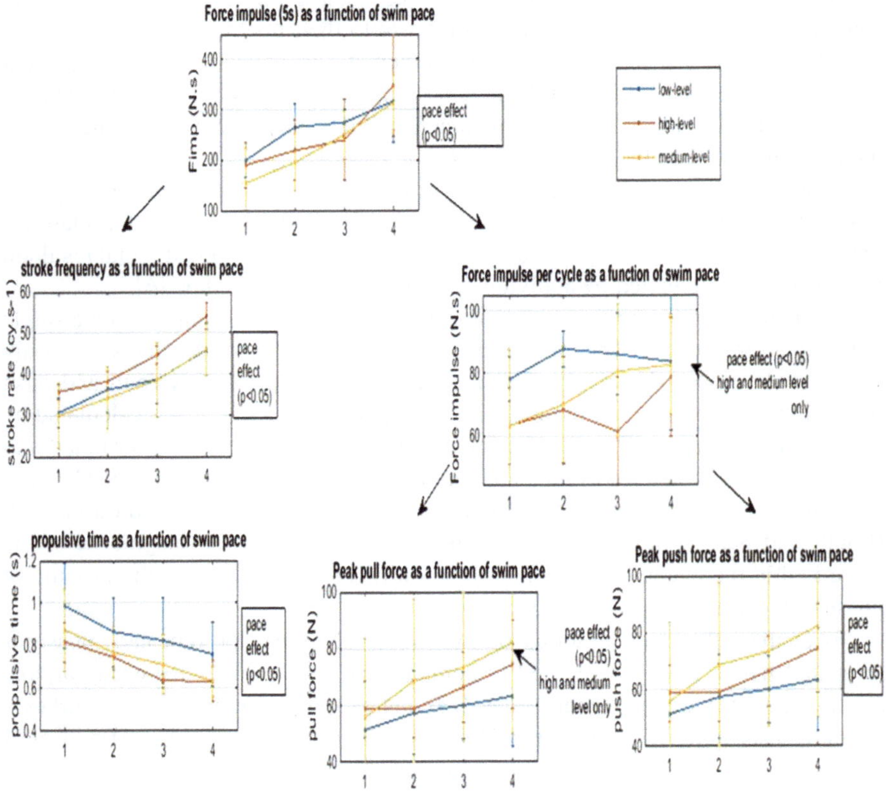

Fig. 1 Force impulse, stroke frequency, force impulse per cycle, propulsive time, peak pull and push forces as a function of pace and skill level

4 Discussion

The objective of this study was to examine and compare the motor adaptive behaviour of swimmers with different skill levels when they increase their swim speed. Main results were twofold: first, the magnitude of the total impulse (I^+) generated during 5 s is not skill dependent. Second, if I^+ increase with swim speed, only medium- and high-level swimmers are capable of increasing $I^+_{/\text{stroke}}$ simultaneously scaling up stroke frequency and discrete impulse.

Propulsive impulses are generated by swimmers to move forward, therefore overcoming drag encountered. In that, better swimmers are characterized by a capacity to produce propulsive impulses of higher magnitude, and/or exhibit lower drag at similar swim speeds [3, 7]. Seifert et al. [8] suggested that lower-level swimmer tended to increase swim speed while increasing SF and IdC but not tangential hand speed, which suggested that their hand tended to "slip" through the water. Our data do not support this hypothesis, since the magnitude of the impulses is no different among skill level. Here, only measurement of the force exerted perpendicular to the hand were taken into account. In other words, we did not measure the efficient component of the force, that is, the component used for propulsion. However, both SL and SI, which both are indicators of swim efficiency, strongly differentiate between skill levels. In accordance with Toussaint et al. [3], those data suggest that this is not the capacity to generate high force impulses (I^+) that differentiate among skill level, but rather the capacity to reach high values of propulsive efficiency.

In what concern kinematical adaptations, all swimmers increase their propulsive impulse (I^+) while swimming faster. This was obtained only by increasing SF in low-level swimmers (strategy 1), whereas middle- and high-expertise level increased both SF and I^+/cyle (strategy 3). So despite the fact that all participants increased their SF, IdC, propulsive peaks, and decreased their PrP(s), only middle- and high-expertise level could scale up the magnitude of the force to allow I^+/cyle to increase despite a decrease in PrP(s). Leblanc et al. [9] already observed such phenomenon in breaststroke: whereas expert swimmers were capable of adapting different speeds by modifying the coordinative mode, non-expert only changed speed by altering SF.

Those data suggest that such situation makes hard for low-level swimmers to produce efficient component at the highest speeds, and confirms that adaptability is a particularity of expert performance [10]. In accordance with past studies, those findings confirm not only that expertise is characterized by swim efficiency, but also that adaptability in itself is a key-element into differentiating among skill levels.

However, due to technical limitation, our study was unable to provide a full picture of swimmer's adaptability. Indeed, only part of the discrete impulses (one hand) could be observed; and as we outlined earlier, the magnitude of the efficient component could only be estimated. But despite those limitations, this study provides a significant breakthrough in the biomechanical measurements in swimming. By using force sensors, we were for the first time able to measure force production in an ecological situation; moreover, combining force and videos helped to determine in a much more precise and objective way the duration of the propulsive time, therefore providing a much more reliable idea of swimmers' behavioural adaptation with speed. Still, confirmatory studies measuring impulses from other part of the body and that takes into consideration efficient component of the force are necessary.

5 Conclusion

The capacity to exhibit adaptive behaviour in highly constrained environment has often been pointed out to be a characteristic of expert performance in sport. In swimming, actions of experts are mostly dedicated to lower active drag and generate propelling action as efficient as possible. The present study shows that expertise could also be characterized in terms of level of adaptability, that is, more capacity to modify coordination and force production to adequately adapt constraints. In that sense, analysis of expert performance in sport seems to be an adequate paradigm to examine how adaptability is important to characterize complex systems.

6 Scientific Validation

This paper has been unanimously validated in a collaborative review mode with the following reviewers:

- Reviewer 1 Ricardo Fernandes, Faculty of Sport, University of Porto
- Reviewer 2: CarlotaTorrents Martins, C University of Lleida (Spain)
- Reviewer 3: Alice Della Penna, CRI, Paris.

References

1. Mitleton-Kelly E (2003) Ten principles of complexity & enabling infrastructures. In: Complex systems and evolutionary perspectives on organisations: the application of complexity theory to organisations. Elsevier
2. Davids K, Button C, Bennett S (2008) Dynamical systems theory: physical constraints on coordination dynamics of skill acquisition: a constraint-led approach. Human Kinetics, Champaign (IL)
3. Toussaint HM, de Groot G, Savelberg HH, Vervoorn K, Hollander AP, van Ingen Schenau GJ (1988) Active drag related to velocity in male and female swimmers. J Biomech 21(5):435–438
4. Robertson D, Caldwell G, Hamill J, Kamen G, Whittlesey S (2004) Research methods in biomechanics. Human Kinetics, Champaign, IL

5. Seifert L, Toussaint HM, Alberty M, Schnitzler C, Chollet D (2010) Arm coordination, power, and swim efficiency in national and regional front crawl swimmers. Hum Movement Sci 29(3):426–439. doi:10.1016/j.humov.2009.11.003
6. Takagi H, Wilson B (1999) Calculating hydrodynamic force by using pressure difference in swimming. In: Keskinen K, Komi P, Hollander AP (eds) Biomechanics and medicine in swimming VIII. Gummerus Printing, Jyvaskyla, pp 101–106)
7. Chatard JC, Lavoie JM, Lacour JR (1990) Analysis of determinants of swimming economy in front crawl. Eur J Appl Physiol Occup Physiol 61(1–2):88–92
8. Seifert L, Toussaint HM, Alberty M, Schnitzler C, Chollet D (2009) Arm coordination, power, and swim efficiency in national and regional front crawl swimmers. Hum Movement Sci 29:426–439
9. Leblanc H, Seifert L, Baudry L (2005) Arm-leg coordination in flat breaststroke: a comparative study between elite and non-elite swimmers. Int J Sports Med 26(9):787–797
10. Seifert L, Button C, Davids K (2013) Key properties of expert movement systems in sport. Sports Med 43(3):167–178

Emerging Collective Shared Behaviors from Individual Exploration in Football Small-Sided Games

A. Ric, C. Torrents, and R. Hristovski

1 Introduction

In team sports, like football, a spontaneous social order emerges from the interaction of two confronting teams, where players try to adapt their behavior to the immediate changing environment. Ecological approach to cognition and action proposes that the exploratory behavior of athletes can maintain or change the nature and amount of available information in sport context been necessary for successful decision-making and action [1]. This permanent flow of player movement configurations due to task heterogeneity (different players have different immediate tasks to solve) and specific personal constraint reveals the hierarchically soft-assembled action landscape [2] which characterizes the metastability of player's behavior attaining to a temporarily stable goal through flexible actions. It has been showed that this metastable dynamic landscape is suitable to be changed under specific task constraints promoting the emergence of specific tactical solutions [3–5].

Small-sided games (SSGs) are common training drill used by football coaches because represent the unstable, dynamic and unpredictable nature of football game [6, 7]. However, by allowing manipulations of some specific variables, different player's behaviors can be elicited. Usually, these kinds of studies were conducted with aim of discovering the effect of task constrains on physical and physiological features (see [8] for review). However, few studies had theoretically explained the

A. Ric (✉) • C. Torrents
National Institute of Physical Education of Catalonia (INEFC), Complex Systems in Sport Research Group, University of Lleida, Lleida, Spain
e-mail: angelric10@outlook.com; carlotat@yahoo.com

R. Hristovski
Faculty of Physical Education, Sport and Health, Skopje, Complex Systems in Sport Research Group, Ss. Cyril and Methodius University, Macedonia
e-mail: robert_hristovski@yahoo.com

P. Bourgine et al. (eds.), *First Complex Systems Digital Campus World E-Conference 2015*, Springer Proceedings in Complexity, DOI 10.1007/978-3-319-45901-1_15

emergence of collective behavior in SSGs [9, 10]. To our knowledge, there were no empirical studies that explicitly show and model how tactical behavior is organized on different space-time scales and how task solutions, shared tactical behaviors, and task constraints are linked between them.

Therefore, the aim of this study was to identify the soft-assembled hierarchical structure of tactical behavior and demonstrate the temporally nested structure of game constraints that shape the emergence of offensive and defensive behaviors on different time scales.

2 Method

Eight male football players (age: 26 ± 4.96 years, playing experience: 19.6 ± 4.9 years; training schedule of five sessions per week) were dividing into two team of four players. Similarly, the opponents were distributed in two different teams of up to seven players. Goalkeepers participated in the protocol but were excluded from the data analysis. All the SSGs were played on an artificial pitch measuring 40×30 m, and in accordance with the official rules of soccer. Three different SSGs were designed changing the magnitude of numerical imbalances and were played in randomized order involving two 3-min periods of play for each condition (360 s):

- Small numerical advantage (4 vs. 3)
- Small numerical disadvantage (4 vs. 5)
- Large numerical disadvantage (4 vs. 7).

Data were gathered through a combination of systematic observation and the use of a 15 Hz non-differential global positioning system (SPI-ProX, GPSports, Canberra, ACT, Australia). The data collected for each player yielded binary data vectors derived from 37 categories, which belonged to the following four categories: tactical actions, inter-player context, pitch zones, and movement speeds (see Table 1); representing the full action configuration during the same time interval (1 s). The value of 1 was ascribed to the active category and a value of 0 to the inactive one. Finally, a binary (Boolean) matrix of size 37×360 s was obtained for each player.

The hierarchical principal component analysis (HPCA) procedure allowed reducing the dimensionality from 360 action configuration vectors sequentially to one final slowest collective variable. The number of significant first-level principal components was determined by identifying those that accounted for $\geq 80\,\%$ of the explained variance [11]. The component correlation matrix of the first-order principal components was then subjected to a further higher-order analysis revealing the hypothesized higher-order structure. The component score matrix was used in order to detect the most salient player action configuration that emerged during the game.

The dynamic overlap $\langle q_d(t) \rangle$ was calculated as an average cosine auto-similarity of the overlap between configurations with increasing time lag to determine the dynamic properties of player's complex tactical patterns. This measure provides information about the time scale on which the exploratory behavior sufficiently

Table 1 Data collected to assess the tactical patterns of each player formed by 37 categories

Tactical actions [13]	Offensive actions	Penetration
		Offensive coverage
		Depth mobility
		Width and length
		Offensive unity
	Defensive actions	Delay
		Defensive coverage
		Balance
		Concentration
		Defensive unity
Inter-player contexts	Rear teammate between advanced opponent and own goal	
	Intermediate teammate between advanced opponent and own goal	
	Advanced teammate between advanced opponent and own goal	
	Rear teammate between advanced and rear opponent	
	Intermediate teammate between advanced and rear opponent	
	Advanced teammate between advanced and rear opponent	
	Rear teammate between rear opponent and the opposing goal	
	Intermediate teammate between rear opponent and the opposing goal	
	Advanced teammate between rare opponent and the opposing goal	
Pitch zones	Deep-defensive right (length <10 m; width <10 m)	
	Deep-defensive center (length <10 m; width 10 < 20 m)	
	Deep-defensive left (length <10 m; width >20 m)	
	Mid-defensive right (length 10 < 20 m; width <10 m)	
	Mid-defensive center (length 10 < 20 m; width 10 < 20 m)	
	Mid-defensive left (length 10 < 20 m; width >20 m)	
	Mid-offensive right (length 20 < 30 m; width <10 m)	
	Mid-offensive center (length 20 < 30 m; width 10 < 20 m)	
	Mid-offensive left (length 20 < 30 m; width >20 m)	
	Deep-offensive right (length >30 m; width <10 m)	
	Deep-offensive center (length >30 m; width 10 < 20 m)	
	Deep-offensive left (length >30 m; width >20 m)	
Movement speeds [14]	<0.7 km/h (stand)	
	0.7–3.6 km/h (walk)	
	3.6–7.2 km/h (jog)	
	7.2–14.4 km/h (medium-intensity running)	
	14.4–19.8 km/h (high-intensity running)	
	>19.8 km/h (sprint)	

Each data vector represented a configuration of the player in a 4D category space

saturates and opens a possibility of defining the scale of short- vs. long-term dynamics. The average dynamic overlap was fitted by the following equation, which is derived for systems with an intricate hierarchical structure (see Eq. 1 [12]):

$$\langle q_d(t)\rangle = (1 - q_{\text{stat}})\, t^{-\alpha} + q_{\text{stat}} \tag{1}$$

where q_{stat} is the asymptotic (i.e., stationary) value of the dynamic overlap, t is the time lag, and α is the dynamic exponent. The average dynamic overlap, $\langle q_d(t)\rangle$, converges to a stationary value, $\langle q_{\text{stat}}\rangle$, detecting the long-term exploratory breadth of the player. However, it initially shows a descending trajectory with a slope defined by the exponent α, corresponding to the short-term (time scale of seconds) exploratory breadth.

3 Results

Under the Kaiser-Guttman criterion, the HPCA initially revealed between 10 and 14 principal components (PCs) on the first-level PC. Each level PC structure was formed by salient correlated clusters of previous order PCs. A significant dimensional reduction was obtained, resulting in between four- or five-order PCs to analyze. Finally was obtained the highest level PCs with only one principal component which always captures the most robust and stable structure of associations within the data, whereas the lower order PCs are more sensitive to more detailed changes in impinging constraints. The projections of action on the lower order PCs yielded a metastable dynamic landscape, where certain action configuration of players attract the system dwelling few seconds inside before switch to another local metastable configuration attractor basin (see Fig. 1, as an example).

Under the influence of numerical imbalance of opponent team, we can detect some differences in the slowest collective variable (highest order of PCs) that contained the dominant long-term persistent action configurations. While under small numerical advantage offensive actions were predominant, playing in numerical inferiority most of player offer to his teammate's offensive coverage in order to maintain the ball and the difficulties to go forward. In 4 vs. 7 it can be seen that some players decided to perform demarks going in depth as unique offensive strategy. Constrained by the increased number of the opponents, the teams played nearer to their goal. Due to the density of opponents, the players performed their actions in contexts surrounded by opponents, placed between advanced and rear opponent, with high risk to lose the ball possession. The speed of action movements decreased with the increasing number of opponents.

The average dynamic overlap $\langle q_d(t)\rangle$ showed a characteristic behavior involving relaxation to an average stationary value $\langle q_{\text{stat}}\rangle$. It can be seen in Fig. 2 thatfor different task conditions a different stationary value was attained in the lag interval from around 15–30 s. The first quickly relaxing part of the curve shows that exploration of different game patterns by individual players exists on a time scale of seconds. Over a scale of a few tens of seconds, however, the exploratory dynamics slows down and attains stationary values. On even larger time scales (hundreds of seconds), the degree of exploration is stationary, partly repeating the

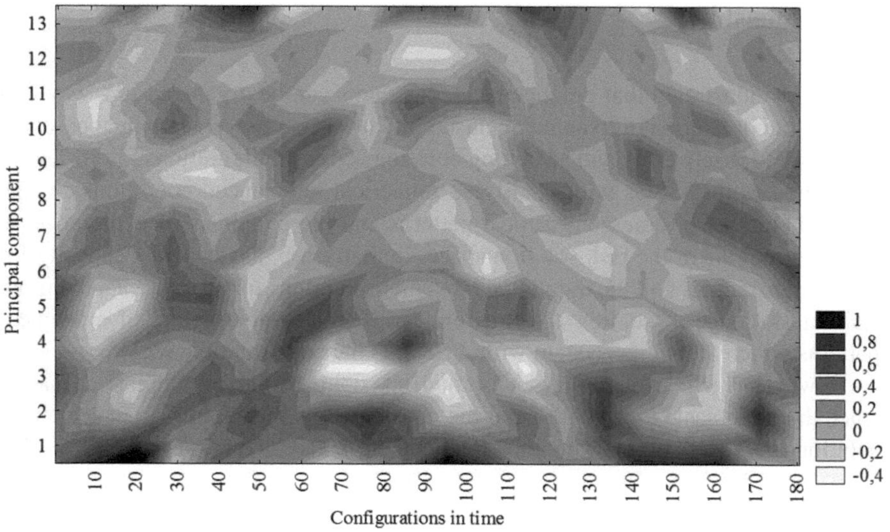

Fig. 1 Typical time evolution of the potential landscape of action configurations of one player in 4 vs. 3. As time passes configurations are attracted to and repelled from certain PC subsets. *Dark areas* are attracting, while *pale* and *white* are repelling regions

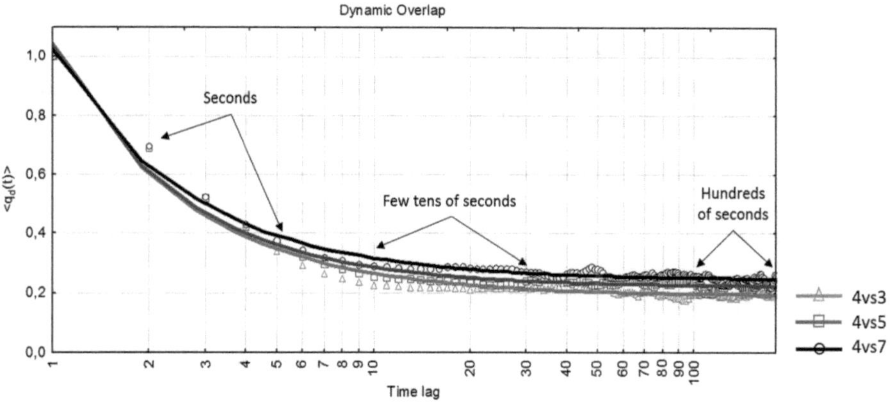

Fig. 2 Profile of the average dynamic overlap $\langle q_d(t) \rangle$ of each condition for different time lags; these dynamics unfold over three time scales (from seconds to several minutes)

already created tactical patterns, showing a consistent increase in the stationary overlap of the tactical patterns with an increasing number of opposing players (4 vs. $3 = 0.202 \pm 0.015$; 4 vs. $5 = 0.230 \pm 0.025$; 4 vs. $7 = 0.254 \pm 0.029$) [15].

4 Discussion

Results demonstrated that practice task constraints can be organized in a way that enables maximization of dynamic metastability affording players maximal ease and flexibility of discovering and switching between action configurations [3–5]. Therefore, constraining SSGs with modifying numerical imbalanced conditions change the probabilities of emergent tactical patterns and the dynamics of exploration in different regions of the task solution space. In that sense, the system's capacity for exploration captures all the individual movement solutions for an immediate or more stable task goal [2] yielding a soft-assembled hierarchical structure of movement patterns which attract the system with different characteristic time scales.

However, each level processes is interlinked in a kind of circular causality, since the large-scale task goal (avoid the goal) governs the shorter-small scale task solution (pass the ball, marking the opponent, etc.), and vice versa [16, 17].

5 Scientific Validation

This paper has been unanimously validated in a collaborative review mode with the following reviewers:

- Paul Rodriguez. UC San Diego, USA
- Ricardo Fernandes. Faculdade de Desporto, Universidade do Porto, Portugal
- Ludovic Seifert. Faculty of Sports Sciences, University of Rouen, France.

References

1. Araujo D, Davids K, Hristovski R (2006) The ecological dynamics of decision making in sport. Psychol Sport Exerc 7:653–676
2. Hristovski R, Davids K, Araújo D, Passos P (2011) Constraints-induced emergence of functional novelty in complex neurobiological systems: a basis for creativity in sport. Nonlinear Dyn Psychol Life Sci 15:175–206
3. Hristovski R, Davids K, Araújo D, Button C (2006) How boxers decide to punch a target: emergent behaviour in nonlinear dynamical movement systems. J Sport Sci Med 5:60–73
4. Pinder RA, Davids K, Renshaw I (2012) Metastability and emergent performance of dynamic interceptive actions. J Sci Med Sport 15:437–443
5. Torrents C, Ric A, Hristovski R (2015) Creativity and emergence of specific dance movements using instructional constraints. Psychol Aesthet Creat Arts 9:65–74
6. Davids K, Araújo D, Correia V, Vilar L (2013) How small-sided and conditioned games enhance acquisition of movement and decision-making skills. Exerc Sport Sci Rev 41:154–161
7. Passos P, Milho J, Fonseca S, Borges J, Araújo D, Davids K (2011) Interpersonal distance regulates functional grouping tendencies of agents in team sports. J Mot Behav 43:155–163
8. Aguiar MVD, Botelho GMA, Gonçalves BSV, Sampaio JE (2013) Physiological responses and activity profiles of football small-sided games. J Strength Cond Res 27:1287–1294

9. Sampaio JE, Lago C, Gonçalves B, Maçãs VM, Leite N (2014) Effects of pacing, status and unbalance in time motion variables, heart rate and tactical behaviour when playing 5-a-side football small-sided games. J Sci Med Sport 17:229–233
10. Frencken W, Lemmink K, Delleman N, Visscher C (2011) Oscillations of centroid position and surface area of soccer teams in small-sided games. Eur J Sport Sci 11:215–223
11. Joliffe IT (2002) Principal component analysis. Springer, New York
12. Sibani P, Dall J (2003) Log-Poisson statistics and full aging in glassy systems. EPL 64:8
13. Costa IT, Garganta J, Greco PJ, Mesquita I, Maia J (2011) System of tactical assessment in soccer (FUT-SAT): development and preliminary validation. System 7:69–83
14. Folgado H, Duarte R, Fernandes O, Sampaio J (2014) Competing with lower level opponents decreases intra-team movement synchronization and time-motion demands during pre-season soccer matches. PLoS One 9:e97145
15. Ric A, Hristovski R, Gonçalves B, Torres L, Sampaio J, Torrents C (2016) Timescales for exploratory tactical behaviour in football small-sided games. J Sports Sci 34:1723
16. Balague N, Torrents C, Hristovski R, Davids K, Araújo D (2013) Overview of complex systems in sport. J Syst Sci Complex 26:4–13
17. Araújo D, Passos P, Esteves P, Duarte R, Lopes J, Hristovski R, Davids K (2015) The micro-macro link is understanding sport tactical behaviours: integrating information and action at different levels of system analysis in sport. Mov Sport Sci 2015:1–32

Matrimonial Patterns and Trans-Ethnic Entities

Philippe Ramirez

Despite the complexity of their object of study, i.e. human society and culture, anthropologists seldom address complexity per se. In this paper I would like to briefly suggest approaches to deal with certain complex issues involved in the modelization of social relations within a naturalistic paradigm. What I want to show primarily is how anthropological knowledge might be markedly enhanced by truly confronting the complexity of anthropological phenomena, i.e. by putting culture and social structures back into the natural/physical world, to understand how "it really works". And the condition for performing such an enterprise is close cooperation with other disciplines dealing with the study of complexity. The topic at hand is modelling the origin of a system of surname equivalences in tribal India. But the work being done might have a much broader range, by helping the studies of matrimonial systems to adopt more naturalistic approaches.

Georg Simmel wrote about "large systems", formed of "immediate interactions that occur among men constantly . . . that have become crystalized . . . attaining their own laws" ([1], p. 10). The present study aims at understanding how one of these systems evolves "by itself", beside or in conjunction with what the actors say and think about it.

Among several tribal societies of North-East India people recognize that some of their neighbours, belonging to altogether different cultures and ethnicities, and speaking different languages, are actually "similar" because of their surname [28, 29]. "Similar" people are identified by their surnames being "the same". We might translate without too much risk "similar" by "synonymous" or "equivalent", as such surnames refer to clans ("descent groups" in the anthropological parlance), the immediate outcome being that equivalent clans are found across different ethnicities. "Similar" people from different tribes are not bound to any mutual

P. Ramirez (✉)
CNRS, Centre for Himalayan Studies, Villejuif Cedex, France
e-mail: pramirez@vjf.cnrs.fr

P. Bourgine et al. (eds.), *First Complex Systems Digital Campus World E-Conference 2015*, Springer Proceedings in Complexity, DOI 10.1007/978-3-319-45901-1_16

obligation other than the absolute ban of intermarriage: being the same they should not marry, which would amount to an incest, the worst crime in those societies. Thus, being different in terms of ethnicity and culture does not preclude being the same in terms of clan, which is a counter-intuitive idea for many anthropologists and for many North-East Indians as well, especially those belonging to the younger generations. Anthropologists have been taught in universities that tribal societies consist of clans who share a common culture and a common social structure. And young North-Easterners are immersed in a neo-essentialist, ethnicized India where different people are different by all means: how could they be both different and similar?

The initial question which the research described here tries to address is the origin of such a system, with the view that in the absence of a centralized authority such relations may have been the product of processes of self-structuration. Systems displaying similar features (clanic equivalence implying exogamy) have been documented in other areas of the world, although they are globally rare: Western Siberia, North America, and North-East Africa [2–6]. Very few systematic studies have been undertaken on the topic, with the exception of Robinne, on the Kachin of Upper Myanmar [30], and Günther Schlee in the Horn of Africa [31]. Schlee suggested that the origin of equivalences among ethnically different—but linguistically related—people was a series of segmentation processes dividing original clans into branches with different titles, but who still remembered being related. This scenario cannot hold for the North-East India case, which involves culturally unrelated people who arrived on the scene at various times and from different directions.

My data pertains to three particular ethno-cultural entities, to three "tribes" (tribe being the established term in local English): the Tiwa, Karbi, and Khasi, and more particularly to the 2500 km^2 or so of low hills on the Meghalaya and Assam states borders where the villages of these tribes are interwoven, each village being monoethnic [29]. The differences among these three tribes are obvious. They speak very dissimilar languages, which either belong to different branches of the Tibeto-Burman family (Karbi, Tiwa) or to the Austroasiatic family (Khasi). The Karbi follow patrilineal descent: clan membership and surnames are inherited on the male line. Khasi follow matrilineal descent. And among the Tiwa, descent is ambilineal: according to where the new couple decides to reside, their children will inherit either their mother's or their father's clan and surname. Social and political structures as well as traditional religions (between 30 % and 60 % are Christians) are less contrasted, but enough to form three (relatively) different cultures. Finally, in the domain of identity, ethnicities are neatly asserted—which may not have been the case in the past. People have no doubt they belong to one particular tribe and that this tribe has "nothing to do" with the next one.

To fully complete this rough picture of the context, two points have to be added. First, before a sharp population growth during the twentieth century, which led to sedentarization, short distance migrations, driven mostly by slash-and-burn agriculture as well as epidemics, were pervasive. Second, interethnic marriages were less than marginal and are now on the rise. Their precise amount is difficult to assess, as the incoming spouses fully adopt the local culture, and even change their surname, actually on the basis of the interethnic surname equivalences. And in interethnic marriages, exogamy between equivalent surnames is, up to my data, strictly respected.

Most elderly persons know about the existence of surname equivalences. In a particular locality, or family, people cannot list the full equivalence relations linking all the clans; they would only cite a few of them; similarly, they most generally cite one to one equivalences relations between one clan of their own tribe and one of the two other tribes. Strikingly, however, when put together, the collected equivalences fully agree and draw a very coherent picture (Fig. 1, below). The lines in the diagram should be read as "Surname A is equivalent to surname B", or "people bearing surname A cannot marry people bearing surname B". The diagram includes exogamic relations internal to each tribe, i.e. the clans within each tribe who are forbidden to marry, each cluster forming what anthropologists call a "phratry". One notices the striking correspondence between "internal exogamies" and "external" ones, particularly between the Karbi and Tiwa. Among these two tribes at least, exogamy is transitive across the ethnic boundaries: e() being the relation of exogamy, T_n a Tiwa clan and K_n a Karbi clan, $(T_a\ e(K_a))\,\hat{}\,(K_a\ e(K_b)) \Rightarrow T_a\ e(K_b)$.

Where did such a coherent system come from? The possibility that, as in the Horn of Africa, equivalences are the links which remained after original descent groups split up, can be discarded, as the tribes concerned differ in all aspects. Intentional creation also comes to mind as a possible origin. However, we're among tribal societies, with no overarching centralized authority. We have no signs that the neighbouring Hindu *raja*s, with whom the tribes had at different times maintained loose alliances, have interfered with their marriage rules—and this seems hardly plausible anyhow. The highest level authorities which could have regulated marriages are the politico-ritual dignitaries responsible for the settlement of disputes and the performance of rituals, but their authority seldom stretched beyond a dozen villages. And even in this case, what would have been the reason for coordinating interethnic marriages?

To me, the most plausible scenario to account for the origin of equivalences is that of self-organization and diffusion. Equivalences would have emerged locally in the context of interethnic marriages, when different matrimonial norms were confronted. They then would have diffused to other localities and/or connected with similar patterns. Presently, among the Khasi and Tiwa, there exists no other matrimonial rule than clan exogamy. On the Karbi side, however, informants describe their traditional but vanished ideal rule as a circulative connubium, i.e. cyclical marriages among their five major clans (A→B→C→D→E→A).[1] Nowadays, the only norm among the Karbi is the avoidance of symmetrical alliances between local lineages, and this is largely adhered to: after a local lineage (i.e. segment of clan) belonging to clan A has "taken a girl" from a lineage belonging to clan B, upcoming marriages between the two lineages will only be allowed in the same direction; B will not "take a girl" from A and the orientation of the relation will

[1]On similar norms, see e.g. Needham [7] and Lehman [8].

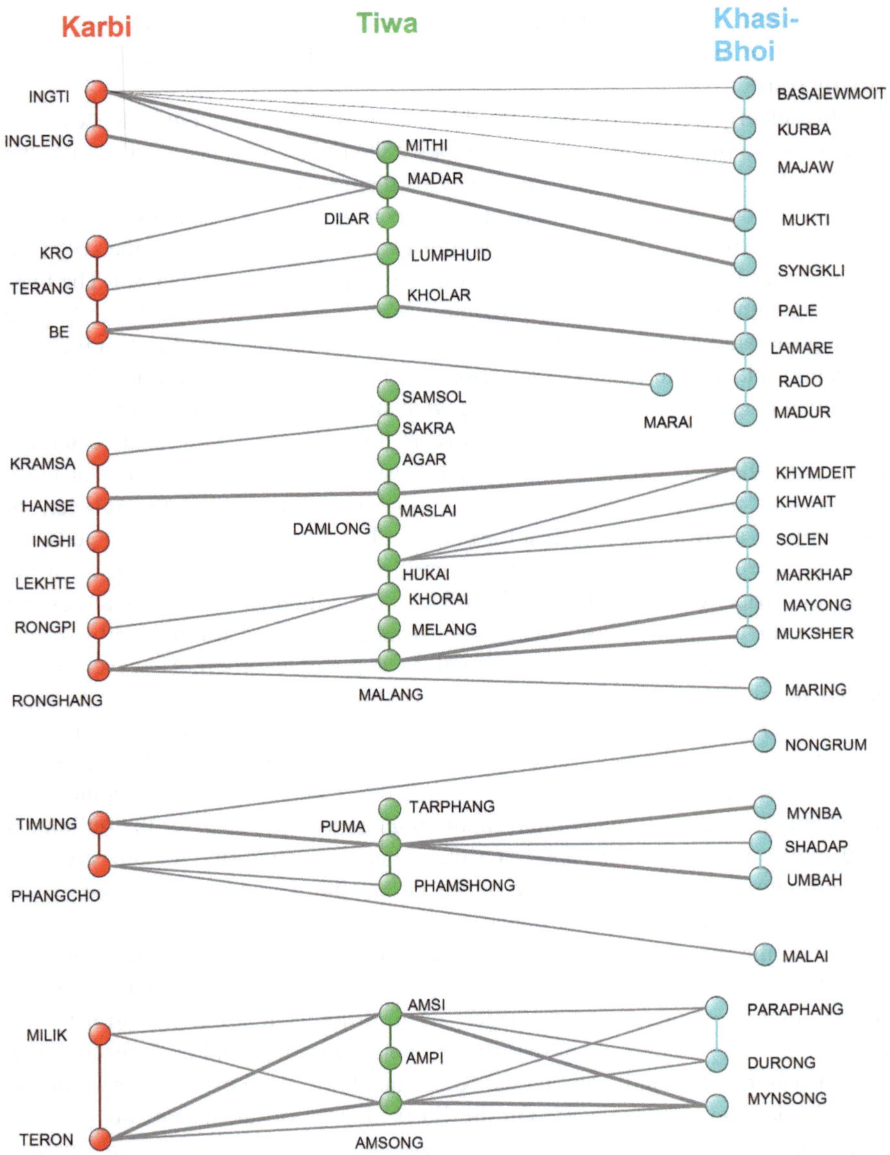

Fig. 1 Endo-ethnic and interethnic exogamies among Karbis, Tiwas, and Khasi-Bhois of Assam–Meghalaya borderlands. *Dots* represent patronymic groups (clans), *lines* represent relations of exogamy. This diagram highlights the convergence of exogamies within ethnic groups (same colour clusters) and across them

not be allowed to reverse for several generations. Any marriage thus creates an asymmetrical relation between "wife-givers" (WG) and "wife-takers" (WT), all men born in the WG lineage being afterwards addressed by their WT as *ong*, indistinctly meaning "father-in-law" and "maternal uncle". Thus, when Karbis consider a new matrimonial alliance, they must ensure that existing WG–WT relations will not be jeopardized.

My hypothesis is the following: after a Karbi marries a member of another tribe, a WG–WT is established between the partner lineages. The foreign lineage becomes structurally equivalent to a Karbi lineage and becomes identified with a Karbi surname. This will be taken into consideration in future marriages. After several such interethnic marriages happen in the neighbourhood, a full set of equivalences becomes established between the Karbi clans and the non-Karbi clans. This set of equivalences then spreads to the neighbouring localities. An alternative, more plausible and more complex scenario, is that different sets of equivalences were born in several distant localities and that after a while they coalesced into a unified regional system.

The question is: how much time would it take for such a scenario to be completed in a particular population/area? The core of this scenario can be qualified as a weak emergence. It is extremely simple to simulate for ideal conditions and a single locality: each new marriage generates a new WG–WT relation and forbids its reversal. After n interethnic marriages have taken place between n distinct Karbi clans and n distinct non-Karbi clans, n equivalences are established. This could be a matter of two generations only.

Modelling the same process in a moderately realistic way would, however, involve a lot of factors, making us enter the realm of complexity (both computational complexity and complex dynamic systems). First, in the real world, there is little chance that three interethnic marriages involving 2×3 distinct clans will happen in a row within one or two generations. The communities involved are villages of a few dozen houses, thus the frequency of matrimonial unions are limited. Considering the time factor brings to the probability that a marriage between two particular lineages takes place, and first of all that it is contemplated by the families. A prospective marriage corresponds to two households situations matching, possibly to the agreement or feelings of the prospective spouses, and to the mere existence of matchable individuals, i.e. at least the availability of a bachelor of the right age: the local population's structure is a critical factor. Furthermore, the probability that a member of a particular clan marries a member of another particular clan depends on the availability of these two clans in the surroundings. Now, considering the small populations at stake, this situation is not obvious. Second, a simulation of the build-up of equivalences in a local isolate does not take into account what occurs outside. Now what would happen if a similar process happens in the next area? There is little chance that the very same equivalence relations will emerge. So, how will contradictions between two sets of equivalence be resolved if a marriage is considered between people living in both areas?

These rather mechanistic issues don't bother most anthropologists. When describing social structures and norms, anthropologists seldom consider their applicability in a realistic context. I believe that the question of the origin of equivalences cannot be settled in a useful manner by considering the social structure and the norms supposed to regulate it in isolation from their biological environment. What would be the point of showing that the asymmetrical marriage rule might generate surname equivalences if there is very little chance that the proper matrimonial configuration appears over several centuries? This is the reason why I decided to solicit the expertise of statisticians, modellers, and ecologists, among others, and set up a working group, ATIMODIS, to decipher the processes at stake in the complex relations between marriage rules, space, and population dynamics.

For the time being, ATIMODIS has tried to identify which factors might be critical in the modelling of the hypothesis in question. To start with, a series of simple simulations (under GAMA[2]) involving only a small population and a simple set of marriage rules have helped to put into light a very critical factor indeed: the fragility of small populations to even seemingly light marriage constraints. Imposing nothing more than the simple rule of clan exogamy ("one should not marry within one own's clan") to a population of 200 individuals divided into three clans, even with optimistic fertility rates, will considerably endanger its survival in less than three centuries. Going a step further by imposing an asymmetrical marriage rule among the clans ("after A has taken from B, it will not be able to give back") will make the population dramatically collapse and disappear within less than three to four generations. As for the circulating connubium, with three clans exchanging, even when a degree of deviance is allowed for, eligible partners become unavailable after a couple of years, preventing any reproduction.

Although the "small population effect" or "demographic stochasticity" is nothing new,[3] realizing how much any pairing constraint amplified its consequences forced us to somehow let the question of equivalence aside for a while, and focus on the relationships between matrimonial rules and population viability. In which conditions could the prescriptive marriage norms formulated by the Karbi have been applied at all? Our simulations showed that the ideal norm of the circulating connubium, what anthropologists after Lévi-Strauss call the "generalized exchange", with three or five clans exchanging spouses in a single direction, seems impossible to put into practice in a small population when a majority follows the rule. This incited me to go back to the anthropological literature on generalized exchange. Lévi-Strauss himself had noticed that the structural model he proposed was hardly applicable in practice, both because of population size effects (on which he did not elaborate much) and because of the inequalities among the exchanging parties ([15], chap. introduction). Indeed, Leach and others had criticized him on the ground that nowhere was generalized exchange statistically dominant [16]. Lévi-Strauss reacted by arguing that generalized exchange is primarily a "model": its "reality" (accuracy) does not consist in its statistical existence but in the fact that it depicts the general trend of all concrete alliances in the long term. As a matter of fact,

---

[2]https://github.com/gama-platform/gama/wiki.

[3]See, for instance, Dyke [9], Gaines and Gaines [10], Legendre et al. [11], Engen et al. [12], and Drake [13]. For human populations, a "Minimum Viable Population" of 500 has been suggested by Livi [14].

some scholars, with the limited tools of the time (1960s–1970s), had started to show through simulation both that the rule was impossible to put fully into practice and conversely that in the absence of any other rule than simple exogamy, there was always a proportion of marriages corresponding to the norm of generalized exchange [17, 18]. If other aspects of the generalized exchange model still arouse new formalizations and modelizations (e.g. graphs[4]), the study of its demographical context has not benefited from the computational and simulation tools now at our disposal. This is one of the tasks ATIMODIS project wishes to take on.

Although generalized exchange in its full cyclic form has not been documented, cases of populations following prescriptive asymmetrical rules are widespread [20–23]. If simulations show that asymmetrical rule is so detrimental to the demographic structure, how then could it be explained that populations practising such marriages were able to survive? The issue was seldom taken up by anthropologists. Recently, however, starting from this very question, a set of studies in Western Australia have brought a very convincing and stimulating answer, and one which compel us to deeply reconsider our methodology, if not our paradigm: a few exchange of partners among local groups could have been sufficient to maintain their demographic viability even while prescriptive alliance remained dominant [24–26]. Small populations are very rarely complete isolates. Their sustainability cannot be explained without taking into account the inputs from outside, even if these look statistically marginal. Statistically marginal exogamy may prove vital for the survival of small populations.[5] "Intermarriage ... should be seen as a component of formal marriage system models that opens up local models to form an explicit, culturally recognized, and realistic continent-wide meta-model" ([26], p. 6).

Thus the need to move from the level of the village, where most marriages happen, to the regional level, at which inter-villages marriages take place from time to time, in order to understand how the two levels are dynamically connected. In other words, shifting from the (false) local isolate to what ecologists call the meta-population. In this regard, the contribution of the tools used by ecological modelling could be essential for the approach of anthropological complexities such as the interactions between marriage norms and demographics.

The "problem" posed by the negative effects of marriage rules on the demographic dynamics of small populations could in fact be completely reversed. In a functionalist perspective, and at least as a working hypothesis, one could consider prescriptive marriage rules not as a constraint but as an incentive to search for mates outside the group—whatever group—which would counter demographic stochasticity as well as genetic inbreeding.

Modelling the dynamics of asymmetrical marriage rules at the level of meta-populations is, however, far from being a trivial task. This is where one enters into hardcore complexity (at least to an anthropologist) and where interdisciplinary becomes compulsory. For instance, how did different local matrimonial patterns

[4]Harary and Hage [19].

[5]This converges with conclusions met by paleo-anthropologists [27].

become a coherent regional pattern? Simulation will obviously be a major component of our approach. The second one will consist in mapping the data available on existing matrimonial links. A set of genealogies I collected in some localities will help to understand the small-scale mechanisms, but hardly any phenomenon of emergence/self-organization. For this purpose, our main corpus will be the Indian electoral rolls, which provide a full size picture of voters with their surnames, including married women and their husband names, and thus of the existing alliances among clans. This set of data has several obvious limitations, among others the inability to show the geographical origin of external spouses, or to show real matrimonial networks. The challenge will be here to devise proper tools that will explore the corpus at successive spatial levels and detect the existence of significant patterns.

For the time being, it is interesting to underline how a very pragmatic problem emerging during simulation, "how to avoid that my population collapses", has led to a deep re-thinking of the general approach. Going back to the literature about the anthropological theory of alliance, I realized that the problem had not been resolved at all, and that other essential issues as the structuration of alliance patterns at higher levels remained almost untouched. An approach in terms of complexity is not an escape out of the classical anthropological debates; indeed it is a constructive return to them.

1 Scientific Validation

This paper has been unanimously validated in a collaborative review mode with the following reviewers:

- Karla de Jesus, University of Porto
- Philippe Collard, Université de Nice.

References

1. Simmel G (1950) In: Wolff KH (ed) The sociology of Georg Simmel
2. Emmons GT, De Laguna F (1991) The Tlingit Indians. University of Washington Press, Seattle
3. Goodwin G (1937) The characteristics and function of clan in a Southern Athapascan Culture. Am Anthropol 39(3):394–407. doi:10.1525/aa.1937.39.3.02a00030
4. Shternberg LIA (Lev Iakovlevich) (1999) The social organization of the Gilyak. Anthropological papers of the AMNH; No. 82. Edited by Bruce Grant. http://digitallibrary.amnh.org/dspace/handle/2246/281
5. Swanton JR (1909) Contributions to the ethnology of the Haida. E.J. Brill/G.E. Stechert, Leiden/New York
6. Verbov (1936) Лесные ненцы. Советская этнография 2:56–70
7. Needham R (1957) Circulating connubium in Eastern Sumba: a literary analysis. Bijdragen Tot de Taal-, Land-En Volkenkunde 113(2):168–178

8. Lehman FK (1963) The structure of Chin Society
9. Dyke B (1971) Potential mates in a small human population. Biodemography Soc Biol 18(1):28–39. doi:10.1080/19485565.1971.9987897
10. Gaines SW, Gaines WM (1997) Simulating success or failure: another look at small-population dynamics. Am Antiq 683–697
11. Legendre S, Clobert J, Møller AP, Sorci G (1999) Demographic stochasticity and social mating system in the process of extinction of small populations: the case of passerines introduced to New Zealand. Am Nat 153(5):449–463
12. Engen S, Lande R, Sæther B-E (2003) Demographic stochasticity and Allee effects in populations with two sexes. Ecology 84(9):2378–2386
13. Drake J (2011) Allee Effects| Nature Education Knowledge 3(10):2, http://www.nature.com/scitable/knowledge/library/allee-effects-19699394
14. Livi L (1949) Considérations Théoriques et Pratiques sur le Concept de 'Minimum de Population.'. Population (French Edition) 4(4):754–756. doi:10.2307/1523777
15. Lévi-Strauss C (1969) The elementary structures of kinship. Beacon, Boston
16. Leach ER (1962) The determinants of differential cross-cousin marriage (correspondence 1). Man 62:153
17. Gilbert JP, Hammel EA (1966) Computer simulation and analysis of problems in kinship and social structure. Am Anthropol 68(1):71–93
18. Kunstadter P, Buhler R, Stephan FF, Westoff CF (1963) Demographic variability and preferential marriage patterns. Am J Phys Anthropol 21(4):511–519
19. Harary F, Hage P (1996) The logical structure of asymmetric marriage. L'Homme 36(139):109–124. doi:10.3406/hom.1996.370120
20. Krige EJ (1975) Asymmetrical matrilateral cross-cousin marriage—the Lovedu case. Afr Stud 34(4):231–258
21. Leach ER (1951) The structural implications of matrilateral cross-cousin marriage. J R Anthropol Inst G B Irel 81(1/2):23–55. doi:10.2307/2844015
22. Löffler LG (1964) Prescriptive matrilateral cross-cousin marriage in asymmetric alliance systems: a fallacy. Southwest J Anthropol 20(2):218–227
23. Moore SF (1963) Oblique and asymmetrical cross-cousin marriage and Crow-Omaha terminology. Am Anthropol 65(2):296–311. doi:10.1525/aa.1963.65.2.02a00060
24. Denham WW (2013) Beyond fictions of closure in Australian Aboriginal Kinship. Math Anthropol Cult Theory 5(1), http://escholarship.org/uc/item/7d69w4sk
25. Dousset L (2005) Assimilating identities: social networks and the diffusion of sections., http://halshs.archives-ouvertes.fr/docs/00/19/33/39/PDF/Dousset_assimilating_identities.pdf
26. White DR, Denham WW (2007) The indigenous Australian marriage paradox: small-world dynamics on a continental scale. In: SASci Annual Meeting, San Antonio. http://www.researchgate.net/publication/227855133_The_Indigenous_Australian_Marriage_Paradox_Small-World_Dynamics_on_a_Continental_Scale/file/d912f4fe4e d0d05ca7.pdf
27. Bocquet-Appel J-P, Masset C (1982) Farewell to paleodemography. J Hum Evol 11(4):321–333. doi:10.1016/S0047-2484(82)80023-7
28. Ramirez P (2013) Ethnic conversions and transethnic descent groups in the Assam-Meghalaya borderlands. Asian Ethnol 72(2):279–97. http://nirc.nanzan-u.ac.jp/nfile/4304
29. Ramirez P (2014) People of the margins: across ethnic boundaries in North-East India. Spectrum, Guwahati
30. Robinne F (2007) Transethnic social space of clans and lineages: a discussion of Leach's concept of common ritual language. In: Robinne F, Sadan M (eds) Social dynamics in the highlands of Southeast Asia. Brill, Leiden/Boston, pp 283–297
31. Schlee G (1985) Interethnic clan identities among Cushitic-speaking pastoralists. Afr J Int Afr Inst 55(1):17–38. doi:10.2307/1159837

Are Innovation Systems Complex Systems?

Emmanuel Muller, Jean-Alain Héraud, and Andrea Zenker

1 Introduction

Basically, complex systems are defined as a large number of autonomous entities in interaction, that create several levels of collective organization leading to emergent (and immergent) behavior. In short, complex systems are characterized by the observation that the whole is more than the sum of the parts (Aristotle). The aim of this paper is to address the issue of innovation systems in order to determine by how much they constitute complex systems.

Innovation systems are generally used by economists interested in depicting innovation processes, technological changes as well as the emergence of new fields in terms of knowledge, services, or business models as an ensemble of resources committed to interrelated activities. Interestingly, only little importance has been given until now by evolutionary economics to the "*complex-systemic*" nature of innovation systems. Too often, such systems are just depicted as networks of actors.

So far, the main assumption was that an innovation system (e.g., a country, a region, specific sectors or technologies, or a private firm) constitutes—taken as a whole—more than just the sum of its elements and that some feedback loops may be at stake related to processes and interrelations between actors. It is, however, important to go further in this analysis in considering the key features of innovation processes and thus identifying the reasons explaining why innovation systems may indeed be complex systems. The coordination between the different types of actors

E. Muller (✉) • A. Zenker
Fraunhofer Institute for Systems and Innovation Research (ISI), Karlsruhe, Germany

E. Muller • J.-A. Héraud
Bureau d'Economie Théorique et Appliquée (BETA), Université de Strasbourg, Strasbourg, France
e-mail: emmanuel.muller@isi.fraunhofer.de

P. Bourgine et al. (eds.), *First Complex Systems Digital Campus World E-Conference 2015*, Springer Proceedings in Complexity, DOI 10.1007/978-3-319-45901-1_17

involved in the system as well as the existence of mechanisms ensuring a form of governance is at the core of the issue addressed. More precisely, the paper is an attempt to identify possible/thinkable systemic characteristics and properties.

The structure of the paper is the following. The first section deals with the concept of innovation as seen by evolutionary economists. The second and third sections consider different hypotheses in particular related to the issues of behavioral patterns and to non-linear and non-predictable choices. The fourth section attempts to identify some emerging properties of innovation systems seen as complex systems whereas the fifth proposes some reflections about policy implications.

2 What Are Innovation Systems According to Evolutionary Economics?

The "father" of the analysis of innovation in economics is without contest Joseph Schumpeter (1883–1950) who put forward the concept of "creative destruction." According to Schumpeter [1] this process rules the historical evolution of capitalism. In this respect, five main types of innovation [2] can be distinguished: (1) new consumptions objects; (2) new production and transport methods; (3) new markets; (4) new sources of production materials; and (5) new market positions (e.g., monopolistic situation).

Following seminal works by Schumpeter the field of evolutionary economics progressively emerged in the 1950s and 1960s. As a consequence, the concept of innovation systems was developed by authors such as Freeman [3] and Lundvall [4]. At first this concerned mainly the so-called National Systems of Innovation (NSI). This in turn inspired reflections dealing with the sub-national level, i.e., the Regional Systems of Innovation (RSI). This last concept was made popular by authors such as Cooke [5]. Therefore, an innovation system can be considered as a complex system in the sense that it is composed of many subsystems that are different in nature while strongly articulated. Evolutionary economics consider innovation systems mainly under the aspect of creation/selection mechanisms and not so much in the framework of real systemic approach. The simplifying representation of neoclassical economics—the market providing all necessary information—is just replaced by another global mechanism, the selection process.

All in all, the main features of innovation systems in the field of innovation economics can be summarized as follows: (1) central role of learning; (2) importance of historical processes; (3) influence of institutions (public actors, legal framework, norms, etc.); (4) existence of feedback loops between non-fully rational actors; and (5) numerous and diverse interrelations between scientific, technological, and organizational innovations.

A very interesting outcome is that those concepts have a real impact on the conception and application of public policies, especially in European Countries (at national and regional levels) as well as on the level of the European Commission (this point will be developed more extensively in Sect. 6). Nevertheless, and despite the wording usually adopted to name such systems, it clearly appears that the systemic nature of innovation systems is not really explored in the corpus of evolutionary economics. The following three sections will try to tackle this issue in presenting arguments in favor of the idea of innovation systems being complex systems.

3 Innovation Systems May Be Complex Systems Since They Are Based on Complex Behavioral Patterns

The crucial issue to be addressed is the following. If innovation systems are complex systems, what are the mechanisms that explain the mobilization and alignment of resources in an innovative direction? According to the literature, innovation systems are based on complementary mechanisms. The most important of such mechanisms are: (1) interactions; (2) trial and error procedures; and (3) selection.

Concerning at first interactions, no firm seems to be able to innovate without interactions with the "outside world." The forms and the actors involved in such interactions may be extremely heterogeneous but the existence of "autarkic innovations" seems to be extremely rare if not impossible. At the same time, an underlying process of "trial and error" can be found in every "innovation story." In other words, a "first shot" success may constitute a very unlikely exception. Moreover, it is helpful to keep in mind that many innovation projects fail. This leads finally to selection phenomena, which concern ideas, production and delivery processes, technologies, marketing approaches, types of interactions and collaboration, and even actors (unsuccessful firms). One should be aware that the average life expectancy of companies is far lower than, for instance, the average human life expectancy.

One may easily recognize the evolutionary character of innovation systems. As a consequence, such systems can be seen as the results of complex behaviors of heterogeneous and numerous actors. Moreover, within such systems, different and variable ensembles of resources are committed to interrelated innovation activities [6]. In other words, innovation systems may be complex systems since some behavioral mechanisms (and not "the market" as in the neoclassical economic approach) govern the alignment of resources, activities, and efforts in innovative directions. These alignments consist in arrangements between actors and result from their—to a certain extent implicit and/or opposite—strategic decisions. This may be seen as a process of stretch and leverage taking place within an environment characterized by co-opetition, connectivity, and selection.

4 Innovation Systems May Be Complex Systems Since They Result from Non-linear and Non-predictable Choices

Considering the heterogeneity of actors of any innovation system (e.g., firms, academic institutions, public actors, intermediaries, individuals, etc.), the issue can be addressed of what is "behind" their behaviors, choices, and decisions and if those behaviors can be seen as corresponding to elements of complex systems.

At least two aspects play a crucial role in this context: First of all, the underlying assumptions of (neo-)classical economics such as perfect competition, full availability of information, atomistic markets, and unbounded rationality are scrutinized and replaced by characteristics like humans' opportunistic behaviors, uncertain context conditions such as complex and unstable production environments, evolving structures, learning processes, and so forth (see, for instance, [7]). Scholars in evolutionary economics assume that complex economic processes—characterized by variety, technology development, firm evolution, etc.—are embedded in and interrelated with dynamic and uncertain contexts. Both parts—in our case innovating actors and their environment—are influencing each other: Innovators' activities can (at least partly) be considered as response to their environment, but the overall environment is in turn also influenced through the activities of embedded innovation actors.

The second aspect to be considered refers to individual (innovation-related) actors and their "views of the world." It cannot be assumed that the totality of "elements" in a system (here: innovation actors) behave in the same way, thus acting as a kind of "uniform crowd," but that every individual has rather built up his/her individual model of the surrounding world. In short: human beings perceive the world in a specific way and create individual mental models of their surrounding which, in turn, strongly guide their behavior (e.g., [8]). Mental models are constructed on the base of memories, experiences, etc., which help to classify and interpret newly perceived information of the external world. Since context conditions including values, rules, and habits are largely homogenous within innovation systems (at least in comparison with other systems), it can be assumed that members of the same system have converging views of the world (again compared with members of other systems, cf., for instance [6]).

5 If Innovation Systems Are Complex Systems What Can Be Deduced From That?

Summarizing the three previous sections, it can be stated that the complexity of innovation systems arises from interaction between actors, existence of feedback loops, and non-linearity of processes. Socioeconomic systems like innovation systems are also extremely difficult to describe because individuals have mental representations of the whole system and play strategic roles on the basis of those representations.

The strategic orientation of an innovation system results from more than just the sum or the aggregation of individual choices. Game theory applied to economics shows the extreme variety and instability of possible outcomes of strategic models. Nevertheless, even if behaviors cannot be predicted at individual level some patterns may exist and moreover the hypothesis can be formulated that some systemic properties may emerge. This is typically the case in physics, where independent particles in interaction may constitute a fluid, that will implement whorls and eddies under turbulent conditions even though Navier–Stokes equations cannot be found anywhere in the description of the particles. Then, in biology, a number of cells working together will implement an organ with properties going beyond those of each cell (the brains is more than a collection of neurons). Hundreds of ants in interaction will show intelligent behavior (finding the shortest path between their nest and a food source) that is beyond the reach of a single ant. On a higher level, is it possible to make the same observations with social networks (such as the Internet or even the stock exchange market)? Or are they complex systems of another type?

Then, observing only a part of a system may not allow drawing conclusions on some other parts. Local robustness is no indication on the robustness of the whole, due to retroactive loops, border effects, and reflexive interactions.

However, collecting and observing data may allow creating a first model of a complex system that can be refined by subsequent simulations (and comparing the simulations to the observed data). The refined model can then be used to predict what can happen (in probability) in order to prevent catastrophes (in the sense of [9]). Having a complex systems model for innovation systems will allow better understanding and predicting what conditions are needed for innovations to emerge.

6 Implications for Policy Making

Over the last decades, considerable efforts were made by public authorities in order to support economic development at different levels (national, but also regional, European, etc.) with the clear goal of ensuring economic growth as base for employment, income, and welfare. Innovation is understood as one means to contribute to economic development and thus wealth creation. On the microlevel of business firms, innovation is a means to maintain or broaden market shares and thus to generate benefits.

Various policies address economic development, both via influencing the business environment and/or via supporting specific economic actors or actor groups. Fostering innovation follows this logic and—often based on the analysis of existing innovation systems—introduces interventions in order to create innovation-friendly framework conditions plus promotional measures for (potentially) innovating actors. With the broadening of knowledge and comprehension of innovation processes, policies also broadened and to an increasing extent included additional actors (e.g., from high-tech manufacturing firms to the inclusion of non-tech firms), their interrelations (e.g., collaborative innovation activities), and also switched to integrative policy approaches instead of unconnected approaches of individual policy fields. Equally, learning, exchange, and gathering of experience received an increasing role (e.g., demonstrator cases, "good examples," etc.).

Several evolutionary aspects of innovation systems should be further considered from a perspective of complex systems:

- the existence of technological trajectories (for instance, in the fields of the automotive or aerospace industries);
- the apparition of hybrid actors and/or mutational effects or even paradigm leaps through the emergence of permissive technologies (such as electricity or ICT);
- the impacts of learning effects and the resulting acceleration of knowledge flows which may be comparable to a certain extent to forms of swarm intelligence.

In consequence, and from a more general point of view, the following assumptions can be formulated in terms of implications for policy making if innovation systems are considered as complex systems. First, innovation-supporting policies should focus on issues related to actors' coordination rather than on optimization and efficiency. Second, only holistic innovation policies may be successful in the long-term and policy instruments must aim at providing good contextual conditions, rather than trying to manipulate individual behaviors. Third, and due to the non-linearity of the considered phenomena, if policies lead to mainly unpredictable events, the incentives mobilized by those policies should target marginal systemic effects rather than huge changes. If innovation systems are complex systems, then such approaches may support a reinvention of the role and legitimacy of public interventions dealing with innovation and economic development.

7 Conclusion

A better understanding of how far innovation systems can be considered as real complex systems would open new possibilities. In particular, some strong implications for public innovation-led strategies may result from sharper complex dynamics analytical abilities.

Nevertheless, the ideas developed in the previous sections are subject to limitations of different nature. First, only a few aspects of the broad spectrum of the concepts developed by academics about innovation systems are taken so far into account. Second, no quantitative data allowing the test of hypothesis are used in this paper. Last but not least, until now no attempt to propose some modeling of innovation systems as complex systems is proposed. Those limitations constitute avenues for future research.

8 Scientific Validation

This paper has been unanimously validated in a collaborative review mode with the following reviewers:

- Pierre Parrend, from ECAM Strasbourg-Europe and University of Strasbourg
- Flavia Mori Sarti.

Acknowledgment The authors wish to thank Pierre Collet (University of Strasbourg) for his valuable support.

References

1. Schumpeter JA (1950) Capitalism, socialism and democracy. Harper and Brothers, New York
2. Schumpeter JA (1935) The analysis of economic change. Rev Econ Stat 17:2–10
3. Freeman C (1995) The 'National System of Innovation' in historical perspective. Camb J Econ 19:5–24
4. Lundvall B-Å (ed) (1992) National systems of innovation. Towards a theory of innovation and interactive learning. Pinter Publishers, London
5. Cooke P (1998) Origins of the concept. In: Braczyk H-J, Cooke P, Heidenreich M (eds) Regional innovation systems - the role of governance in a globalized world. UCL Press, London, pp 2–25
6. Stamboulis Y (2008) Exploring the system dynamics of innovation systems. System Dynamics Society, Athens, Online: http://www.systemdynamics.org/conferences/2008/proceed/papers/STAMB454
7. Simon H (1957) Models of man. Wiley, New York
8. Johnson-Laird PN (1983) Mental models. Towards a cognitive science of language, inference and consciousness. Cambridge University Press, Cambridge
9. Thom R (1989) Structural stability and morphogenesis: an outline of a general theory of models. Addison-Wesley, Reading

Dynamic Emissions Reduction from Vehicles with Technical and Behavioral Approach

Mohammad Syafrizal, Estelle Bretagne, Nadia Hamani, Bambang Sugiarto, and Setyo Sarwanto Moersidik

1 Introduction

Jakarta, the capital of Indonesia, is among the most populous, fastest-growing, and pollutant cities in the world. Air pollution from traffic-related pollutant has become a very serious problem. It is caused by the traffic congestion, the two-stroke engines, the lack of catalytic converters, the fuel quality, etc. It is predicted that Jakarta would experience total gridlock in 2014. Transjakarta is the name of the bus rapid transit (BRT) system, which started in Jakarta in 2004. Transjakarta buses carry 315,000 people per day; the number is expected to double if new buses hit the streets by the end of 2013 [1].

Exhaust emissions and fuel consumption depend on various factors: driving pattern, gradient road, type of vehicle, vehicle's technology, maintenance, load, traffic, fuel, etc. A driving cycle represents the traffic behavior and is representative of a given region. There are many standards of driving cycles. The development of a real driving cycle is very important to make arrangements in transportation management, measurement and control pollutant emissions, and also to study energy and fuel savings [2–4]. The European cycle (UN-ECE cycle R-83 for short), seven modes of California cycle, and Japan cycle are made of several phases of constant acceleration and constant speed. Worldwide Motorcycle Test Cycle

M. Syafrizal
Faculty of Engineering, University of Indonesia, Depok, 16424 Indonesia

University of Picardie Jules Verne, Amiens, France

B. Sugiarto • S.S. Moersidik
Faculty of Engineering, University of Indonesia, Depok, 16424 Indonesia

E. Bretagne • N. Hamani (✉)
University of Picardie Jules Verne, Amiens, France
e-mail: nadia.hamani@u-picardie.fr

P. Bourgine et al. (eds.), *First Complex Systems Digital Campus World E-Conference 2015*, Springer Proceedings in Complexity, DOI 10.1007/978-3-319-45901-1_18

(WMTC) and Common Artemis Driving Cycle (CADC) driving cycles are more dynamic [5]. These cycles are models of driving characteristics in an area considered in each driving cycle [4–7]. However, these driving cycles cannot represent the actual condition of Jakarta where the average traffic speed is about 8 km/h. Jakarta driving cycle was developed by taking the patterns of actual driving on the road [8]. Sampling enough vehicles provided more real data. Moreover fuel usage and selection of widely used vehicles were also a major consideration. This real driving cycle is very suitable to represent traffic conditions with sharp accelerations and decelerations. Jakarta driving cycle is the first step for determining real emissions in order to decrease pollution and to influence vehicle choice in Jakarta. The second step deals with chassis dynamometer test, where vehicle is placed on a chassis dynamometer. The purpose of the chassis dynamometer test is to measure exhaust emissions. So exhaust emissions and emissions factors can be obtained in function of average speed in case of Jakarta. Then these emissions factors are used to determine exhaust emissions in a road named "Semanggi" where traffic assessment is known hourly by video and manual counter. Semanggi intersection is a very important road in Jakarta because about 70 % vehicles crossing on this road. Daily traffic volume might be greater than 385,000 vehicles [9].

Besides the technical approach, the behavioral approach is also considered. The theory of planned behavior [10] is an integrating framework for psychological factors underlying public transportation use. The approach of behavior intervention is designed to improve the intentions of private car users to switch to public transport; in this case it is the BRT system [11, 12]. The aim of behavioral approach consists in understanding, which beliefs or norms contribute to the change of transportation mode choices. An application of some behavioral approach results is presented in the case of Semanggi intersection. This modeling part aims to investigate the impact of car-use reduction for Semanggi intersection based on both statistical methods and psychological factors associated with the change in transportation mode choices. So analysis of feasible policy will be done to reduce air pollution from transport sector in the future.

2 Results and Discussion

2.1 Jakarta Driving Cycle and Chassis Dynamometer Test

The objective was the development of driving cycle that can represent the traffic conditions in Jakarta. After obtaining Jakarta driving cycle, the measurement of exhaust emissions and fuel consumption of vehicles can be done by performing a test on the chassis dynamometer. In a previous work [3] we have shown the following results. Test results on the chassis dynamometer show that the average daily speed and real emissions from parameters of CO-HC with the method of Jakarta driving cycles are very different from test results with UN-ECE cycle

R-83. For the parameters of NOx and CO_2 the opposite observation is obtained. The European ECE R-83 test results show that a non-Euro vehicle produces always more emissions and fuel consumption than a Euro-II vehicle. But it is wrong with the Jakarta driving cycle test results and this fact demonstrates that local driving cycle is needed to understand local driving behavior and real emissions. Development of Jakarta driving cycle still needs to be improved. Chassis dynamometer testing is required for more samples of vehicles, so that emission factors can be determined for both heavy- and light-duty trucks and both heavy- and light-duty buses. Vehicles are divided into six types, namely: passenger car, light-duty truck, heavy-duty truck, heavy-duty bus, light-duty bus, and motorcycle. Inbound is vehicles coming in Semanggi intersection and outbound is vehicles coming out Semanggi intersection. The traffic volume survey was presented in previous work [13, 14].

2.2 Emissions Analysis with Improved Intention

At first, weekday and holiday emissions (CO, HC, NOx, PM, and CO_2) were calculated (unit: ton, distances 5 km). Then, emission calculation is based on a study of behavioral intention of car owners to use BRT in Jakarta [9, 14]. In intention, the respondents were asked, "In the future, I intend to use bus as my travel mode to go to office to substitute my private car." Majority of car users tended to have positive attitude toward BRT Transjakarta using, weak subjective norms, but strong behavioral intentions to use BRT Transjakarta. Their intentions were even stronger if BRT Transjakarta was totally convenient. About 68.7 % of the respondents say that they would go to office using BRT Transjakarta if convenient [9]. The respondents that were interviewed have only 1 passenger car (70 % of total passenger car users in the corridor 1 and Blok M-Kota). We also assume that one Transjakarta bus is equivalent to 85 passenger cars. We take into account the possibility to have one bus not completely filled. Weekday emissions with improved intentions are showed in Figs. 1, 2, 3, 4, and 5. Weekday emissions without (a) and with (b) improved intention are compared by category of vehicle and by ratio. The highest result of CO emission comes from four-stroke Euro-II motorcycles and the result is 5.195 ton/day. The total emission of CO is 7.792 ton/day. The highest results of HC emission come from four-stroke Euro-II motorcycles and two-stroke non-Euro motorcycles with 0.803 ton/day and 0.263 ton/day, respectively. The total emission of HC is 1.300 ton/day. The highest results of NOx emission come from non-Euro heavy-duty trucks gasoline Euro-II passenger cars and non-Euro heavy-duty buses with 0.325 ton/day, 0.247 ton/day, and 0.227 ton/day, respectively. The total emission of NOx is 1.393 ton/day. The highest result of PM emission comes from non-Euro heavy-duty trucks and the result is 97.071 ton/day. The total emission of PM is 158.265 ton/day. The highest results of CO_2 emission come from gasoline Euro-II passenger cars, non-Euro heavy-duty trucks, and four-stroke Euro-II motorcycles with 50.129 ton/day, 34.153 ton/day, and 20.530 ton/day, respectively. Total emission of CO_2 is 200.027 ton/day.

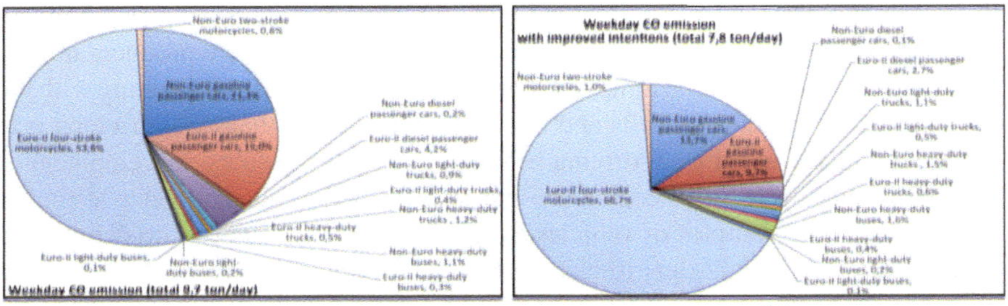

(a) (b)

Fig. 1 Weekday emissions of CO without (**a**) and with (**b**) improved intentions

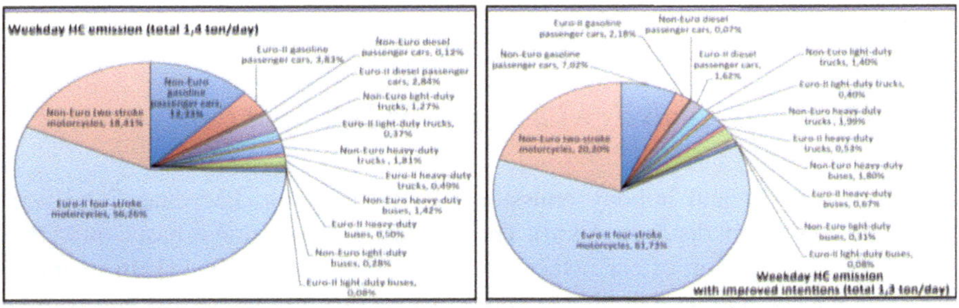

(a) (b)

Fig. 2 Weekday emission of HC without (**a**) and with (**b**) improved intentions

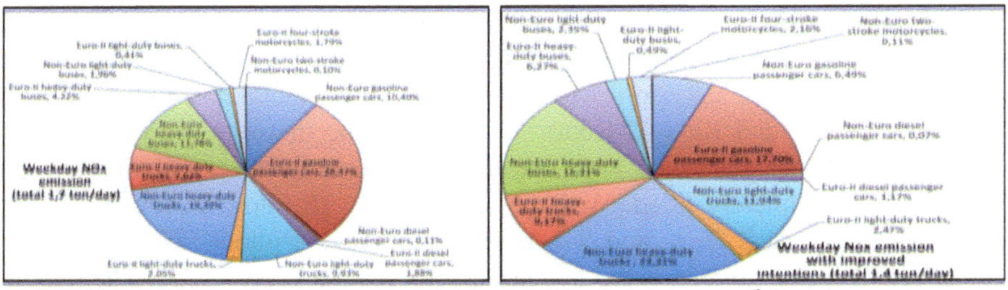

(a) (b)

Fig. 3 Weekday emission of NOx without (**a**) and with (**b**) improved intentions

Motorcycle's emissions are very important for CO and HC. NOx emissions come majority from cars and trucks. PM emissions come majority from non-Euro heavy-duty trucks. CO_2 emissions come majority from passenger cars. Except for PM, with improved intentions, a decrease of emission is observed (Figs. 1, 2, 3, 4, and 5). This decrease of pollutant is attributable to modal shift to the BRT. Bigger emission reductions can be expected from both Jakarta driving cycle for the BRT buses and the lower emissions of the new buses compared with those of the old buses. Insufficient data exist to quantify precisely BRT buses emissions.

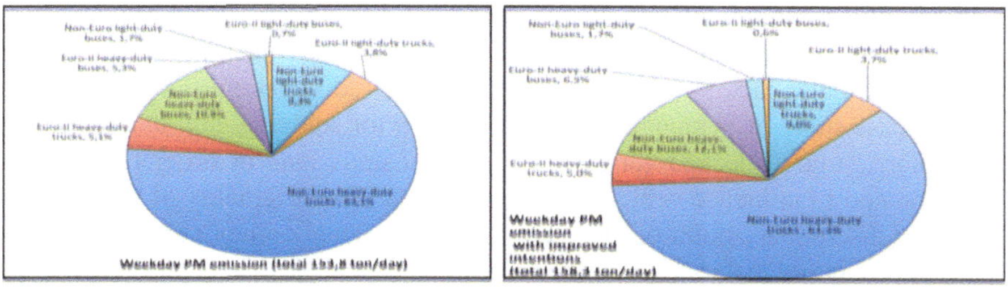

(a) (b)

Fig. 4 Weekday emission of PM without (**a**) and with (**b**) improved intentions

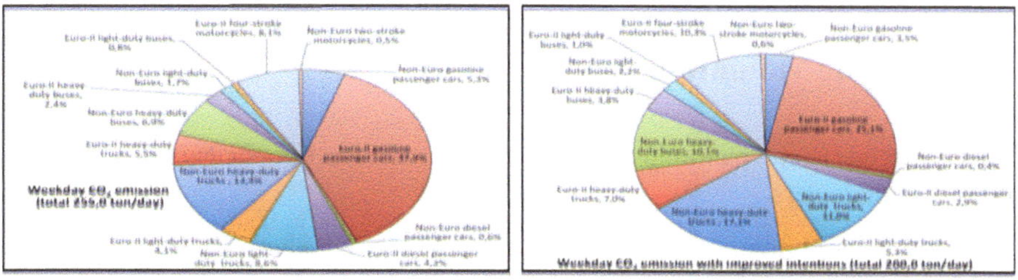

(a) (b)

Fig. 5 Weekday emission of CO_2 without (**a**) and with (**b**) improved intentions

We also assume that one Transjakarta bus is equivalent to 40 and 60 passenger cars. Bigger emission reductions can be expected from both Jakarta driving cycle for the BRT buses and the lower emissions of the new buses compared with those of the old buses. Insufficient data exist to quantify precisely BRT buses emissions. The results for trucks, cars, and motorcycles are the same as presented above.

3 Conclusion

The objective of this research was the understanding of emission reduction from the traffic in Jakarta. After obtaining Jakarta driving cycle, the measurement of exhaust emissions of vehicles can be done by performing a test on the chassis dynamometer. In the development of Jakarta driving cycle, the collection of data of the selected roads and traffic characteristics has been investigated. Development of Jakarta driving cycle still needs to be improved. Chassis dynamometer testing is required for more samples of vehicles, so that emission factors can be determined for each vehicle category. Jakarta driving cycle is the first step for determining real emissions in order to decrease pollution and to influence vehicle choice in Jakarta. Our article deals with a case study in real conditions. Some results [15] of the theory of planned behavior (TPB) are used so as to estimate possible reduction of emission. As a conclusion of this reference [9] 10 behavioral beliefs about transportation mode, 11 control beliefs, and 2 normative beliefs were identified: if these points were improved, car users would change to Transjakarta bus. Nevertheless we precise that first intention to change is not enough to ensure loyalty [16]: perceived service quality is a major factor of loyalty. Assessment of reduction emission is made in the real case study of Semanggi. It is the first step for studying emission reduction on a more large scale and for determining real emissions in order to decrease pollution and to influence vehicle choice in Jakarta. The originality of this work is the multi-disciplinary point of view and the transnational collaboration. Moreover real conditions and real traffic were taking into account so as to understand precisely Jakarta case without cultural bias. Future research will deal with incentives for car owners to use BRT system in Jakarta and with parameter influences.

4 Scientific Validation

This paper has been unanimously validated in a collaborative review mode with the following reviewers:

- Rafaela Carolina da Silva, from Universidade Estadual Paulista "Julio de Mesquita Filho"—Brazil
- Maria Carmen Lemos, from University of Seville
- Abdellatif Benabdelhafid, from Université du Havre.

References

1. Editorial, The Jakarta Post, 06 April 2013
2. Tamsanya N, Chungpaibulpatana S (2009) Influence of driving cycles on exhaust emissions and fuel consumption of gasoline passenger car in Bangkok. J Environ Sci 21:604–611
3. Syafrizal M, Sugiarto B, Moersidik SS, Hamani N, Bretagne E (2013) Jakarta driving cycle for passenger cars: analysis of chassis dynamometer test. In: The 13th international conference on QiR (quality in research), Yogyakarta, Indonesia, 25–28 June 2013
4. Internet website (2013) United nations environment program. http://www.unep.org/, transport section. Accessed 26 Feb 2013
5. Fontaras G, Franco V, Dilara P, Martini G, Manfredi U (2014) Development and review of Euro 5 passenger car emission factors based on experimental results over various driving cycles. Sci Total Environ 468–469(15):1034–1042
6. Nesamani KS, Subramanian KP (2011) Development of a driving cycle for intra-city buses in Chennai, India. Atmos Environ 45:5469–5476
7. Tamasanya S, Chunpaibulpattana S, Attajariyakul S (2006) Development of automobile Bangkok driving cycle for emissions and fuel consumption assessment. In: The 2nd joint international conference on sustainable energy and environment (SEE 2006), Bangkok, Thailand, 2006
8. Kanari S (2009) Development of driving cycle (practice version), Japan Automobile Research Institute (JARI), internal report (unpublished)
9. Hafiyah N, Soesetyo SR (2011) Explaining commuters' travel behaviour from psychological perspective: studies of behavioral intention of car owners to use bus rapid transit in Jakarta. In: The 12th International conference on quality in research (QiR), Bali, Indonesia, 2011

10. Ajzen I (1991) The theory of planned behaviour. Organ Behav Hum Decis Processes 50:179–211
11. Bamberg S, Fujii S, Friman M, Gärling T (2011) Behaviour theory and soft transport policy measures. Trans Policy 18:228–235
12. Heath Y, Gifford R (2002) Extending the theory of planned behaviour: predicting the use of public transportation. J Appl Soc Psychol 32(10):2154–2189
13. BTMP-BPPT, Data pengujian kendaraan, IJ-EPA Partnership, internal data, unpublished, 2009–2011
14. Sugiarto B, Soesetyo SR, Hafiyah N, Syafrizal M, Mauludi F, Darsono D (2010) Studi polusi udara Jakarta dengan pendekatan intensi masyarakat pengguna kendaraan pribadi untuk berpindah ke bus rapid transit, final report, Hibah riset multidisiplin, Universitas Indonesia, 2010
15. Ernst JP (2005) Initiating bus rapid transit in Jakarta, Indonesia. Trans Res Record J Trans Res Board 1903:20–26
16. Ningtyas DU, Joewono TB (2009) The relations of the image and fact of the service quality of busway with user loyalty. J Trans 9(1):15–24

Frontal Systems as Mechanisms of Fish Aggregation

Alberto Baudena, Francesco d'Ovidio, Guido Boffetta, and Silvia De Monte

1 Introduction

The oceans cover the 70 % of the planet's surface, and are the place in which earth life was born and later has colonized the lands. Nevertheless, terrestrial ecosystems are slightly different from oceanic ones [5]. The differences of marine environment are various: e.g., the phytoplankton, the major responsible of primary production in the oceans, have a biomass several-fold less than that of their dependents, the gravity is not of central importance, the new lineages tend to coexist with the older ones instead of replacing them, etc. A marine life peculiar characteristic is that it has developed in an environment whose timescale dynamics overlap with the ones of its ecological processes.

In the past years, it has been shown that horizontal stirring at the (sub-) mesoscale (10–100s of km, days to several weeks) plays a central role in organizing the phytoplanktonic landscape along physical fronts [2]. These fronts enclose water patches with contrasted water properties ("fluid dynamical niches") long enough for types locally adapted to become predominant. Eventually, these niches are stretched into thinner filaments that are mixed with waters far apart from their origin. This mechanism allows to build a highly diverse and widely distributed background of plankton types at the basin scale: this is a central issue in the dynamic and maintenance of planktonic biodiversity.

This dynamics has been described for primary producers. However, it is not clear how it affects the whole ecosystem, in particular because while plankton are passively advected, higher trophic elements can move actively.

A. Baudena (✉) • F. d'Ovidio • G. Boffetta • S. De Monte
Locean - Université Pierre et Marie CURIE, 4 Place Jussieu, 75005 Paris, France
e-mail: alberto.baudena@hotmail.it

P. Bourgine et al. (eds.), *First Complex Systems Digital Campus World E-Conference 2015*, Springer Proceedings in Complexity, DOI 10.1007/978-3-319-45901-1_19

On the other end of the trophic chain, recent studies about the behavior of top predators have shown their tendency to fish in structures within intense (sub-) mesoscale activity, which correspond to waters which have supported planktonic blooms during spring [1, 3].

This suggest that (sub-) mesoscale features could have strong ecological implications because they can host high fish densities.

2 Results and Discussion

Here we study the evolution of an idealized patch of tracer (an unspecified cue for fish, for instance, zooplankton concentration) initially distributed in a large area, exposed to a dynamic of diffusion and stretching, which is typically observed in the frontal regions visited by top predator during their foraging trips.

Starting assumption is that fish patches can react to the concentration of the cue by climbing the gradient. Exposed to diffusion and stirring dynamic, the initial patch of cue will tend to modify its shape and become smaller. If the fish are able to follow the gradient , they will tend to aggregate into a gradually smaller region.

It is useful to represent the surface currents in terms of phase space of a 2D dynamical system: in this way it is possible to represent a frontal system as a hyperbolic point, with a stable and an unstable manifold.

A first Lagrangian 1D approach is considered, and it is assumed, along the stable manifold, equilibrium between the diffusion and the stirring.

Along the front, the unstable manifold, we consider the speed as $u = \lambda x$, with λ Lyapunov exponent and a initially normal distribution of the cue. It is possible to write an analytical expression for the position and the speed of the point of a specific concentration, that can be chosen as the boundary of the optimal niche.

The second approach considers the evolution of the concentration of a 2D patch of cue, following the advection–diffusion equation:

$$\frac{\partial C}{\partial t} + u\frac{\partial C}{\partial x} + v\frac{\partial C}{\partial y} = K_H\left(\frac{\partial^2 C}{\partial x^2} + \frac{\partial^2 C}{\partial y^2}\right) \tag{1}$$

in which $u = \lambda x$ e $v = -\lambda y$.

We realized a numerical simulation of a 2D patch, using physical parameters representative of the Southern Ocean ($\lambda = 0.05\,\mathrm{d}^{-1}, k = 5\,\mathrm{m}^2\,\mathrm{s}^{-1}$), with a starting normal distribution of $\sigma_x = 10\,\mathrm{km}$ and $\sigma_y = 4\,\mathrm{km}$, $dx = dy = 1000\,\mathrm{m}$, $dt = 6\,\mathrm{h}$.

The simultaneous action of stirring and the diffusion compresses the cue toward the front and stretches it in the unstable direction, making the filament thinner and with lower concentrations.

Supposing that fish prefer to live in areas in which the cue is higher than a certain threshold (e.g., the 60 % of the maximum value at $t = 0$), we study the position and

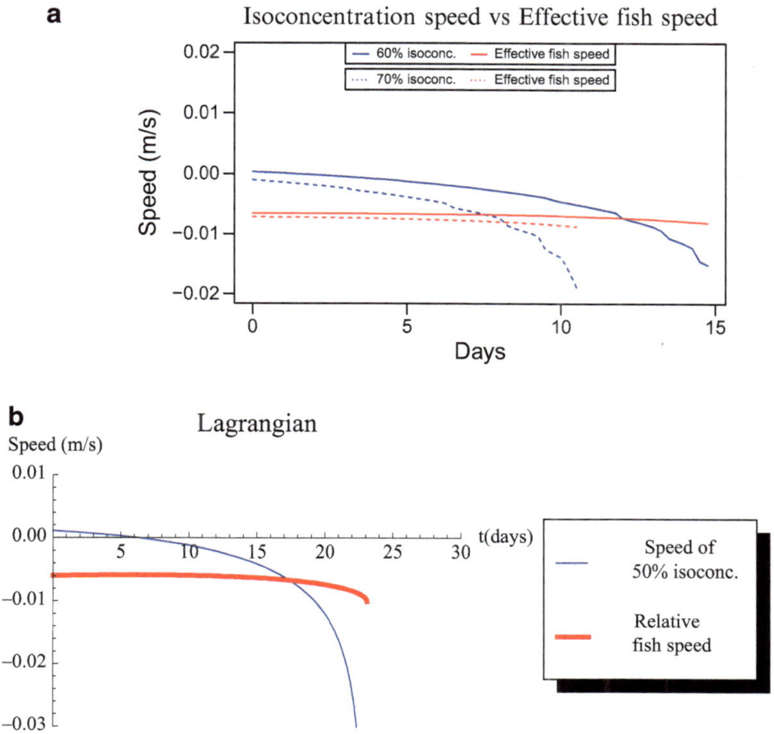

Fig. 1 Speed of the line of isoconcentration along the front (*blue line*) compared to the effective myctophid speed (order of magnitude). The first is considered positive if the patch is expanding, otherwise negative. (**a**) represents the speed of the isoconcentration of the cue (e.g., zooplankton concentration) at the 60 % (*continuous line*) and 70 % (*dotted line*) of the initial maximum value, while (**b**) refers to the 50 %. (**a**) is obtained with a Eulerian model, (**b**) with a Lagrangian one

the velocity of the optimal niche boundaries, to verify if they are able to follow the gradient of with their swimming speed. In particular, we concentrate on the unstable manifold, where fish have more difficulties to go after the patch.

We define so an *effective fish speed* as:

$$u_{\text{eff}} = u_M - u_C(x) \tag{2}$$

where u_M is the average fish speed (that we consider as $0,01\frac{\text{m}}{\text{s}}$) and $u_C(x)$ the velocity of the current in the point in which the fish are, that is so dependent from the position.

The value that we have chosen for the fish speed is taken from swimming speed of the myctophid, a mesopelagic fish which constitute a large component of the diet of many predators of the open ocean [4]. As it is possible to see from Fig. 1a, b we found that the isoconcentration lines speed are compatible with the myctophids one, suggesting that the fish may be able to follow the optimal niche for about 10 days, increasing so their local density.

3 Scientific Validation

This paper has been unanimously validated in a collaborative review mode with the following reviewers:

- Francesco Nencioli, Plymouth Marine Laboratory
- Yohav Lehah, Department of Earth and Planetary Sciences.

References

1. Cotté C, d'Ovidio F, Dragon A-C, Guinet C, Lévy M (2015) Flexible preference of southern elephant seals for distinct mesoscale features within the antarctic circumpolar current. Prog Oceanogr 131:46–58
2. d'Ovidio F, De Monte S, Alvain S, Dandonneau Y, Lévy M (2010) Fluid dynamical niches of phytoplankton types. Proc Natl Acad Sci 107(43):18366–18370
3. Godø OR, Samuelsen A, Macaulay GJ, Patel R, Hjøllo SS, Horne J, Kaartvedt S, Johannessen JA (2012) Mesoscale eddies are oases for higher trophic marine life. PLoS ONE 7(1):e30161, 01
4. Irigoien X, Klevjer TA, Røstad A, Martinez U, Boyra G, Acuña JL, Bode A, Echevarria F, Gonzalez-Gordillo JI, Hernandez-Leon S et al (2014) Large mesopelagic fishes biomass and trophic efficiency in the open ocean. Nat Commun 5:1–10
5. Smetacek V (2012) Making sense of ocean biota: how evolution and biodiversity of land organisms differ from that of the plankton. J Biosci 37(4):589–607

Science Information and the Complexity: Are We Oriented to a Transdisciplinary Science?

Rafaela Carolina da Silva and Rosângela Formentini Caldas

The concept of transdisciplinarity is an extent of complex expression in society. Currently, the human, social, and phychological contexts are connected and the science needs an answer about this phenomenon. This article conceptualizes the transdisciplinarity as a strategic application of complexity, which is a methodological concept that bases the transdisciplinarity.

Then, the concepts of transdisciplinarity and complexity connected, since we understand transdisciplinary as a fundamental question related to complex structures inherent in the world. This idea also leads to the thinking of a design that considers a new perception of reality. According to Morin [1, p. 138, our translation], understand the complexity is "[. . .] at the same time, separate, associate [. . .] the emergency of levels reality without reducing them to elementary units and general laws."

In this context, "Transdisciplinarity as the prefix 'trans' indicates, indicates that is a relation between the disciplines, across different disciplines and beyond any discipline. This goal is the understanding of this world for which one of the imperatives is the unity of knowledge." ([2], s.n, our translation).

So, the question is: "Is the Information Science oriented to a transdisciplinarity science?" It depends on the object that we are studying. If we study the object and think that the object of Information Science is the information, it can be a transdisciplinary object. We can see that the literature of Information Science hasn't a conceptualization of what information would be. On the other hand,

R.C. da Silva (✉)
Information Science, UNESP, Marília, Brazil
e-mail: rafaelacarolinasilva@gmail.com

R.F. Caldas
Information Science Department, UNESP, Marília, Brazil
e-mail: rcaldas@marilia.unesp.br

P. Bourgine et al. (eds.), *First Complex Systems Digital Campus World E-Conference 2015*, Springer Proceedings in Complexity, DOI 10.1007/978-3-319-45901-1_20

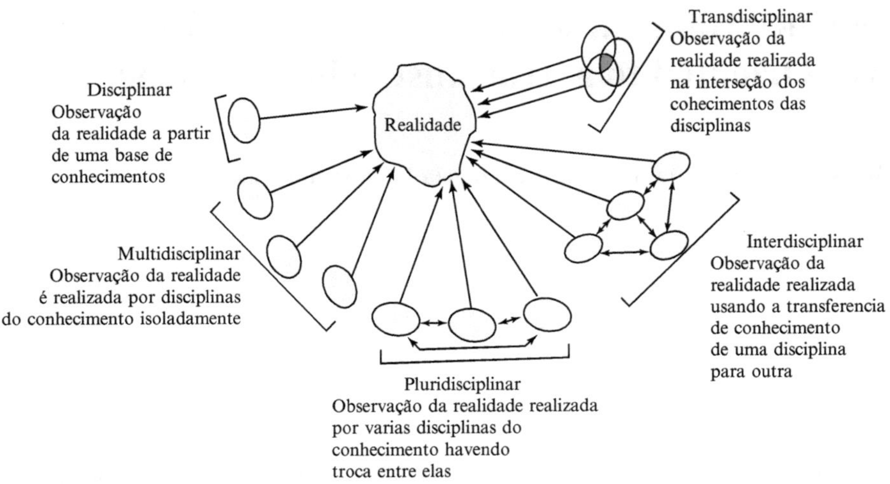

Source: larozinski Neto; Leite (2010. p. 4).

Fig. 1 Interdisciplinarity, pluridisciplinarity, multidisciplinarity, and transdisciplinarity. Source: Iarozinski Neto ([3], p. 4)

if you concept that your object of study is the various types of applications that information professionals realize in the library, we can discuss other concepts, such as multidisciplinarity, pluridisciplinarity, and interdisciplinarity.

We can observe that Fig. 1 relates the interdisciplinarity, pluridisciplinarity, multidisciplinarity, and transdisciplinarity concept. The transdisciplinarity is multireferential and multidimensional, taking concepts of time and history in a transhistorical horizon [4].

For Information Science can be considered as transdisciplinary science, there must be a change in the perception of phenomenon, society, reality, ethics, etc. In this perspective, it is how the professional exercises his/her profession that designates if a science is multi/pluri/inter or transdisciplinary.

Finally, this paper aims to propose a discussion about the theme instead of trying to conceptualize transdisciplinarity. It's a question that impacts all the world in a complexity system of emotions, rights, rationality, humans relations, and contexts.

References

1. Morin E (2001) Ciência com consciência, 5 th edn. Bertrand, Rio de Janeiro
2. Nicolescu B (1997) A evolução transdisciplinar a universidade: condição para o desenvolvimento sustentável. In: Congresso Internacional "A Responsabilidade da Universidade para com a Sociedade", 1, 1997, Bangkok. Anais... Bangkok: International Association of Universities; Chulalongkorn University, 1997. Disponível em: http://ciret-transdisciplinarity.org/bulletin/b12c8por.php. Acesso em: 05 jun. 2015

3. Iarozinski Neto A, Leite MS (2010) A abordagem sistêmica na pesquisa em engenharia de produção. Produção 20:1–14
4. Carta de Transdisciplinaridade (1994) In: Primeiro Congresso Mundial de Transdisciplinaridade. Convente de Arrábida, Portugal, 2–6 novembro 1994

Flexible Perception-Action Strategies: The Influence of Segmental and Global Motion Information on Follow-the-Leader Coordination

Laurentius A. Meerhoff, Harjo J. De Poel, and Chris Button

Humans can coordinate their movements with each other efficiently in complex and dynamic situations. For example, on a daily basis humans can navigate effortlessly through crowds of people, drive a car through traffic or play sports. In many situations, there are no (obvious) rules or constraints that determine how inter-personal coordination emerges, some less tangible construct (e.g. social norms, mood or skill) may dictate how behaviour unfolds. In sports, the emergent interaction can arguably be linked to an individual's skill to interact with others effectively—their 'interact-ability' [1]. Through visual and motor experience humans learn to attune and calibrate their actions with information presented by others to satisfy mutual goals. The present study explores how such information may be available along a proposed spectrum of global and local sources of motion information.

The study employs a follow-the-leader task where followers have to maintain their initial distance with the leader (cf. [2]). The leader is a pre-recorded video display of a back- and forwards moving leader shape. A variety of shapes were presented to manipulate the availability of available motion information (see Fig. 1).

The spectrum is suggested to have segmental motion information on the one end, requiring more local perception of parts of the object in relation to each other (i.e. the limbs). The other end is suggested to consist of a global perception of the optical size change of an object [3]. In-between cadence information, referring to

L.A. Meerhoff (✉) • C. Button
School of Physical Education, Sport and Exercise Science, University of Otago, Dunedin, New Zealand
e-mail: laurentius.meerhoff@otago.ac.nz

H.J. De Poel
Center for Human Movement Sciences, University Medical Center Groningen, University of Groningen, Groningen, The Netherlands

P. Bourgine et al. (eds.), *First Complex Systems Digital Campus World E-Conference 2015*, Springer Proceedings in Complexity, DOI 10.1007/978-3-319-45901-1_21

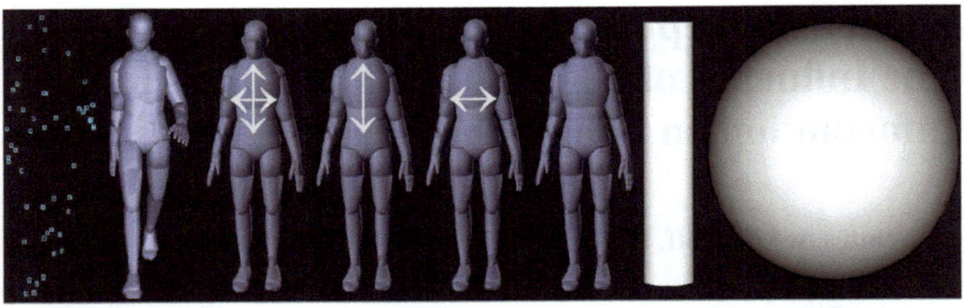

Fig. 1 Leader shapes as presented to the follower participants; the *arrows* indicate the cadence movement of the static displays. From *left* to *right*: point-light, avatar, cadence XY, cadence Y, cadence X, fixed, cylinder and sphere

the up-and-down movements that occur during gait, is suggested to play a role. In the reference condition, a leader was presented as a back-and-forth walking animated avatar. It was based on a 15 segment model of a pre-recorded leader participant. In the other conditions, various aspects of motion information were either suppressed or emphasized. We emphasized movement of body segments using a point-light display. In other conditions we showed a mannequin (i.e. an avatar without moving limbs) with various combinations of the subtle, but potentially pertinent, lateral and vertical displacements that occur during gait. Another set of conditions showed different shapes (a mannequin, cylinder and sphere) that only displayed the optical size change corresponding to the back-and-forth displacement.

Synchrony between follower and leader was analysed using a point-estimate relative phase and an estimate of the virtual interpersonal distance. The first informs about the temporal synchrony and to what extent there was a lead–lag relationship. A negative value indicates a phase-lead of the leader, scaled to 2π. The second informs about a visual angle based measure of spatial alignment. A negative value indicates an overall decrease of distance between follower and leader. It was revealed that local information led to a tighter temporal synchrony (Fig. 2, left panel), whereas global information to a closer spatial alignment (Fig. 2, right panel). These findings contribute towards understanding how individuals can coordinate their movements in dynamic environments. It suggests that in order to act most efficiently it is important to be able to flexibly switch and prioritize different information sources. Such flexible behaviour to pertinent motion information could be the underlying process that construes interact-ability.

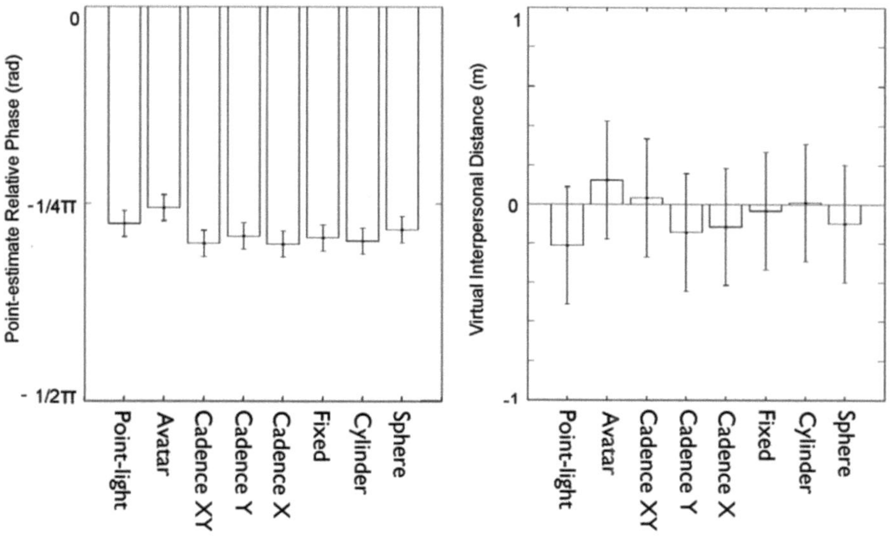

Fig. 2 Temporal synchronization (*left*) reflected by the relative phase and spatial alignment (*right*) represented by virtual interpersonal distance. Averages and 95 % confidence intervals are displayed

1 Scientific Validation

This paper has been unanimously validated in a collaborative review mode with the following reviewers:

- Rafaela Carolina da Silva, from Universidade Estadual Paulista "Julio de Mesquita Filho"—Brazil
- Laura Melinda Stan, from Faculty of Economics and Business Administration, West University of Timisoara, Romania
- Angel Ric, from Complex Systems and Sport Research Group, National Institute of Physical Education of Catalonia (INEFC), University of Lleida, Spain.

References

1. Meerhoff LA, De Poel HJ (2014) Asymmetric interpersonal coupling in a cyclic sports-related movement task. Hum Mov Sci 35:66–79
2. Meerhoff LA, De Poel HJ, Button C (2014) How visual information influences coordination dynamics when following the leader. Neurosci Lett 582:12–15
3. Rio KW, Rhea CK, Warren WH (2014) Follow the leader: visual control of speed in pedestrian following. J Vis 14(2):1–16

Dynamic Process of Pulmonary Data Analysis: From the Athlete Mouth to the Coach Hands

A. Sousa, K. de Jesus, L. Machado, J.P. Vilas-Boas, and R.J. Fernandes

1 Introduction

The dynamic behaviour of pulmonary gas exchange variables (particularly oxygen uptake—VO_2) during and after exercise has been studied throughout different exercise modes [1]. However, from the time VO_2 is measured (usually at the mouth using a breath-by-breath gas analyser—e.g. K4b^2, Cosmed, Italy), until it is available for coaches (for training control and athletes evaluation), a complex system involving multi-level dynamics experimental steps must occur. Contrarily to this, the VO_2 data analysis is traditionally conducted considering only the smoothing and averaging procedures of the breath-by-breath VO_2 data itself. Therefore, extreme importance should be given to this dynamic process; otherwise the physiological significance of the data obtained could be misunderstood, not reflecting the truth of the physiological mechanisms underlined [2].

A. Sousa • L. Machado • J.P. Vilas-Boas • R.J. Fernandes
CIFI2D, Faculty of Sport, University of Porto, Porto, Portugal

LABIOMEP, University of Porto, Porto, Portugal

K. de Jesus (✉)
CIFI2D, Faculty of Sport, University of Porto, Porto, Portugal

LABIOMEP, University of Porto, Porto, Portugal

Human Performance Laboratory, Faculty of Physical Education and Physiotherapy, Federal University of Amazon, Manaus, Brazil
e-mail: kellydejesus@ufam.edu.br

P. Bourgine et al. (eds.), *First Complex Systems Digital Campus World E-Conference 2015*, Springer Proceedings in Complexity, DOI 10.1007/978-3-319-45901-1_22

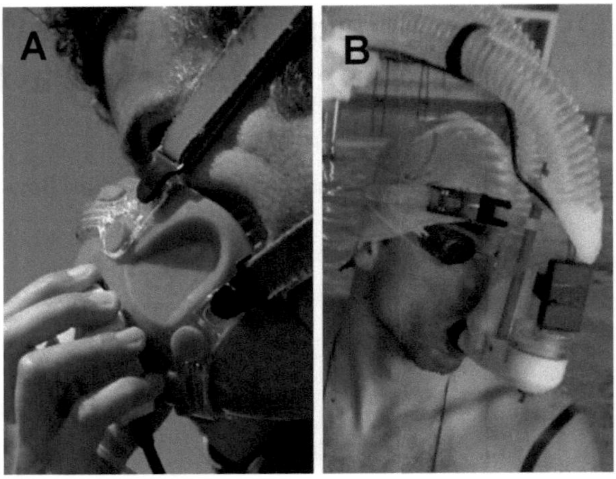

Fig. 1 Traditional facemask (**a**) and respiratory snorkel and valve system (**b**) used for breath-by-breath pulmonary data assessment

2 Development

Firstly, individual breath-by-breath VO_2 data that typically arises as a result of some constraints caused by the traditional facemask or respiratory snorkel and valve system used (Fig. 1), like swallowing, coughing, sighing, signal interruptions and so forth, should be omitted.

For that, respiratory frequency (Rf—b/min), tidal volume (VT—l), oxygen expiratory fraction (FeO_2—%) and carbon dioxide expiratory fraction ($FeCO_2$—%) breath-by-breath values greater and lower than ± 4 SD from the local mean are not considered for further analysis (Fig. 2a). In fact, these ventilatory parameters are the most affected by data variability, and consequently, are the ones that influence more VO_2 data [3, 4]. Secondly, individual breath-by-breath VO_2 responses are smoothed using a 3-breath moving average. After this process, and considering that breath-by-breath data might contribute to a great imprecision of the physiological VO_2 response, multiple analysis strategies, fundamentally by averaging across breaths and discrete time intervals, are used (Fig. 2b). We recommend the 10 and 15 s time-averaging intervals strategies to more precisely assess the VO_2 values. All data analysis is conducted by using the time-averaging function of the Cosmed analysis software (Cosmed, Italy).

An example of the dynamic process of pulmonary data analysis with raw breath-by-breath VO_2 data and after smoothing and averaging processes is presented in Fig. 3.

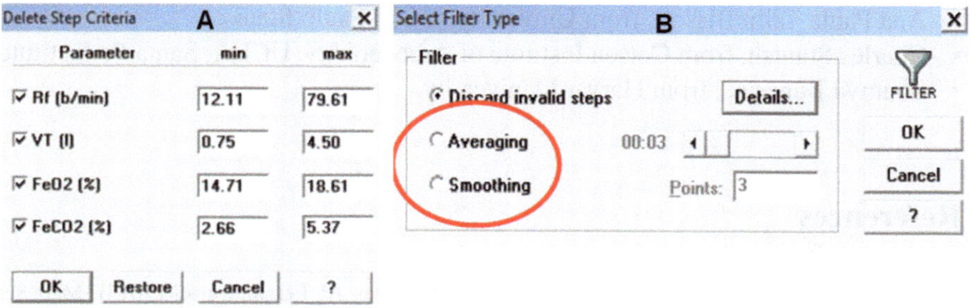

Fig. 2 Breath-by-breath values (Rf, VT, FeO_2 and $FeCO_2$) greater and lower than ±4 SD from the local mean omission (**a**) and averaging across breaths and discrete time intervals (**b**) processes conducted for data analysis using the Cosmed analysis software

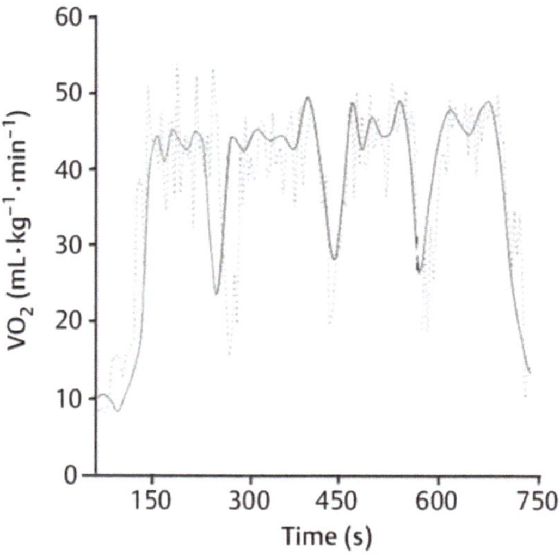

Fig. 3 VO_2 response of a representative subject during 4 sets of 200 m swimming moderate exercise where raw breath-by-breath (*dashed line*) and smoothed and averaged processes (*continuous line*) are evident

3 Conclusion

The emergence of breath-by-breath gas analysers enabled the acquisition of pulmonary data with greatest temporal resolution [2]. However, attention should be given to this dynamic process, which should provide the best compromise between the accuracy and the reliability in pulmonary assessment.

4 Scientific Validation

This paper has been unanimously validated in a collaborative review mode with the following reviewers:

- Ana Paula Talin Bissoli, from University of São Paulo State
- Charles Stanish, from Cotsen Institute of Archaeology, UCLA, Santa Fe Institute
- Soumya Banerjee, from Harvard University.

References

1. Sousa A, Figueiredo P, Zamparo P, Pyne DB, Vilas-Boas JP, Fernandes RJ (2015). Med Sci Sports Exerc 47(8):1705–13
2. de Jesus K, Guidetti L, de Jesus K, Vilas Boas JP, Baldari C, Fernandes R (2014) Which are the best VO2 sampling intervals to characterize low to severe swimming intensities? Int J Sports Physiol Perform 35(12):1030–1036
3. Özyener F, Rossiter H, Ward S, Whipp B (2001) Influence of exercise intensity on the on and off-transient kinetics of pulmonary oxygen uptake in humans. J Physiol 533(3):891–902
4. Lamarra N, Whipp BJ, Ward SA, Wasserman K (1987) Effect of interbreath fluctuations on characterizing exercise gas exchange kinetics. J Appl Physiol 62(5):2003–2012

Lyapunov Exponents and Oceanic Fronts

Francesco Toselli, Francesco d'Ovidio, Marina Lévy, Francesco Nencioli, and Olivier Titaud

The ocean is a turbulent system where its physical and biogeochemical tracers (like heat, salinity, and phytoplankton) present strong inhomogeneities that are structured over a large range of spatiotemporal scales by features like vortices (eddies) and fronts. Several methods have been proposed to analyze the surface currents and track the physical features that constrain tracer distributions through the horizontal transport. In particular, Lagrangian methods allow to mimic the transport dynamics by creating synthetic particle trajectories which are obtained by integrating the velocity field and then analyzed.

One powerful diagnostic which has been used to identify frontal structures, i.e., lines where discontinuities or strong gradients are expected to occur in the ocean, is the calculation of the local Lyapunov exponent. In general the Lyapunov exponents are used in a dynamical system approach in order to detect chaotic behavior for an invariant system by measuring the growth of the perturbations occurring along particle trajectories. For geophysical systems, the calculation of the Lyapunov exponent is usually performed at finite time and finite space. When computed in this way along backward trajectories, this calculation has been shown to provide exponent maxima (ridges) along frontal structures. Intuitively, this is due to the fact that one common mechanism for producing a front in the ocean is by confluence of water masses originally located far apart, so that the trajectories of nearby particles in a frontal region diverge fast when observed evolving backward in time [1–3, 7]. Correspondingly, the Lyapunov vector associated with the leading exponent contains information on the front orientation [6]. Many studies have been conducted in the past [4, 5, 8].

F. Toselli (✉) • F. d'Ovidio • M. Lévy • F. Nencioli • O. Titaud
LOCEAN Laboratory, Sorbonne Universités (UPMC, Univ Paris 06), 4 place Jussieu, 75005 Paris, France
e-mail: francesco.toselli@edu.unito.it

P. Bourgine et al. (eds.), *First Complex Systems Digital Campus World E-Conference 2015*, Springer Proceedings in Complexity, DOI 10.1007/978-3-319-45901-1_23

Here we explore the reliability of the finite size Lyapunov technique applied to satellite-derived ocean currents, in estimating front orientation. In order to achieve this we compute systematically Lyapunov exponent and vectors and we compare to the direction of sea surface temperature images observed from microwave satellites. We study this comparison by focusing to the North Atlantic basin where we analyze and calculate the angles between Lyapunov vectors, SST gradient direction, and surface velocity fields vectors, in order to measure how much better Lyapunov predicts fronts in respect to the direction of the flow. Firstly we compute two angular difference tests: one between velocity fields and SST gradient and one between Lyapunov vectors and SST gradient. After that we calculate the difference between these two tests, showed in Fig. 1, in order to assess the skill of the Lyapunov analysis for predicting SST gradient orientation in respect to velocity directions.

Figure 1 presents a preliminary results where we can see differences between zones of high kinetic energy, like the Gulf Stream Current, where Lyapunov vectors mostly align with the direction of currents. On the other hand, in lower energetic regions, like the subtropical gyre, Lyapunov vectors appear as a better predictor of SST gradient than current direction.

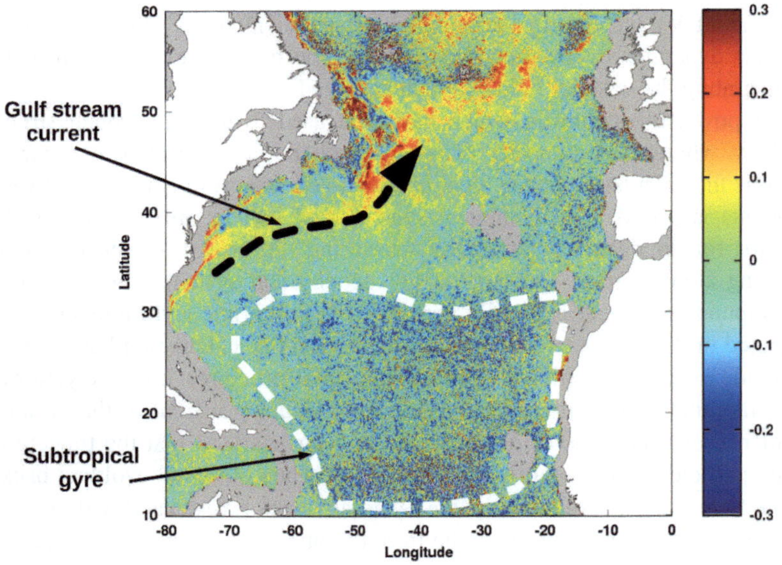

Fig. 1 Assessment of the capabilities of the Lyapunov analysis in predicting the orientation of SST gradients in the North Atlantic. Regions in *blue* (in particular the subtropical gyre) correspond to areas where the Lyapunov analysis predicts SST gradient orientations better than current direction. *Red regions* (in particular higher energetic features like the Gulf Stream) are areas where current direction performs better. See text for details. Altimetry data were provided by AVISO/CLS with support from CNES. The map shows 2003–2007 mean (degrees) computed in frontal regions (defined as Lyapunov exponents larger than 0.1 day^{-1})

1 Scientific Validation

This paper has been unanimously validated in a collaborative review mode with the following reviewers:

- Rens Meerhof
- Laure Resplandy, LOCEAN-Paris
- Maria Carmen Lemos, School of Natural Resources and Environment-University of Michigan.

References

1. Abraham ER, Bowen MM (2002) Chaotic stirring by a mesoscale surface-ocean flow. Chaos 12(2):373–381
2. Beron-Vera FJ, Olascoaga MJ, Goni GJ (2008) Oceanic mesoscale eddies as revealed by Lagrangian coherent structures. Geophys Res Lett 35(12):L12603
3. Boffetta G, Lacorata G, Redaelli G, Vulpiani A (2001) Detecting barriers to transport: a review of different techniques. Physica D 159(1–2):58–70
4. d'Ovidio F, De Monte S, Alvain S, Dandonneau Y, Lévy M (2010) Fluid dynamical niches of phytoplankton types. Proc Natl Acad Sci 107(43):18366–18370
5. d'Ovidio F, De Monte S, Della Penna A, Cotté C, Guinet C (2013) Ecological implications of eddy retention in the open ocean: a Lagrangian approach. J Phys A Math Theor 46(25):254023
6. Lehahn Y, d'Ovidio F, Lévy M, Heifetz E (2007) Stirring of the northeast atlantic spring bloom: a Lagrangian analysis based on multisatellite data. J Geophys Res Oceans 112(C8):C08005
7. Shadden SC, Lekien F, Marsden JE (2005) Definition and properties of Lagrangian coherent structures from finite-time Lyapunov exponents in two-dimensional aperiodic flows. Physica D 212(3–4):271–304
8. Waugh DW, Abraham ER (2008) Stirring in the global surface ocean. Geophys Res Lett 35(20):L20605

Software is Not Fragile

William B. Langdon and Justyna Petke

1 Introduction

It is often assumed that computer programs are fragile and any single change will destroy them totally. Figures 1, 2 and 3 show this is not true. We automatically make changes like those a human programmer might make. That is, delete a line of a program, replace a line with another and insert a copy of a line. Figure 1 shows the impact of all possible changes to the part of the program which is used. 63 % do not compile and 14 % abort when run. However, most of the modified programs which compile and run normally produce exactly the same result as the original program source code. Indeed a few produce identical answers but are slightly faster. We suggest often wholesome changes can be quickly found.

Software engineering continues to produce some of the most complex human artefacts on the planet. Mostly it has succeeded in its goal of describing in complete detail the computer systems people create, maintain and use. Nonetheless, despite extensive tools, large software systems are beyond comprehension. They cannot be fully understood by anyone no matter how clever, nor can they be understood by groups of people, not even the team of experts who may have been working on them for years. Nonetheless they continue to yield enormous economic advantage which has led to the world economy being addicted to software.

Genetic Improvement (GI) [1–3] applies search-based optimisation techniques (SBSE) [4], such as genetic algorithms (GA) [5], genetic programming (GP) [6] and hill climbing to make measurable improvement in existing human-written code. Although GAs [7] and GP [8] have been analysed for sometime, very little theory is available which can predict how the performance of real software systems will

W.B. Langdon (✉) • J. Petke
CREST, Department of Computer Science, University College London, Gower Street, London WC1E 6BT, UK
e-mail: w.langdon@cs.ucl.ac.uk

P. Bourgine et al. (eds.), *First Complex Systems Digital Campus World E-Conference 2015*, Springer Proceedings in Complexity, DOI 10.1007/978-3-319-45901-1_24

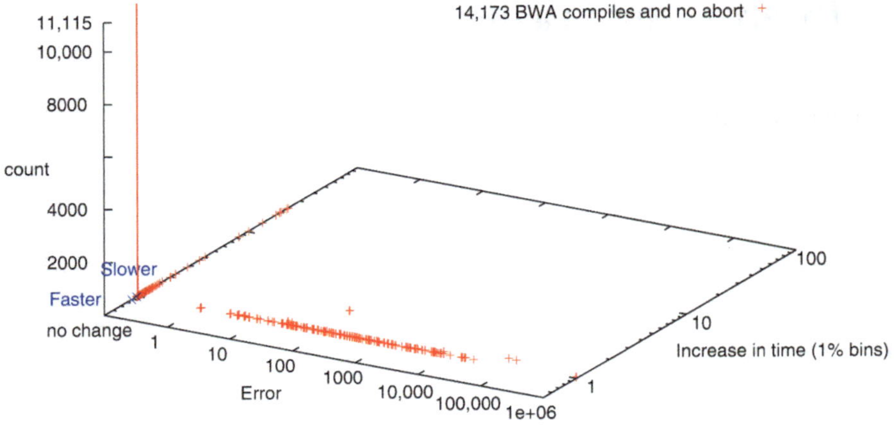

Fig. 1 Impact of all possible executable changes to example program (Sect. 2.1). Of the 23 % which compile and run normally 89 % produce the same answer as the original code ("no change"). Indeed three of them are faster (×)

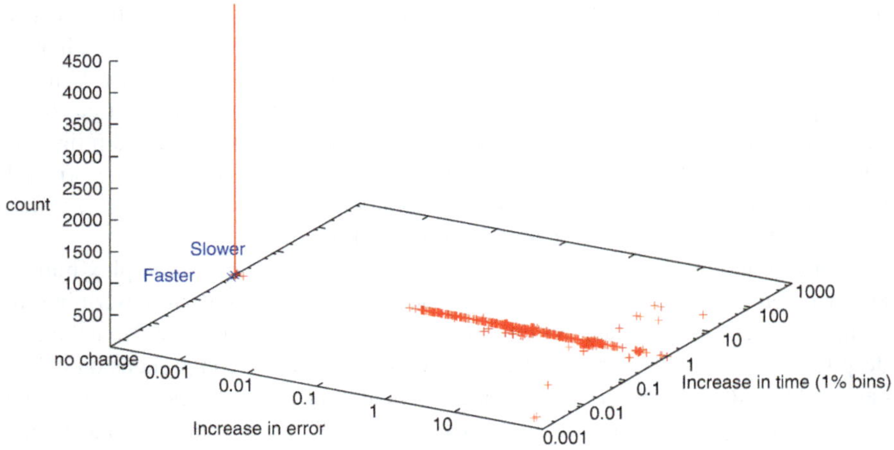

Fig. 2 Impact of all (7079) one change mutations on StereoCamera kernel [16] running on an nVidia GeForce GT 730 graphics card. Nine mutants failed to compile due to a bug in the nVidia CUDA 6.0.1 compiler (fixed in CUDA 7.0). 16 caused infinite loops, 318 others failed at run-time. 4400 (62 %) mutants do not change the output at all ("no change"), indeed at least 41 of them are faster (×)

change in response to changes to their source code. In SBSE this is known as the fitness landscape, which depends both on the program and the mechanisms available to change it.

Section 3 shows software, like other human endeavours, can, both in terms of source code [9] and numeric values, follow Zipf's law [10]. But we start in Sect. 2 with two leading open source Bioinformatics tools (one in C, the other in C++)

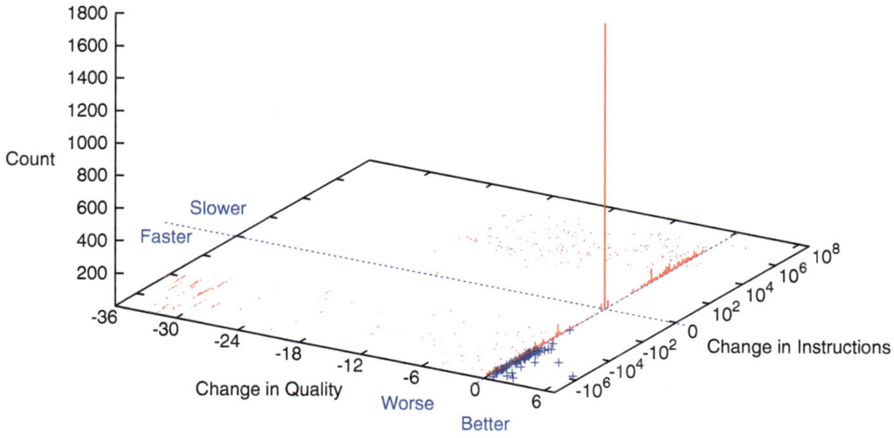

Fig. 3 Impact of single mutations to Bowtie2 [14]. Non-linear scales [19]

and an open source parallel CUDA GPGPU kernel and show, for these examples, software is not as fragile as is often assumed. For these diverse source code examples, we start to map their software fitness landscapes. This initial mapping suggests that software is much more resilient than is commonly assumed. This in turn supports the idea of artificial evolution of programs via mutating existing code (GI).

2 Mutating Useful Software

The following three sections describe the impact of making simple changes (of the three types described at the start of Sect. 1) to three programs written in C, C++ and CUDA. We call these changes mutations. We only make one change at a time and so these are known as first order mutations. Multiple changes are known as higher order mutants [11]. In Figs. 1, 2 and 3 we report the change in output (excluding logging information) and the change in run-time. In each case the mutations target code which is used by the test case and yet many mutations do not change the output at all (cf. equivalent mutants [12]).

2.1 BWA

BWA [13] is a leading open source Bioinformatics tool written in the C programming language for matching short next generation DNA sequences to the human genome. It comprises 33 .c and 20 .h files containing 13,000 lines of code. We down-loaded the complete installation kit (version 0.7.12) from GitHub and automatically converted all the .c C source files into a specialised BNF grammar of 18,621 rules. As with our earlier work [14], we use the grammar to control the mutations [11]. It ensures all mutations are syntactically correct (in that brackets match, there are semi-colons where required, etc.), however, due to variable out-of-scope and other errors, it need not ensure a mutant compiles.

We also obtained a test sequence from http://fg.cns.utexas.edu/fg/course_notebook_chapter_seventeen.html. We used the GNU gcov test analysis tool to discover all the lines of C source code executed. (gcov shows 532 lines of the BWA program are used by this test case.)

We use the three types of mutation: delete a line of source code, replace it with a copy of another line of code and insert a copy of another line of code. As with our previous work [14], we restrict replacements and inserts to be of the same type and from the same .c source file. Notice mutations are limited to re-using existing source code. They do not create new code from scratch. To ensure all mutations are executed, all mutations either: delete or replace a line which is executed or insert a source line immediately before it. There are 61,775 possible mutations of BWA. We generate and test them all (see Fig. 1).

To catch indefinite loops, we impose a CPU limit. The limit is about 40 times longer than normal operation. We also automatically remove computer jobs (known as processes to the Unix operating system) which failed to terminate normally. Our zombie.awk killer removed nine such zombies.

2.2 *Stereo Pair CUDA Image Processing*

StereoCamera is open source CUDA code written by nVidia's vision processing expert [15]. It takes two stereo images and from the parallax differences between them it infers distances between objects in the images and the camera. We had previously used our BNF grammar approach to automatically evolve considerable speedups [16]. Here we reuse our automatically generated grammar to start to map the fitness landscape for CUDA software. As a test case we use a stereo image pair taken inside a typical modern office (provided by Microsoft's I2I database). In order to get a somewhat different example, we concentrate on the parallel CUDA code written for the graphics card, (known as a GPU kernel) and ignore the program code which runs on a normal personal computer. The kernel contains 276 lines of CUDA code.

The StereoCamera GPGPU [17] kernel is more constrained than normal C program code. Therefore firstly all of it will be used, Secondly we can readily force indefinite loops to stop running and finally (excluding a compiler bug) the grammar will ensure mutants always compile.

We deliberately exclude changes to tuning parameters, CUDA specific pragma and kernel parameter changes (which were used in [16]) as by design they cannot break the existing CUDA kernel.

We tried all possible single change mutants (i.e. first order mutants). Figure 2 shows that about five in eight source code mutations do not change the kernel's output on the randomly chosen test image pair.

2.3 Bowtie2

Bowtie2 [18], like BWA, is a state-of-the-art next generation DNA analysis program. We made random source code mutations to Bowtie2 and measured their impact on random test cases, see Fig. 3. Although many mutations cause Bowtie2 to fail (not plotted) and others cause it to produce very poor solutions (e.g. reducing quality by 36, left) others have less dramatic impact. Some slow down Bowtie2 and others make it faster. However, many changes have no impact on quality (although they may change Bowtie2's speed, plotted along x=0). Indeed a large number do not change its speed either (note spike at the origin). There are even a few mutations which give better quality solutions and even 139 which are both better on a random test and faster (plotted in Fig. 3 with +). It is from these a seventyfold speed up can be evolved [14].

3 Zipf's Law

George Zipf (USA, 1902–1950) proposed a "universal" law of human behaviour [10] in which the frequency of repeated items is inversely proportional to their rank (or order). Thus when plotted on log–log scale we get an inverse power law with a slope of -1. This has been shown to hold for the frequency with which words in many languages are used, the population of cities and many other human endeavours, including software engineering [9].

Using the grammars described in Sect. 2, Fig. 4 shows the distributions of exactly repeated BWA C and Bowtie2 C++ lines of source code both approximately follow Zipf's law.

3.1 What's My Favourite Number?

The GNU C library, version 2.22 released 14 August 2015 (excluding its test suite), contains 845,360 lines of C code, which contain in total 1,203,104 integer constants (see Fig. 5). Many glibc integers are associated with mapping non-ASCII character sets, e.g. Chinese.

nVidia's CUDA 7.0 comes with an extensive set of examples (including test code) totalling 85,711 lines of C++ or CUDA code. These contain 73,620 integer constants (see Fig. 6). Many numbers in the CUDA 7.0 samples are taken from the OpenGL

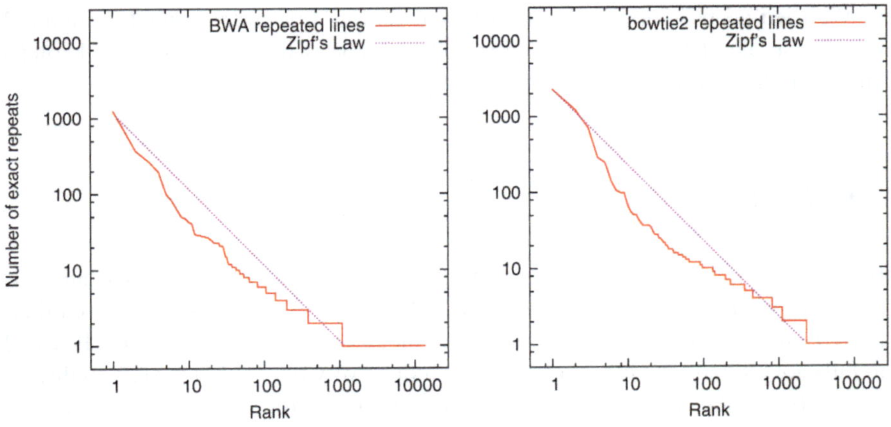

Fig. 4 Distribution of exactly repeated BWA C program source code (*left*) and Bowtie2 C++ source code (after macro expansion) [14, Fig. 5] (*right*). Zipf's law [10] predicts a straight line with slope of −1

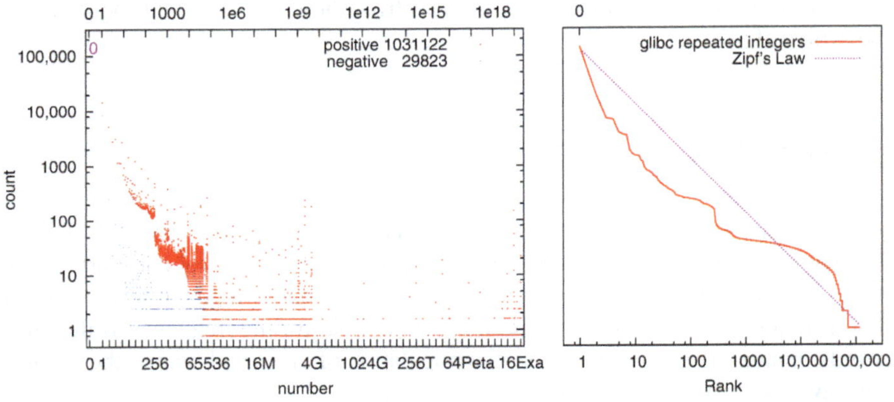

Fig. 5 Distribution of integer constants in GNU C library. Zero is the most common number, occurring in various formats a total of 142,159 times, followed by 1 (18,642) and −1 (6907). Every integer between −28 and 40,957 occurs at least once. There are 116,685 distinct integer constants

package. For example, the OpenGL source code includes integers for use as bit masks which are used to turn on image effects, such as stippling. Nevertheless we do not see pronounced clustering around the powers of 2, visible in our earlier work on CUDA 5.0 samples which included macro expansion.

In both the GNU C library and nVidia's samples, as with the whole lines of source code in Sect. 3, numeric values approximately follow Zipf's law.

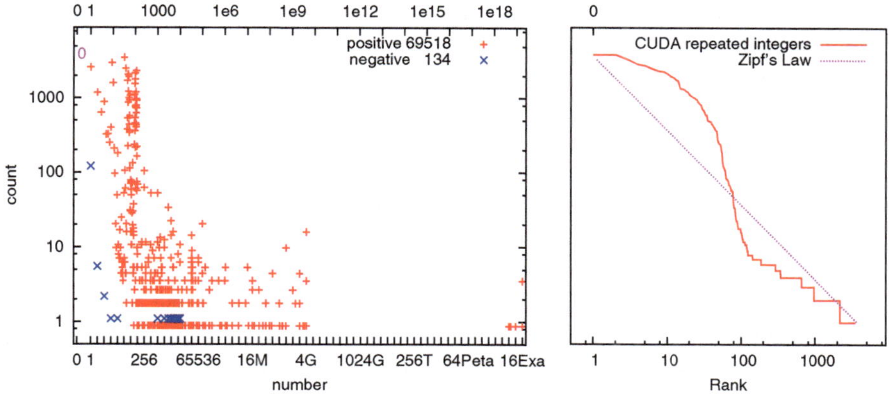

Fig. 6 Distribution of integer constants in CUDA 7.0 sample source code. Zero is again the most common number, occurring in various formats 3968 times. Surprisingly the second most popular number is 32 (3962), which probably reflects the nVidia GPU architecture. There are even fewer negative numbers than in the GNU C library (Fig. 5). Every integer between −2 and 60 occurs at least once. There are 3490 distinct integer constants

4 Conclusions

The existence of power laws in software has been previously reported. For example, Louridas et al. [9] reported power laws in both the patterns with which functions or classes are used and in the frequency with which machine code instructions are used. Nevertheless the prevalence of Zipf's law in human-written program code does appear to be under-appreciated. Although Louridas et al. caution against getting carried away with the supposed universality of power laws, it is tempting to suggest that the Zipf law we see with numbers might make it a good candidate for either numeric values to be used when creating test suites. Or indeed (although this is not yet done) numbers to be embedded into GI-created code might be drawn from a Zipfian distribution. Actual numerical values (see left-hand sides of Figs. 5 and 6) might be drawn with a preference for low entropy decimal, hexadecimal or octal strings. (In genetic programming [6, 20] floating point numbers, sometimes known as ephemeral random constants, ERCs, are typically drawn uniformly from a small range [21], although we have had good results from a tangent distribution [22].)

In Sect. 2 we saw in three different (albeit similar) industrial strength programming languages examples of high quality software (including serial and parallel programs) written by experts where it was comparatively easy to find simple source code changes which on a random test make no external difference. Indeed, if the new program code can avoid the trap of introducing variable out-of-scope errors (i.e. avoid compiler errors) most mutants will not break the program immediately. (In a few cases [16, 23–26] we have set up the GI system to carefully track the scope of variables and thereby avoided compilation errors.) Of course other programming languages are much more forgiving and will assume newly introduced variables should have been declared and will do this automatically for the lazy programmer.

It may be countered that we have not tested the changes sufficiently. This is deliberate. We know how much we have tested each change (exactly once). This gives us a solid benchmark from which to do comparisons. In real GI work, post evolution code maybe subjected to much more rigorous validation. (In [23] in addition to manual validation, we tested the new code back-to-back with the old more than a million times. No difference was ever found.) Although great strides have been made with automatic testing [27], it remains true that devising tests to execute newly written program code is difficult. Indeed one often gets the impression that simply running the new code is considered sufficient, without caring if indeed it calculated the right answer. We have set up our experiment so that we know our mutated program code is executed. We can tell if it calculated the right answer, or at least we can tell if it calculated the same answer as the human-written program.

Our purpose is to put to bed the myth that any random change will destroy human-written programs. Instead we have given persuasive evidence that whilst a random changes might be bad, if you are prepared to try multiple times, perhaps adapting a population approach, you can quickly find equivalent mutants and you may, if guided by a suitable fitness function, find improvements to your software.

References

1. Langdon WB (2015) Genetically improved software. In: Gandomi AH, Alavi AH, Ryan C (eds) Handbook of genetic programming applications, chap 8. Springer, pp 181–220. Forthcoming
2. Langdon WB (2015) Genetic improvement of software for multiple objectives. In: Labiche Y, Barros M (eds) SSBSE, Bergamo, Italy. LNCS, vol 9275. Springer, Cham, pp 12–28; Invited keynote
3. Langdon WB, Petke J, White DR (2015) Genetic improvement 2015 chairs' welcome. In: Langdon WB, Petke J, White DR (eds) Genetic improvement 2015 workshop, Madrid. ACM, New York, pp 791–792
4. Harman M, Jones BF (2001) Search based software engineering. Inf Softw Technol 43(14):833–839
5. Holland JH (1992) Adaptation in natural and artificial systems: an introductory analysis with applications to biology, control and artificial intelligence. MIT Press, Cambridge; First Published by University of Michigan Press (1975)
6. Poli R, Langdon WB, McPhee NF (2008) A field guide to genetic programming. Published via http://lulu.com and freely available at http://www.gp-field-guide.org.uk (With contributions by J.R. Koza)
7. Beyer HG, Langdon W (eds) (2011) Foundations of genetic algorithms, Schwarzenberg, Austria. ACM, New York
8. Langdon WB, Poli R (2002) Foundations of genetic programming. Springer, Berlin
9. Louridas P, Spinellis D, Vlachos V (2008) Power laws in software. ACM Trans Softw Eng Methodol 18(1):2:1–2:26
10. Zipf GK (1949) Human behavior and the principle of least effort: an introduction to human ecology. Addison-Wesley Press Inc., Cambridge

11. Langdon WB, Harman M, Jia Y (2010) Efficient multi-objective higher order mutation testing with genetic programming. J Syst Softw 83(12):2416–2430
12. Yao X, Harman M, Jia Y (2014) A study of equivalent and stubborn mutation operators using human analysis of equivalence. In: Proceedings of the 36th international conference on software engineering, ICSE 2014, Hyderabad. ACM, New York, pp 919–930
13. Li H, Durbin R (2010) Fast and accurate long-read alignment with Burrows-Wheeler transform. Bioinformatics 26(5):589–595
14. Langdon WB, Harman M (2015) Optimising existing software with genetic programming. IEEE Trans Evol Comput 19(1):118–135
15. Stam J (2008) Stereo imaging with CUDA. Technical Report, nVidia
16. Langdon WB, Harman M (2014) Genetically improved CUDA C++ software. In: Nicolau M, Krawiec K, Heywood MI, Castelli M, Garcia-Sanchez P, Merelo JJ, Rivas Santos VM, Sim K (eds) 17th European conference on genetic programming, Granada, Spain. LNCS, vol 8599. Springer, Berlin, pp 87–99
17. Owens JD, Houston M, Luebke D, Green S, Stone JE, Phillips JC (2008) GPU computing. Proc IEEE 96(5):879–899; Invited paper
18. Langmead B, Salzberg SL (2012) Fast gapped-read alignment with Bowtie 2. Nat Methods 9(4):357–359
19. Langdon WB (2014) Genetic improvement of programs. In: Winkler F, Negru V, Ida T, Jebelean T, Petcu D, Watt S, Zaharie D (eds) 16th international symposium on symbolic and numeric algorithms for scientific computing (SYNASC 2014), Timisoara. IEEE, New York, pp 14–19; Keynote
20. Koza JR (1992) Genetic programming: on the programming of computers by natural selection. MIT Press,
21. Daida JM, Bertram RR, Stanhope SA, Khoo JC, Chaudhary SA, Chaudhri OA, Polito II JA (2001) What makes a problem GP-hard? Analysis of a tunably difficult problem in genetic programming. Genet Program Evolvable Mach 2(2):165–191
22. Langdon WB (1998) Genetic programming and data structures: genetic programming + data structures = automatic programming!. Genetic programming, vol 1. Kluwer, Boston
23. Langdon WB, Harman M (2010) Evolving a CUDA Kernel from an nVidia template. In: Sobrevilla P (ed) 2010 IEEE world congress on computational intelligence, Barcelona. IEEE, New York, pp 2376–2383
24. Langdon WB, Modat M, Petke J, Harman M (2014) Improving 3D medical image registration CUDA software with genetic programming. In: Igel C, Arnold DV, Gagne C, Popovici E, Auger A, Bacardit J, Brockhoff D, Cagnoni S, Deb K, Doerr B, Foster J, Glasmachers T, Hart E, Heywood MI, Iba H, Jacob C. Jansen T, Jin Y, Kessentini M, Knowles JD, Langdon WB, Larranaga P, Luke S, Luque G, McCall JAW, Montes de Oca MA, Motsinger-Reif A, Ong YS, Palmer M, Parsopoulos KE, Raidl G, Risi S, Ruhe G, Schaul T, Schmickl T, Sendhoff B, Stanley KO, Stuetzle T, Thierens D, Togelius J, Witt C, Zarges C (eds) GECCO '14: proceeding of the sixteenth annual conference on genetic and evolutionary computation conference, Vancouver, ACM, New York, pp 951–958
25. Langdon WB, Lam BYH, Petke J, Harman M (2015) Improving CUDA DNA analysis software with genetic programming. In: Silva S, Esparcia-Alcazar AI, Lopez-Ibanez M, Mostaghim S, Timmis J, Zarges C, Correia L, Soule T, Giacobini M, Urbanowicz R, Akimoto Y, Glasmachers T, Fernandez de Vega F, Hoover A, Larranaga P, Soto M, Cotta C, Pereira FB, Handl J, Koutnik J, Gaspar-Cunha A, Trautmann H, Mouret JB, Risi S, Costa E, Schuetze O, Krawiec K, Moraglio A, Miller JF, Widera P, Cagnoni S, Merelo J, Hart E, Trujillo L, Kessentini M, Ochoa G, Chicano F, Doerr C (eds) GECCO '15: proceedings of the 2015 on genetic and evolutionary computation conference, Madrid. ACM, New York, pp 1063–1070
26. Langdon WB, Harman M (2015) Grow and graft a better CUDA pknotsRG for RNA pseudoknot free energy calculation. In: Langdon WB, Petke J, White DR (eds) Genetic improvement 2015 workshop, Madrid. ACM, New York, pp 805–810
27. Fraser G, Arcuri A (2011) EvoSuite: automatic test suite generation for object-oriented software. In: 8*th* European software engineering conference and the ACM SIGSOFT symposium on the foundations of software engineering (ESEC/FSE '11). ACM, New York, pp 416–419

Open Systems Science: A Challenge to Open Systems Problems

Mario Tokoro

1 Introduction

Science and technology burgeoned in the sixteenth and seventeenth centuries. Quintessential examples of the achievements of this age are the physics and astronomy of Copernicus and Galileo, as well as Newton's physics and mathematics. The seventeenth century saw a modern scientific approach develop on a Cartesian basis. This methodology contributed enormously to scientific advances from the eighteenth century onwards, and to technological progress that accelerated in the nineteenth century. Thus, the methodology of modern science can be largely credited with the industrial prosperity and economic development that the world has achieved today. It has also advanced medicine and improved standards of living.

2 The Contribution of Descartes

For all this, we are greatly indebted to René Descartes, the father of modern science. In his famous book, "Discourse on the Method," published in 1637, Descartes proposed the scientific method consisting of four steps. Reflecting on these steps, he wrote

1. The first was to never accept anything as true which I could not accept as obviously true.
2. The second was to divide each of the problems I was examining in as many parts as I could, as many as should be necessary to solve them.

M. Tokoro (✉)
Sony Computer Science Laboratories, Inc., Tokyo, Japan
e-mail: mario.tokoro@csl.sony.co.jp

P. Bourgine et al. (eds.), *First Complex Systems Digital Campus World E-Conference 2015*, Springer Proceedings in Complexity, DOI 10.1007/978-3-319-45901-1_25

3. The third was to develop my thoughts in order, beginning with the simplest and easiest to understand matters, in order to reach by degrees, little by little, to the most complex knowledge, and
4. The last was to make my enumerations so complete and my reviews so general that I could be assured that I had not omitted anything.

We can say without hesitation that modern sciences and technologies in its entirety have been built on a methodology comprising these four steps.

3 Drawbacks

Nonetheless, there are still plenty of complex issues that are hard to resolve. One such example is sustainability of the earth. This issue involves energy, climate, population, food, biodiversity, poverty and inequality, safety assurance, and many other mutually dependent factors that cannot be solved independently from one another.

Another example is the matter of life and health. Although medical science has settled almost every issue, there remain diseases such as cancer, metabolic disorder, and immunodeficiency that develop only through the interrelation of complex factors.

Yet another category of examples concerns instability in the global economy, and safety in food and huge social infrastructures, all of which consist of networks of people, materials, and information. Although constituting elements and network topologies are changing every day, our expectation is that—even in the event of unforeseen incidents—they will continue to deliver services without causing any critical impact on everyday life.

These challenges that we wish to address seem to have two common characteristics. The first is that all of these challenges are related to the problem resolution of integrated systems comprising numerous ever-changing, interrelated subsystems. In many cases, we cannot even identify the boundaries of the systems we are examining, what subsystems are involved, and how they interact in the problem systems.

The second characteristic is that these challenges require that we somehow predict the future and take action even though we know that our prediction is imperfect. In other words, these challenges are not reproducible in reality as we cannot stop or reverse the systems in order to retry—a fact that we are reminded of by earth sustainability, life and health, and safety.

The question, therefore, is how can we really solve these issues? Through segmentation into specialized field, science has given us an extremely precise understanding of fundamental principles. With its help, many challenges have now been resolved, but these were static in nature or both regular and reproducible. In the case of challenges with dynamic, intricately interrelated factors, and non-reproducible events, the steps of separation into specialized fields and reconstitution cannot be applied in the same way as before.

More than a few philosophers have reflected on the limits of Cartesian-style reductionism. Karl Popper [1], for example, considered "falsifiability" is the base for science. He held that a scientific theory is valid until it is falsified and replaced by a new theory. Thomas Kuhn [2], meanwhile, pointed to the "paradigm shifts" in that normal state and revolutionary state occurs in the development of science.

Certain scientists have proposed a new perspective of the "system." For example, in the years leading up to the 1960s, Bertalanffy proposed General System Theory [3] and Ilya Prigogine and his collaborators proposed the concept of dissipative systems [4]; Francisco Varela and his colleagues proposed the notion of autopoiesis [5] in the 1970s. Yet, none of these led to a specific methodology that could replace the conventional reductionism. Eventually, I came to believe that we were in need of a new scientific method to address "open systems"—systems that interact with the world outside and that include elements which themselves interact in an intricate fashion and that change over time.

4 Open Systems

As the name suggests, open systems are the opposite of closed systems. A closed system is a system that has no interaction at all, or a fixed interaction with the outside world. This type of system may consist recursively of closed subsystems. A closed system can be characterized as having its boundary, structure, and functions fixed. A closed-system problem is a problem that resides in a closed system.

An open system, on the other hand, is a system that interacts with the outside world, similar to what Bertalanffy defined in his General System Theory, so that it can change its boundary, internal structure, and functions over time. An open system can consist recursively of open subsystems. An Open System Problem is a problem that resides in an open system (see Fig. 1).

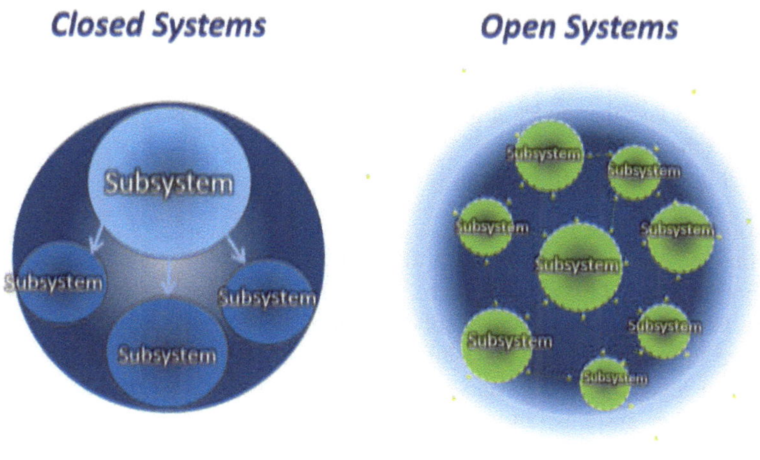

Fig. 1 Closed systems vs. open systems

Table 1 Characteristics of closed and open systems

Closed systems		Open systems
Fixed and definable	Boundary	Indefinable
Fixed	Function	Change in time
Fixed	Structure	Change in time
Equilibrium	Nature	Temporal development
Yes	Reversible	No
Yes	Reproducible	No
Yes	Divisible	No
External observer	Viewpoint	Internal observer

In reality, everything interacts with the outer world, even though it doesn't look like. Nevertheless, there are many problems that are simple in nature, meaning that they can be treated as closed-system problems, and therefore, reductionism works perfectly with the corresponding problems. However, it is not always true, especially in regard to those issues that we face in the twenty-first century. Table 1 shows a comparison of the characteristics of closed systems and open systems.

5 Open Systems Science

I wondered whether we could actually solve an Open System Problem. It appeared to be impossible in the sense that we could give a strong, complete solution to a closed system. However, it did appear to be possible from the perspective of supplying a means to make the entire situation better, not worse, through our best effort. Confident in this belief, I proposed Open Systems Science [6, 7] as an approach to solving the problems of open systems.

The method can be described as follows:

1. Provisionally define the system in which the problem is considered to reside.
2. Model the problem in the system.
3. Investigate whether the behavior of the model over time is self-consistent and consistent with the actual system's behavior.
4. If not, (a) revise the model and (b) if necessary, redefine the problem system, and remodel the problem.
5. Repeat until a satisfactory result is obtained.

Please also see Fig. 2.

As may be clear, Step 1 to Step 4(a) as corresponding to reductionism. A small but important new addition is Step 4(b), and this is the characteristic feature of Open Systems Science. In reductionism, subdivision of a given system into subsystems is performed to precisely understand the problem system in detail. In Open Systems Science, however, more emphasis is placed on identifying the problem system in relation to other systems, and understanding relationships among its subsystems

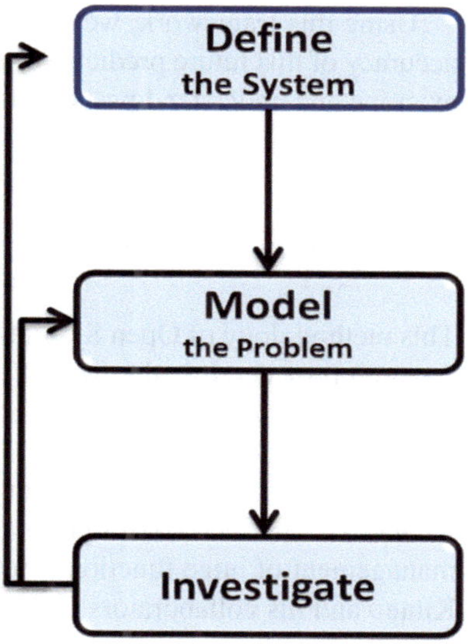

Fig. 2 Methodology of Open Systems Science

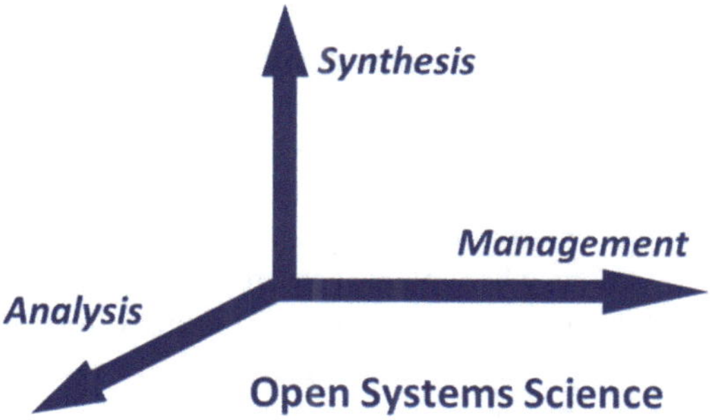

Fig. 3 Three perspectives of Open Systems Science

to ascertain a system's true nature. Digging deep into each subsystem while also mapping how they are interconnected leads to a better understanding of each problem, and therefore, to better solutions.

Another important point is that Open Systems Science introduces the new perspective of "management" into science, thereby augmenting the perspectives of "analysis" and "synthesis" that are provided by reductionism (see Fig. 3). It explores the most appropriate problem system so that the degree of contradiction or divergence from the actual system's behavior does not exceed acceptable levels. Such levels are determined initially by researchers, but a consensus must ultimately be reached by stakeholders and society. Such a consensus is not made on the basis of whether the solutions are true or false, but whether they are acceptable or not for each stakeholder, based on sufficient evidence to assure, with various trade-offs, including cost, also taken into consideration. Hence, this is not a matter of truth (or understanding the principle) but value (finding a solution), and as such, it exists in the domain of management.

Using this framework, we can predict our future in an explainable way. And the accuracy of this future prediction can be gradually improved. Of course, we can also examine and understand past events better.

6 A Few Examples

This methodology of Open Systems Science has been put to practical use in various research projects. It is the fruit of long and diverse discussions with researchers at Sony Computer Science Laboratories [8] and others through the investigation of various concrete research topics.

Systems biology [9] is a new method of biological study established by Hiroaki Kitano. He has shed fresh light on the essence of life by defining it in terms of the management of huge functional networks called pathway networks. Based on this, Kitano and his collaborators proposed long-tail drugs and personalized medication [10]. Systems biology is being extended by Kazuhiro Sakurada, from the viewpoint of historicity. Sakurada took internal, irreversible structural changes caused by an individual's development into consideration. This has provided another base for personalized medication [11, 12].

Masatoshi Funabashi has proposed a new system of agriculture called synecoculture [13]. Based on symbiotic associations of edible species, this system allows the natural environment to be powerfully recovered and reconstructed under any climate conditions where plants can grow. Kaoru Yoshida has integrated molecular biology with healthcare, food, and agriculture to form a new field of food science [14, 15]. Takahiro Sasaki is investigating the co-evolution of influenza viruses and human society, by incorporating multiple factors such as infection mechanisms, carrier animals, geology, climate, and social systems, all interrelated on diverse scales along temporal and spatial axes [16].

One final example is the DEOS project in which we aim to ensure the dependability of massive, man-made, continually changing systems. For such a system, we defined Open Systems Dependability [17]. We have shown that an iterative improvement process, called the DEOS process, with its embedding architecture can achieve Open Systems Dependability.

7 Discussion

Science as it has historically been practiced—adopting a reductionist standpoint in the pursuit of pure research and disconnected from society—represents a tacit license for scientists to be indifferent to the potential applications of their discoveries. This way of thinking has long been the touchstone of science. Great value has been placed on pure research, thereby stoking relentless competition among researchers to lay claim to advances.

Table 2 Properties of reductionism vs. Open Systems Science

Reductionism		Open Systems Science
Equilibrium systems	Object	Temporal development systems
Understanding principles	Purpose	Making consensus
Decomposition	Means	Identification of problem systems
Inward	Direction	Outward
In vitro	Treatment	In vivo
Strong/Perfect	Outcome	Action for the better
Observer	Subject	Actor
Indifferent	Consequence	Responsible
Justice	Value system	Humanity
Profession	Attitude	Citizen

However, in reality, the challenges we need to address—such as those suggested by the global agenda—are all woven into the fabric of society. These issues are products of vast, irreducible, interconnected systems, and feature a bewildering diversity of stakeholders. As such, solutions devised for sub-problems in isolation from the overall system routinely fail to address the broader scope of the issues they seek to resolve. Proposals to address such challenges must win the approval of diverse stakeholders, and achieve a social consensus. This is the point at which scientists, as researchers and as experts, can no longer act as external observers: they must become actors. And this is where Open Systems Science comes into its own, by defining scientists not as people of narrow expertise, but as fully rounded individuals with the capacity to make holistic value judgments, while accepting responsibility for the outcomes and applications of their research and acting selflessly to accelerate the formation of a consensus.

In such situations, it must be recognized that there is often no single correct solution; and furthermore, that decisions made at a given point in time can prove in retrospect to be mistaken. Stakeholders, accordingly, must work toward a consensus about the solution to a given problem, with science providing both the methodology and evidence for achieving that end.

Table 2 shows a comparison of how I see the properties of reductionism and Open Systems Science. I would greatly appreciate your comments and suggestions.

8 Conclusion

In this paper, I suggested that current challenges needing to be addressed urgently are mostly Open System Problems. As an appropriate methodology to address Open System Problems, I proposed Open Systems Science and gave the definition of that method. In Open Systems Science, a new perspective of management becomes important in addition to the conventional perspectives of analysis and synthesis, since we need to achieve consensus among stakeholders. I showed some examples of the methodology being applied in establishing new areas of research. Finally, I extended my perspective toward the future of science.

My path to this new scientific methodology was first made clear in an address in 2008 at the Sony CSL Symposium. The content of that address later appeared in a book [6] but my thinking was still embryonic. I then gave a few lectures and wrote an article [7] as my thinking became clearer. And I hope this current paper presents Open Systems Science in a concrete and understandable way.

Open System Science is a scientific methodology suited to the modern day in the sense that we can fully utilize high performance computers, databases, and the Internet to solve complex problems. I believe that Open Systems Science can be applicable to a wide range of real problems that must be addressed for the sake of this and future generations, and my greatest wish is that it can contribute to the progress of humanity and peace in our world.

About the Author

Dr. Mario Tokoro is Founder and Executive Advisor of Sony Computer Science Laboratories, Inc., Japan. He is a former Professor of Computer Science at Keio University and a former Senior Vice President and Chief Technology Officer of Sony Corporation. In 1988, he established Sony Computer Science Laboratories, Inc. He served as Director of Research, President and CEO, and later as Chairman and CEO. He is an expert in Computer Science and Engineering, Philosophy of Science and Technology, and Research Management. He is currently leading Sony CSL's Open Energy Systems initiative.

Acknowledgement I would like to thank Hiroaki Kitano, Kazuhiro Sakurada, Masatoshi Funabashi, Kaoru Yoshida, Takahiro Sasaki, and all the members of Sony Computer Science Laboratories, Inc., as well as all the members of the DEOS project for their invaluable comments and suggestions in the course of our work together.

References

1. Popper K (1959) The logic of science discovery. Hutchinson & Co
2. Kuhn TS (1962) The structure of scientific evolution. The University of Chicago Press, Chicago
3. Bertalanffy L (1968) General system theory: foundations, development, applications. G. Graziller, New York
4. Nicolis G, Prigogine I (1977) Self-organization in nonequilibrium systems: from dissipative structures to order through fluctuations. Wiley
5. Maturana HR, Varela FJ (1980) Autopoiesis and cognition: the realization of the living. D. Reidel/Springer, Boston
6. Tokoro M (ed) (2010) Open systems science – from understanding principles to solving problems. IOS Press, Amsterdam
7. Tokoro M (2015) Une nouvelle method pour de nouveaux enjeux, Pour la Science-Supplement innovation n° 477, Janvier 2015; A New Method for New Challenges, Scientific American, Innovation Special Issue Supplement of the French edition n° 477, January 2015

8. Sony CSL, https://www.sonycsl.co.jp/en/
9. Kitano H (2002) Systems biology: a brief overview. Science 295:1662–1664
10. Hsin KY, Ghosh S, Kitano H (2013) Combining machine learning systems and multiple docking simulation packages to improve docking prediction reliability for network pharmacology. PLoS One 8(12), e83922. doi:10.1371/journal.pone.0083922
11. Sakurada K (2014) Personalized and proactive medicine based on life-course data. Int J Rheum Dis 17(Suppl 2):6–29
12. Sakurada K (2015) Realization of preemptive medicine by data-driven research. Exp Med (Japan) 33(7):1182–1189
13. Funabashi M (2013) IT-mediated development of sustainable agriculture systems: toward a data-driven citizen science. J Inf Technol Appl Educ 2(4):179–182
14. Yoshida K, Funabashi M (2014) Taste analysis on conventionally, organically and naturally grown cabbage. In: Proceedings of 1st IMEKOFOODS metrology promoting objective and measurable food quality and safety, Rome, Italy, 12–15 October 2014, p 115
15. Yoshida K, Shimizu-Yoshida Y, Funabashi M (2014) Functional and compositional characteristics of conventionally, organically and naturally grown cabbage and carrots. In: Proceedings of fourth international conference on food studies, Prato, Italy, 20–21 October 2014, p 41
16. Sasaki T (2013) Hierarchical multi-agent-based model for simulating the prevalence and evolution of influenza virus. In: Proceedings of SpringSim 2013: spring simulation multi-conference, 2013
17. Tokoro M (ed) (2015) Open systems dependability – dependability engineering for ever-changing systems, 2nd edn. CRC Press, Boca Raton

Open Systems Exploration: An Example with Ecosystems Management

Masatoshi Funabashi

1 Introduction

1.1 Open Complex Systems and One-Time-Only Events

Complex systems science has been applied in various domains where theory and experiment meets with a medium of computation (e.g., [1]). Complex systems science with external observation drastically advanced laboratory measurements, and in some confined conditions succeeded to analyze the living phenomena as an augmented phenomenology, without reducing the whole process into the parts (e.g., [2]).

On the other hand, complex systems in real world cannot be fully simulated when the observation is limited from inside of the systems. When the system scale is larger than a controlled laboratory, when the sensor resolution is not sufficient to reconstruct a predictable model, and when inherent dynamics such as chaos produces principal unpredictability, we are forced to handle internal observation (e.g., [3–6]). Internal observation is not only a compromise of conventional scientific methodology but also a subjective strategy to yield an effective description of the system in dynamical functioning, where characteristic measures can only be defined on the transient configuration of many-to-many bodies systems [7].

This working hypothesis becomes especially informative when a system is open to external environment. In open complex systems (open systems in short), we cannot fully define a system's boundary as it interacts with external environment through time line. The configuration of subsystems is also fuzzy and change temporally. The systems may not be possible to model with parent-slave relation

M. Funabashi (✉)
Sony Computer Science Laboratories, Inc. 3-14-13, Higahi-Gotanda,
Shinagawa-Ku, Tokyo, Japan
e-mail: masa_funabashi@csl.sony.co.jp; masa314159265358979@gmail.com

P. Bourgine et al. (eds.), *First Complex Systems Digital Campus World E-Conference 2015*, Springer Proceedings in Complexity, DOI 10.1007/978-3-319-45901-1_26

in closed environment. The systems are basically unpredictable in a long term by internal observers, uncontrollable depending on the fragility to external disturbance and complexity of interactions, and manifest one-time-only events that are neither fully predictable by modelling nor reproducible by the real phenomenon itself [8, 9]. Whether it be technological innovation, social order reformation, natural disasters, etc., transition of history in open systems has been always triggered by a new event of unpredictable scale [10–12].

In such open systems lie greatest challenges of complex systems science, especially those concerned with the sustainability of our civilization, that is left behind as negative legacies of the modern scientific achievement. For example, environmental problems, epidemic outbreak, life-course chronic diseases, technology-inherent breakdown of social infrastructure, climate change, and associated social-ecological transitions are predominant examples of one-time-only events that require open systems approach [13–17]. These tasks require the application of effective measure by internal observers during the operation, as it cannot be halted, analyzed, and experimented separately from the real world.

In coping with the needs of such global agenda, we need to explore novel scientific methodologies that can be applied in open complex systems. Based on the past achievement of rigorous science with external observation, we need further extend the effectiveness of internal observers' science in an open environment, where real-world problems remain untouched. In contrast to the perfection myth of science seeking the control of the system as a dominant objective, we rather need to struggle in the real-world operation where the prediction and control is not always valid. How much can we attain with incomplete observation, heterogeneous database, in unpredictable environment, with lots of new events that have never happened, but with the aid of fine mathematical theory, ubiquitous sensors, social networks of citizens, and massive computation power? What should we explore during the time-limited operation of open complex systems, in order to survive and create sustainability options in various forms?

In this article, we investigate a conceptual framework of scientific exploration in open complex systems and develop a framework of exploration interfaces taking an example in ecosystems management.

2 Open Systems and Closed Systems Approximation

Most of the natural systems can be described as open systems, and open systems science includes a proto-scientific description ranging between phenomenology and science. In a broad term, conventional science or closed systems science is an approximation of originally open systems with an artificial boundary definition that prohibits open interaction with further external environment. We need, however, to clarify what is common with conventional scientific methodology and what is new or explorable with the conception of open systems. For that purpose, we formalize the comparison between open systems and its closed systems approximation that already has specified examples in conventional science.

Table 1 Conception of open systems in contrast to closed systems approximation in dynamical systems

	Closed systems approximation	Open systems
Nature	Reproducible events	One-time-only events
Control objective	Resilient feedback to controlled state	Active transition to alternative state
Information requirement	Information quantity	Information generation
Methodology	Modelling and simulation	Exploration of management

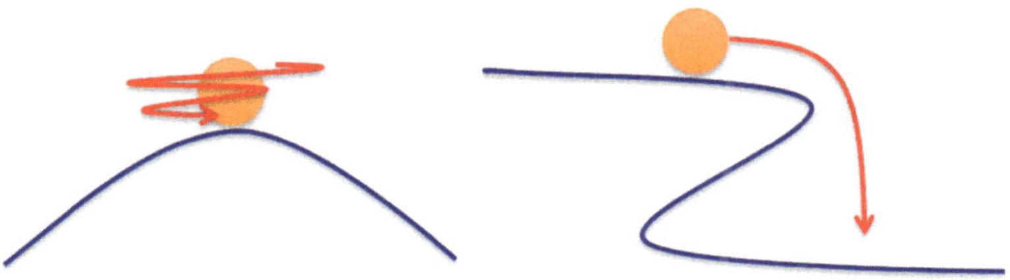

Fig. 1 Conception of open systems in contrast to closed systems approximation in dynamical systems. *Left*: Resilient feedback to controlled state in closed systems approximation. *Right:* Active transition to alternative state in open systems. *Blue lines* indicate the potential of the environment, in which systems depicted with *orange circles* are controlled and managed with *red* trajectories

2.1 *Open Systems with Respect to Dynamical Systems*

Dynamical systems modelling is one of the primary methods in complex systems science [18]. Table 1 and Fig. 1 compare open systems with closed systems approximation in dynamical systems perspective. Dynamical systems, when used in closed systems application, usually treat isolated systems with finite boundary conditions, in which control of reproducible events with a feedback to a desired state is the object of analysis. For such purpose, high-resolution modelling and simulation with external observation is efficient, and controlling the phenomena requires the information quantity in terms of information theory defined on a closed environment without the dynamic exchange of components with external environment.

On the other hand, open systems as it is in real world contain important dynamics in one-time-only events. Such phenomena cannot be externally controlled nor can be finely predicted from past data. Instead, we need to cope with the emerging phenomena and seek for an active transition to an alternative state with strategic adaptation that resolves the conflict. This is not a resilient feedback with a fixed definition of systems, but rather an expansion of the systems including outer environment that leads to the redefinition of the boundary with transition phenomena, in which effective information measures should be redefined. This process is associated with both the exploration and management from inside of the systems that precede modelling and simulation. The importance of exploration is not to gain the information quantity with a fixed framework of observation, but to explore an extended definition of the systems that could encapsulate necessary information for the management as a result. We call this process of extending the systems definition and evaluate the information within to cope with irreproducible events as *information generation.*

2.2 Open Systems with Respect to Machine Learning

Machine learning incorporates a wide forms of statistical modelling in complex systems [19]. Theoretically, non-linear statistical measures can classify any kind of statistical dependency within the effective dimensions of feature space [20]. However, basic frameworks of machine learning are mostly based on closed systems approximation.

Table 2 and Fig. 2 compare open systems with closed systems approximation in machine learning perspective. While standard closed systems approaches define the format of database and observation methods, open systems reality do not always guarantee the continuity both in the definition of data items and its quality. Ubiquitous sensor network and citizen observation, for example, inevitably contain biases in various scales. This situation has a common challenge with the frame problem in artificial intelligence [21]. In the open systems reality where we do not sufficiently know how to assume the effective boundary of the systems,

Table 2 Conception of open systems in contrast to closed systems approximation in machine learning

	Closed systems approximation	Open systems
Framework of database	Fixed	Dynamically change
Protocol	Single algorithm	Workflow of algorithms
	Optimization	Exploration and optimization
Implication	Evaluation	Ontogenesis

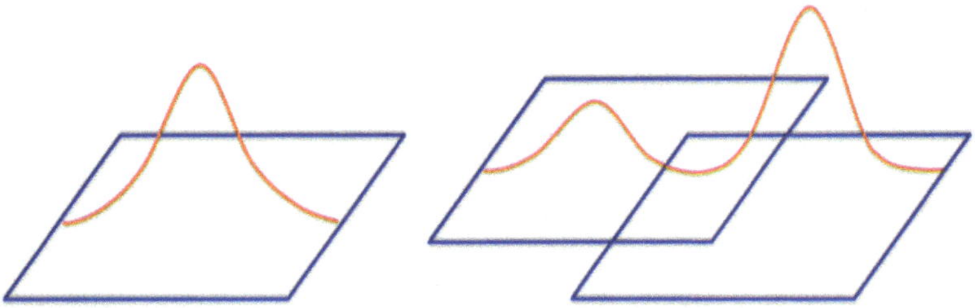

Fig. 2 Conception of open systems in contrast to closed systems approximation in machine learning. *Left:* Single algorithm optimization on a fixed database framework in closed systems approximation. *Right:* Exploration and optimization with a workflow of algorithms in open systems. *Blue rectangles* correspond to the framework of databases or observation, in which algorithmic optimizations are performed with information criteria depicted as *orange* distributions

evaluation with a single algorithm can be a blind measure with respect to the global management goal including future utility. We need to prepare a portfolio of various evaluations within available resources, with respect to a conceivable range of future scenarios, in order to set up a try-and-error workflow that can maximally avoid the operation to fail. This process is not a mere evaluation with an external algorithmic measure, but a creation of novel suitable measures for future transition, in which sense it can be represented as ontogenesis associated with *information generation.*

3 Open Systems Exploration: An Example with Ecosystems Management

Based on the above conception of open systems exploration, we develop a concrete example of the interfaces for the management of ecosystems as open systems.

3.1 Towards Dynamical Assessment of Ecosystems

Ecosystems functions and the services they provide are major sources of social-ecological sustainability [22]. Although an increasing number of literatures reveal general positive relation between biodiversity and ecosystems functions, local assessment and its utilization depend highly on local initiative and industrial inertia that devoid of appropriate scientific support [23]. We try to convert the conventional environment assessment protocol with the use of open systems science methodology in order to achieve a dynamical assessment of ecosystems.

Figure 3 and Table 3 show the comparison between typical environmental assessment and possible open systems extension. Usually, environmental assessment is performed on a basis of static, fixed scoring framework that is derived from past empirical studies [24, 25]. Current environmental studies are based on sensing parameters and index species whose score in relation to environmental quality is defined with past experience [26]. There is, however, little consideration of possible future change of base-line ecosystems, especially regime shifts in response to climate change and human perturbation in a global scale [27]: The number of index species is pre-defined and limited. Observation methods are specified that often require training by professional to assure the quality of data. By respecting the quality of reproducible observation based on the past statistics, therefore limiting the target systems in space and time, conventional assessment lacks in some aspects the accessibility to a wide public and adaptability to abrupt environmental changes where redefinition of the systems, descriptive index, and future insight should be renewed on time.

To cope with an ever-changing open systems that lies in the nature of ecosystems and associated human activities, we need to extend assessment protocols to an

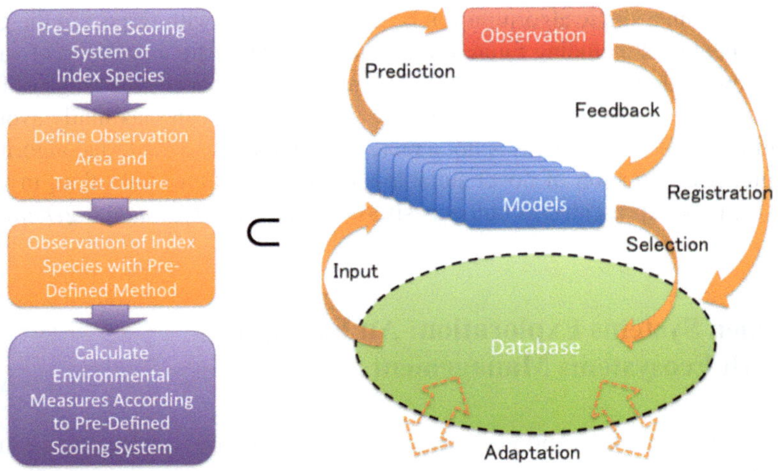

Fig. 3 Environmental assessment protocols in closed systems approximation and open systems exploration. *Left:* Typical conventional protocol with closed systems perspective (based on [24–26]). To ensure the objectivity and reproducibility of observation, *violet* processes are usually fixed based on the past assessment data. *Orange* processes need to respect pre-defined methods that usually call for training by professionals. *Right:* Dynamical assessment as a process of open systems exploration applied in ecosystems management. Hence the right protocol can include the left one by fixing the corresponding parameters

Table 3 Characteristics of environmental assessment protocols in closed systems approximation (current environmental assessment) and open systems exploration (dynamical assessment)

	Current environmental assessment	Dynamical assessment
Interface	Static, fixed scoring framework	Interactive, dynamical, on-the-fly ICT
Index	Pre-defined and limited	Can be expanded and renewed by observation
Observation method	Fixed	Can be modified, various
Accessibility	Mainly for trained professionals	Open to wide public without training
Evaluation	Based on the past experience	By renewal of the observation scheme according to the focused change

interactive and dynamical interface that can treat on-the-fly modification of the protocol itself. The acceleration of information sharing, processing, and augmentation of interactivity can further modify the way of environmental assessment, and contribute to the readiness of the management. Information communication technology (ICT) is expected to bring more dynamic and reflexive dimension in citizen science, allowing to fill the gap between crude, diverse data, and refined governance on multifunctional ecosystems [28]. Since model-based prediction from physical to biological diversity still confronts complexity of ecological response [4], direct biodiversity measurement with human observation still plays an essential role. The distributed measurement of biodiversity with interactive ICT has a potential to shift the modality of indexing and scoring of species, from stable qualitative description to dynamic quantitative data-driven assessment in real time. This approach will expand current assessment in its observation network, data quantity, and analytic tools on an integrated design of distributed ICT. By means of the on-the-fly observation, reflexive redefinition of index species and its environmental score becomes possible. Such dynamic reconfiguration of assessment criteria would introduce more flexibility for rapidly adapting to changing situation. For that purpose, we propose an iterative framework that comprises database, models, and observation that can modify its relationship according to the actual change of situation.

Observation of multi-scale systems such as society and ecosystems is internal observation in principle. We cannot rely on empirical external measurement in terms of data efficiency and analytical predictability [4]. Rather, we need to assure a diversity of strategies to allow multiple actors to explore possible scenarios that are rich enough to mitigate unpredictable change. Open systems exploration in ecosystems management may not realize the reproducibility or predictability on what will happen, but should seek for the capacity of exploration on what could happen for a flexible planning of strategy portfolio. In short, we may fail to predict rigorously but should succeed to survive in any possible situation. This is a common principle with ICT-mediated citizen science in the roadmap of complex systems science [28].

With this respect, structural design of exploratory simulation tools should have emphasis on the diversity of the models, their parameters, and reflexive evaluation of substantive variables for dynamic adaptability. Figure 3 (right) shows conceptual framework of open systems exploration in ecosystems management: Distributed measurements including the sensing of ecosystem agents collect massive data with multiple and fluctuating criteria. A copious combination of analytical and numerical simulators produce possible predictions in the background, which are given feedback by the on-going measurement to evaluate the efficacy of each model and weight the data variable in a reflexive workflow with multiple time scale. Not only the effect of single variable but also synergetic effects between variables can be explored with a variety of model functions. The observation network should be reconfigured according to the efficiency of the actual management, in order to assure sufficient diversity of substantive variables by eliminating useless ones and investing for novel exploration. Here, the frame problem of determining sufficiently diverse and effective subset of variables is a consistent task to resolve. Cloud computing resource and parallel-processed simulators would play essential role for the on-site implementation.

For the example, data-driven assessment of biodiversity and associated environmental quality can be realized with this framework. Taking environmental variables and biodiversity as a database, a wide range of possible definitions of index species and their environmental scores can be generated from simulators, which will be selected to extract high-resolution assessment scheme as actual measurement continues. Steep change of biodiversity, environment, and observation network can be immediately reflected to the assessment protocol by producing new possibilities of scoring system with new inputs. We develop basic interfaces and models of such protocol in the next chapters.

3.2 Example of Data Interface: Multi-partite Graph Exploration

We develop prototypical interfaces for open systems exploration applied in ecosystems management. As a testbed we use an ecological database developed in Synecoculture project [29]. The database comprises biodiversity observation in various Synecoculture farms and surrounding environment in Japan. To assess these environments in open systems perspective, one needs to diversify the observation until it can attain the saturation of the biodiversity measures related to the management principles.

For this purpose, extensive link of data and related information is useful as an initial hands-on interface. Figure 4 shows a multi-partite graph visualization of biodiversity records. The observation of plants and insects species are linked according to the geographical cooccurrence with taxonomical relationships and observation places. The users can explore on this graph to seek concurrent and/or allied species, that could extend their observation activity and learn related ecological information. This model can support extensive search for data registration within the framework of cumulative past experience. It represents a simplest model for prediction in which all past cooccurrences are superimposed.

Management requires wider choice in response to a change. Ecosystems dynamics under human perturbation is especially irregular and difficult to harness [30]. By combining further information source such as climate data and ecological literature, multi-partite links can provide wider choice triggered by actual observation when a new data is recorded and connected in the web of multi-partite relations. The real-time development of complex network of observation with automated link to relevant information is a primary interface that complex systems science can offer to open systems exploration. The evolution of complex network autonomously combines observation and related knowledge, and extends the framework of possible observation to provide collective suggestion between users.

3.3 Example of Suggestion Tool: Integration of Environmental and Biodiversity Data with Symbolic Dynamics Analysis

Integration of biodiversity and sensor data is a fundamental task in data-driven environmental management. While current studies try to integrate biodiversity

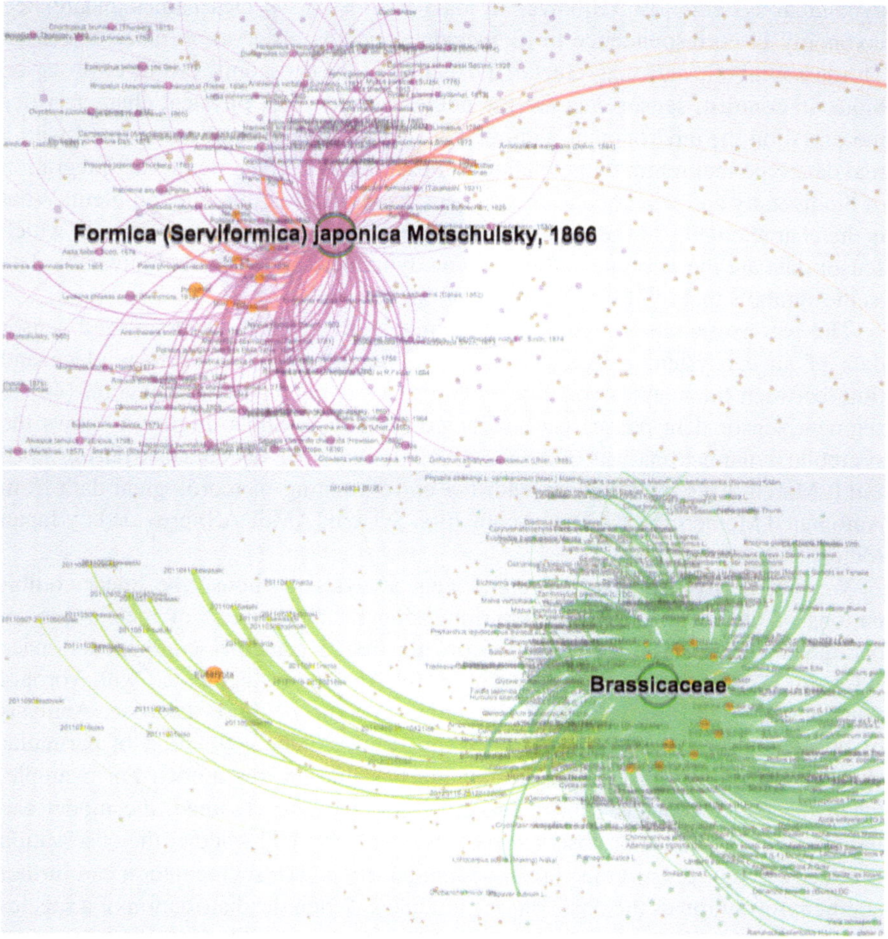

Fig. 4 Snapshots of multi-partite graph between plants (*green*), insects (*magenta*), biological taxonomy (*orange*), and observation place (*yellow*). The links represent the total cooccurrence in the database (Synecoculture CMS [29])

records with remote sensing databases [31, 32], little has been investigated on a local scale under direct effect of management. For example, in agricultural land, sensor-based measurement and control of precision agriculture [33] is not connected with local biodiversity observation. Natural farming practices based on local biodiversity, on the other hand, rely merely on human observation and have little introduced sensor technology [34, 35]. In actual management of farmland with both yield and biodiversity promotion, one needs to consider the integrated aspects of biodiversity and environmental conditions [36–38].

We propose a general framework to integrate biodiversity data based on human observation and sensor data in general with the use of symbolic dynamics in dynamical systems [39]. Biodiversity data is a list of species names and related taxonomy in correspondence to its metadata such as observation place and time. This is a symbolic data that refers to the quality of the taxonomic profile of observed biota. In contrast, sensor data are the numerical values of physical characteristics measured on the environment with metadata. This is in general represented with a real data type that refers to the quantity of each measurement item. The integration of biodiversity and sensor data can be generalized into the following problem: What is the characteristics of the symbolic dynamics of a measured ecosystem, in which sensor data are the estimates of underlying dynamical system and biodiversity data as the symbols that represent the states of the systems?

The reconstruction of symbolic dynamics with given biodiversity and sensor data of an ecosystem is possible by matching the metadata such as place and time between them. As a concrete example, we employ Voronoi diagram [40] to segment sensor data phase space with biodiversity symbols. Figure 5 shows the symbolic dynamics analysis of the Synecoculture biodiversity database during April 2011–March 2013 by matching with the corresponding meteorological data from Automated Meteorological Data Acquisition System (AMeDAS) provided by Japan Meteorological Agency [41].

We first performed principal component analysis to choose the linear combinations of the most distinctive two-dimensional feature space of meteorological parameters (Fig. 5 Top Left). Based on the first 2 principal components space (PC1–PC2), 30 previous days mean of AMeDAS data is segmented with Voronoi diagram for each observation date recorded in Synecoculture database. Analysis of observable species diversity (Fig. 5 Top Right), niche estimation of particular species (Fig. 5 Bottom Left and Right) are possible on this model. For example, when the meteorological sensor data of a new day are obtained, the model can indicate what is the list of observed species in the past, and whether the observation is already rich or poor in the corresponding partition. The segmentation can further augment resolution as the observation cumulates. When the distribution of a species is confined in a subspace of the Voronoi diagram, it is possible to estimate its niche boundary by an interpolation. Significant correlation between estimated niches (e.g., order-wise correlation [42]) can provide suggestions that there might be underlying ecological dependence between those species.

Theoretically, infinite sequence of finite biodiversity symbols can specify any arbitrary trajectory of meteorological data with real-value precision, if the system is deterministic and the partition is "generating" in terms of symbolic dynamics [43]. To enrich the suggestion based on the spatio-temporal structure, this model is further accessible to mathematical analysis of symbolic dynamics that can treat complex trajectories in dynamical systems including chaos.

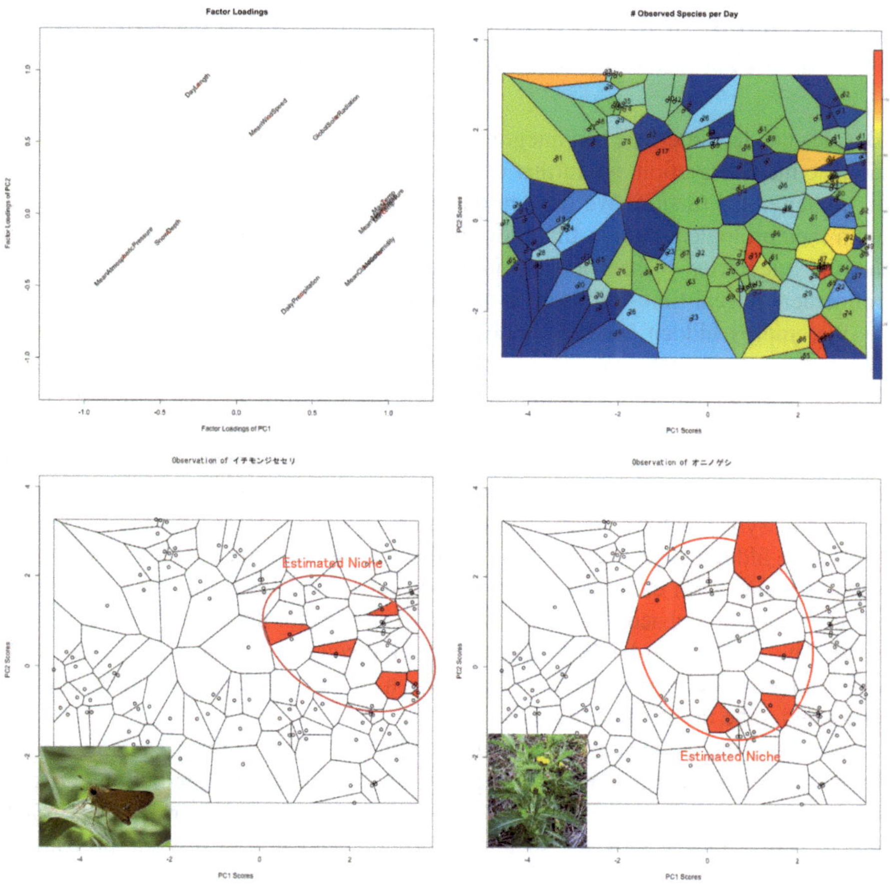

Fig. 5 Example of symbolic dynamics analysis of biodiversity and meteorological data. *Top Left*: Factor loading of principal components analysis (PC1 and PC2) of 11 daily meteorological parameters (mean/maximum/minimum temperature, daily precipitation, day length, global solar radiation, mean wind speed, mean vapour pressure, mean atmospheric pressure, mean humidity, mean cloud cover, and snow depth) in AMeDAS data. *Top Right*: Voronoi segmentation of AMeDAS data PC1–PC2 space with Synecoculture biodiversity database for each 30 days mean. The *color* represents the number of species observed in the same partition. *Bottom Left*: Example of niche estimation of *Parnara guttata guttata (Bremer et Grey, 1852)* (*in picture*) on the symbolic dynamics analysis. *Bottom Right*: Example of niche estimation of *Sonchus asper (L.) Hill* (*in picture*) on the symbolic dynamics analysis. Partitions where the species appeared are filled with *red*

3.4 Example of Model Selection: Seasonal Segmentation and Prediction of Biodiversity Observation

Besides the data interface and integration model that can provide interactive suggestions to the observation, we further consider how to select a better predictive model in a changing situation. We take an example of biodiversity prediction combined with meteorological data in time development. This is again a prototypical model for the integration of sensor and biodiversity data, but with consideration to the refinement of real-time feedback on observation based on the model selection.

We employ hidden Markov model (HMM) as a primitive example of seasonal segmentation of meteorological data [44]. We applied the standard forward–backward algorithm for the inference of hidden states from the past AMeDAS data, and the Viterbi algorithm to inversely infer hidden states with new data for each observation. Figure 6 Top shows an example of seasonal segmentation of AMeDAS data. Hidden states with the highest probability was chosen to associate the observed species in Synecoculture database in the same day. The species diversity associated with each hidden state is expressed as a discrete distribution on a set of observed species name, with cumulative occurrence probability. Each time new species is observed, the model acquires additional list of species for the corresponding hidden state. The discrete probability distribution of species occurrence associated with each hidden state can be used as a prediction model, when a new observation is estimated to be in the same hidden state.

Based on the estimated models with the hidden states number ranging from 2 to 10, we performed a numerical experiment to evaluate the prediction capacity of each HMM with respect to each 30 observations mean (Fig. 6 Middle). Each model was evaluated with the standard likelihood function of discrete probability distribution with respect to the observed species. The results show a dynamical trend in the number of hidden states that gives the best prediction model. For example, in Fig. 6 Bottom, the initial phase during April 2011–January 2012 shows an increase of the number of hidden state for the best model, which implies an increase of model resolution for seasonal segmentation. Observation of new species also tends to saturate as it is in winter time. Between February 2012 and October 2012, as the summer time reactivates the ecosystems, new species records become more frequent which leads to the decrease of the model resolution (hidden states number of the best model). The models go through a heuristic learning process of biodiversity change with low likelihood for estimation, until it regains the resolution and relative likelihood in the next winter time around November 2012–March 2013. Since likelihood of the models monotonously decreases as the list of observed species expands, relative increase/decrease of likelihood is important to characterize the model resolution. When the relative increase of likelihood is associated with the increase of the number of hidden states in the best model, model resolution is considered to increase. During the observation, the diversity of observation is maintained sufficiently high without producing statistical bias on new species

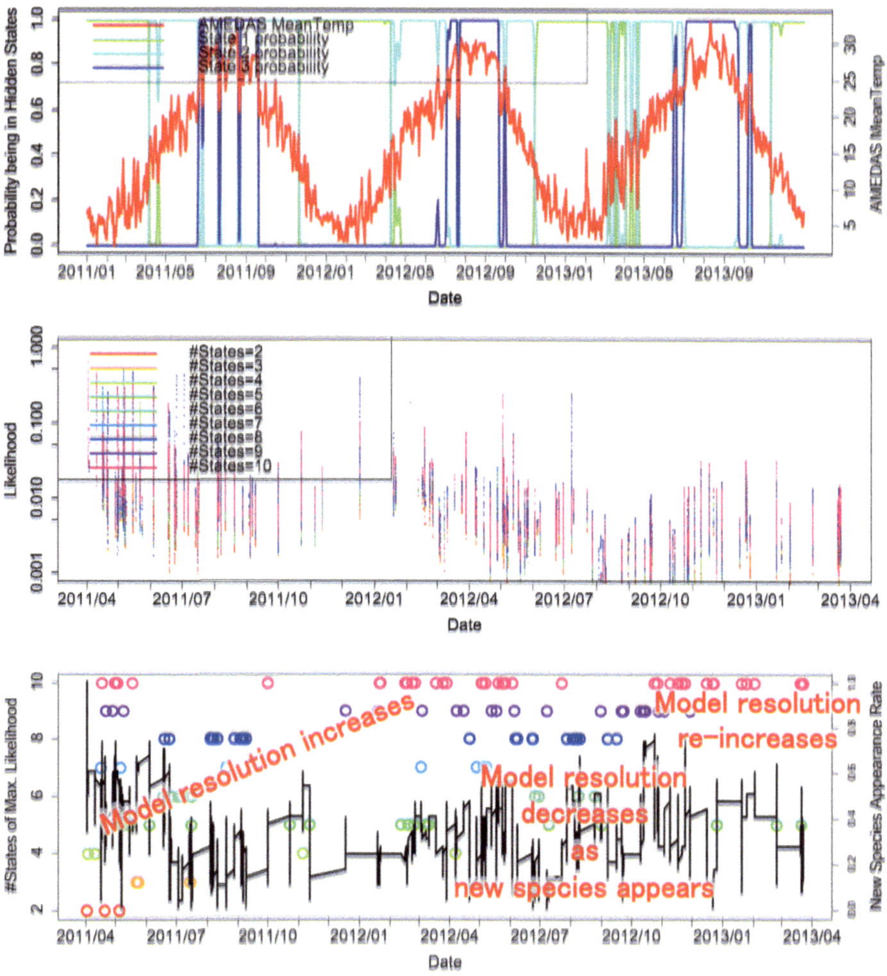

Fig. 6 Example of model selection on an integrated model of biodiversity and sensor data with hidden Markov model (HMM). *Top*: Example of seasonal segmentation of AMeDAS daily mean temperature data with 3 hidden states. Estimated probability of each state is plotted with corresponding color. *Middle*: Numerical experiment of model selection based on the likelihood of HMMs with hidden states 2 to 10, for each 30 observations mean. Likelihood of each HMM is depicted as dots with colors that corresponds to the number of hidden states. *Bottom*: Time development of new species appearance rate for each 30 observations and number of hidden states of selected HMM giving maximum likelihood for prediction. Dynamic trend of model selection and learning occur with the real-time feedback of observation

appearance rate (data not shown). Therefore, the numerical experiments imply a dynamic model selection process during the real-time learning, in which the system manages to select the best-in-time prediction model by compromising between the adaptability to new observation and reproducibility of past statistics.

4 Conclusion

4.1 Components Evaluation of Dynamical Assessment

We have conceptualized the methodology of open systems exploration based on open systems science, and developed prototypical interfaces with models taking an example in ecosystems management, namely dynamical assessment. Basic properties of the example systems in view of incorporation into dynamical assessment are summarized in Table 4. By generalizing these properties such as data processing mode from batch to real-time, parameter segmentation type from simple superposition to spatio-temporal segmentation, and model selection range from single to group selection, respectively, these systems can be further developed and integrated to augment a whole cycle of dynamical assessment.

The correspondences between the processes in Fig. 3 (Right) of dynamical assessment and the utilization of each example model are summarized in Table 5.

The *information generation* proposed as the essential dynamics of open systems exploration in Table 1 can further be explored in the following contexts:

Table 4 Achievement of basic system properties of three example models, multi-partite graph (MPG), symbolic dynamics model (SDM), and hidden Markov model (HMM) for the integration in dynamical assessment

Properties	MPG	SDM	HMM
Ex.1 Data processing mode	Batch	Batch	Real-time
Ex.2 Parameter segmentation type	None	Spatial	Temporal
Ex.3 Model selection range	Single	Single	Group

Table 5 Correspondence between dynamical assessment process in Fig. 3 (Right) and multi-partite graph (MPG), symbolic dynamics model (SDM), and hidden Markov model (HMM)

Process in Fig. 3 (Right)	Process in example models
Input	AMeDAS and Synecoculture database
Prediction	Links in MPG
	Suggestion from SDM
	Prediction with HMM
Feedback	Selection of effective information in MPG
	Selection of time window in SDM
	Parameters selection in HMM
Selection	Selection of AMeDAS variables in SDM and HMM
	Geographical and time window selection of Synecoculture database
Registration	Modification of actual observation
	Introduction of new observation method
	Setting of new sensors

- Multi-partite graph: Exploration of links and validation by observation
- Symbolic dynamics model: Field exploration of suggested species diversity, niche condition, and its validation
- Hidden Markov model: Exploration of wider parameter spaces, model selection with a real-time observation likelihood during operation

4.2 Example of Assessment Result: Generative Index Species Scoring Systems

By gradually introducing the suggestion from prototypical models, environmental assessment in Synecoculture project started to operate the initial steps of dynamical assessment. Data-driven lists of index species candidates are obtained from the field practice between August 2014 and July 2015 as in Table 6. These generative index species, when connected with other database that refers to the quality of environment such as yield, will serve as timely reconfigurable measures of environmental quality in an ever-changing open systems surrounding the practice and management.

Table 7 gives the list of observed species in Table 6. As an example of scoring system generation, the environmental score of these species is calculated from the

Table 6 Numbers of generative candidates of index species extracted from dynamical assessment in Synecoculture project

Date	Place	Suggestion	Observation	Consistent index	Past index	Novel index
2014/8/7	Todoroki (Tokyo)	16	23	12	4	11
2014/9/13	Todoroki (Tokyo)	36	35	22	14	13
2014/9/14	Oiso (Kanagawa)	33	31	23	10	8
2014/11/22	Oiso (Kanagawa)	17	16	8	9	8
2014/12/21	Oiso (Kanagawa)	21	14	3	18	11
2015/3/28	Todoroki (Tokyo)	22	21	4	18	17
2015/4/25	Todoroki (Tokyo)	22	18	6	16	12
2015/5/2	Oiso (Kanagawa)	62	26	22	40	4
2015/5/30	Todoroki (Tokyo)	16	25	5	11	20
2015/6/13–14	Ise (Mie)	96	64	23	73	41
2015/6/27	Todoroki (Tokyo)	37	21	9	28	12
2015/7/19	Todoroki (Tokyo)	37	24	11	26	13
2015/7/25	Oiso (Kanagawa)	36	21	14	22	7
2014/8–2015/7	Total	229	147	80	190	99

The numbers indicate the number of species that were suggested from the prototypical models, observed on field, and classified as consistent/past/novel index species according to the inclusion and exclusion relationships between suggestion and observation: Consistent index species commonly appeared in both suggestion and observation, while past and novel index species only appeared in either suggestion or observation, respectively

Table 7 List of observed species and its environmental score based on the edible species diversity during the observations between August 2014 and July 2015

Academic name	Score	Category
Morella rubra Lour.	37	C
Ficus carica L.	37	C
Zanthoxylum ailanthoides Siebold et Zucc.	37	C
Megacopta punctatissima (Montandon, 1894)	37	C
Popillia japonica Newman, 1844	37	C
Graphosoma rubrolineatum (Westwood, 1873)	37	C
Mimela splendens (Gyllenhaal, 1817)	37	C
Microcerasus tomentosa (Thunb.) G.V.Eremin et Yushev	37	C
Ipomoea batatas (L.) Poir.	37	C
Ficus erecta Thunb. var. erecta	37	C
Lycaena phlaeas daimio (Matsumura, 1919)	37	C
Orthetrum albistylum speciosum (Uhler, 1858)	37	N
Rubus fruticosus	37	N
Ziziphus jujuba Mill. var. inermis (Bunge) Rehder	37	N
Cyanococcus	37	N
Hyla japonica	37	N
Papilio protenor	37	N
Locusta migratoria Linnaeus, 1758	37	N
Uroleucon nigrotuberculatum	37	N
Aronia melanocarpa	37	N
Eumeta japonica Heylaerts, 1884	37	N
Actinidia polygama (Siebold et Zucc.) Planch. ex Maxim.	37	N
Hydrangea serrata (Thunb.) Ser. var. thunbergii (Siebold) H.Ohba	37	N
Camellia sinensis (L.) Kuntze	37	N
Citrus limon (L.) Osbeck	37	N
Oleandraceae	37	N
Eurema hecabe (Linnaeus, 1758)	37	N
Allium chinense G. Don (variant Shimarakkyo)	37	N
Elaeagnaceae	37	N
Prunus avium	37	N
Fragaria x ananassa Duchesne ex Rozier	37	N
Epilachna vigintioctomaculata Motschulsky, 1857	37	N
Diptera Linnaeus, 1758	37	N
Metaplexis japonica (Thunb.) Makino	37	N
Neoscona adianta (Walckenaer, 1802)	37	N
Vitis spp	30.66666667	N
Paederia scandens (Lour.) Merr.	29.5	C/N
Aralia cordata	27.5	N
Acca sellowiana (O.Berg) Burret	27.5	C/N
Trifolium repens L.	26.5	N
Rosa multiflora Thunb.	24.5	C/N

(continued)

Table 7 (continued)

Academic name	Score	Category
Vitis ficifolia Bunge	23.66666667	C/N
Lycopersicon esculentum Mill.	23.33333333	C/N
Acrida cinerea (Thunberg, 1815)	23.33333333	C/N
Solidago altissima L.	23	C
Morus	22.66666667	N
Perilla frutescens (L.) Britton var. crispa (Thunb.) H.Deane	22	N
Smilax china L.	22	C
Ginkgo biloba L.	22	C
Rubus hirsutus Thunb.	22	N
Angelica keiskei (Miq.) Koidz.	22	C
Polistes rothneyi iwatai van der Vecht, 1968	22	N
Gonista bicolor (de Haan, 1842)	22	C
Ampelopsis glandulosa (Wall.) Momiy. var. heterophylla (Thunb.) Momiy.	22	C
Scolia (Scolia) histrionica japonica Smith, 1873	22	C
Lycoris radiata (L'Hér.) Herb.	22	N
Momordica charantia var. pavel	22	N
Artemisia indica Willd. var. maximowiczii (Nakai) H.Hara	22	C/N
Diaea subdola	21.66666667	N
Houttuynia cordata Thunb.	21.4	C/N
Asteraceae	19.83333333	C/N
Colocasia esculenta (L.) Schott	19.8	C/N
Commelina communis L.	19.5	C
Formica (Serviformica) japonica Motschulsky, 1866	19.5	C/N
Dioscorea japonica Thunb.	19.25	C/N
Allium fistulosum L.	18.5	C/N
Allium tuberosum Rottler ex Spreng.	18.42857143	N
Coccinella septempunctata Linnaeus, 1758	18.25	C/N
Pieris (Artogeia) rapae crucivora Boisduval, 1836	18	C/N
Daucus carota L. subsp. sativus (Hoffm.) Arcang.	18	C/N
Eurydema rugosa Motschulsky, 1861	17.66666667	C/N
Brassicaceae	17.54545455	C/N
Cucumis sativus L.	17.5	C/N
Poaceae	17.5	C/N
Equisetum arvense L.	17.4	C/N
Ericaceae	17	C
Parnara guttata guttata (Bremer et Grey, 1852)	17	C
Nonarthra cyanea Baly, 1874	17	C
Portulaca oleracea L.	17	C
Eurydema dominulus (Scopoli, 1763)	17	C
Arctium lappa L.	17	N

(continued)

Table 7 (continued)

Academic name	Score	Category
Menochilus sexmaculatus (Fabricius, 1781)	17	N
Solanum tuberosum L.	17	C/N
Eriobotrya japonica (Thunb.) Lindl.	17	C
Amygdalus persica L.	17	C
Cichorium intybus	17	N
Eucalyptus globula Labill.	17	N
Formicidae	16.8	C/N
Brassica oleracea L. var. capitata L.	16.6	C/N
Glycine max (L.) Merr. subsp. max	16.5	C
Rubus trifidus Thunb.	16.25	C/N
Aedes (Stegomyia) albopictus (Skuse, 1894)	16	C/N
Apis mellifera Linnaeus, 1758	16	C/N
Capsicum annuum "grossum"	16	N
Nerium oleander L. var. indicum (Mill.) O.Deg. et Greenwell	16	C
Armeniaca mume (Siebold et Zucc.) de Vriese	16	N
Promachus yesonicus Bigot, 1887	16	C
Setaria viridis (L.) P.Beauv.	16	N
Cynara scolymus L.	15.66666667	N
Papilio machaon hippocrates C. et R.Felder, 1864	15.66666667	C/N
Dolycoris baccalum (Linnaeus, 1758)	15.66666667	C
A. officinalis	15.6	N
Atractomorpha lata (Motschulsky, 1866)	15.5	C
Aphididae	15.5	C/N
Polistes jadwigae jadwigae Dalla Torre, 1904	15.5	C
Mentha suaveolens	15.5	N
Cornus controversa Hemsl. ex Prain	15.5	N
Akebia quinata (Houtt.) Decne.	15.5	N
Solanum nigrum L.	15.33333333	C
Rutaceae	15.33333333	C
Mentha canadensis L. var. piperascens (Malinv. ex Holmes) H.Hara	15.2	N
Helianthus annuus L.	15	C
Capsicum annuum L.	15	N
Nephotettix cincticeps (Uhler, 1896)	15	N
Lavandula officinalis Chaix.	15	N
Colias erate poliographus Motschulsky, 1860	15	N
Melissa officinalis	15	N
M. pumila	14.75	N
Nysius plebejus Distant, 1883	14.66666667	C
Brassica oleracea L. var. italica Plenck	14.66666667	C/N
Solanum melongena L.	14.66666667	N

(continued)

Table 7 (continued)

Academic name	Score	Category
Petroselinum neapolitanum	15	N
Rosmarinus officinalis	14.66666667	N
Pisum sativum L.	14.6	C/N
Zingiber mioga (Thunb.) Roscoe	14.5	C
Raphanus sativus L.	14	C
Aphis craccivora craccivora Koch, 1854	14	C
Vicia faba L.	14	C
Dolerus similis japonicus Kirby, 1882	14	C
Coccinellidae	14	C
Fabaceae	14	C
Canna	14	N
Phytomyza horticola (Goureau, 1851)	13.66666667	C
Eruca vesicaria	13.5	N
Aulacophora femoralis (Motschulsky, 1857)	13	C
Nephila clavata	13	N
Gryllidae	13	C
Xanthophthalmum coronarium (L.) P.D.Sell	13	C
Diospyros kaki Thunb.	13	C/N
Brassica rapa L. var. perviridis L.H.Bailey	13	C
Illeis koebelei koebelei Timberlake, 1943	13	N
Veronica persica Poir.	12	N
Takydromus tachydromoides (Schlegel, 1838)	12	N
Armadillidium vulgare	10	N
Cycas revoluta Thunb.	10	N
Camellia japonica	10	N
Citrus japonica Thunb.	10	N

The category refers to *C* consistent index, *N* novel index, and *C/N* consistent or novel index depending on the observation place in Table 6

number of edible species observed in the same date and place as an indicator of the productivity. The environmental score of each species was calculated as follows:

1. Calculate the observation-wise environmental score of each species as the number of edible species for each observation.
2. Take mean value of all observation to obtain the overall environmental score of each species.

These environmental scores will evolve as the observation continues and can serve as a data-driven predictor of edible species diversity. Although the scores are not yet fine-grained due to the limit of observation numbers, future observations can be evaluated using the generated scoring systems of index species, further refine the scores and expand the list. The conditions such as time scale of the database that generates a better scoring system can then be selected to optimize the predictability at that moment. The scoring systems can also enrich exploration process since species with similar scores are susceptible of efficient exploration to entail more comprehensive observation. When sufficient diversity of observation is assured in the loop of dynamical assessment, the scoring systems are expected to yield an effective measure with available means, timely reflecting ever-changing conditions of open systems. *Information generation*, a crucial requirement for open systems exploration, can therefore be evaluated by the dynamical reconfiguration of the generative index species scoring system in response to environmental change.

Acknowledgements The author acknowledges Hidemori Yazaki, Kousaku Ohta, Tatsuya Kawaoka, Kazuhiro Takimoto, and Shuntaro Aotake who worked as research assistant.

References

1. Funabashi M (2010) Dynamical system and information geometry - a complementary approach to complex systems. Ph.D. thesis, General Physics, Ecole Polytechnique X
2. Olivier N et al (2010) Cell lineage reconstruction of early zebrafish embryos using label-free nonlinear microscopy. Science 329(5994):967–971
3. Lorenz EN (1963) Deterministic nonperiodic flow. J Atmos Sci 20(2):130–141
4. Xu L et al (2013) Temperature and vegetation seasonality diminishment over northern lands. Nat Clim Chang 3:581–586
5. Kaneko K, Tsuda I (2000) Complex systems: chaos and beyond, a constructive approach with applications in life sciences. Springer, Berlin
6. Funabashi M (2015) Synthetic modeling of autonomous learning with a chaotic neural network. Int J Bifurcation Chaos 25(4):1550054
7. Kaneko K, Tsuda I (1994) Constructive complexity and artificial reality: an introduction. Physica D 75:1–10
8. Tokoro M (2010) What is open systems science? In: Tokoro M (ed) Open systems science: from understanding principles to solving problems (the future of learning). IOS Press, Amsterdam
9. Tokoro M (2015) A new method for new challenges. Sci Am, vol 477, 64–66; Innovation Special Issue
10. Taleb NN (2010) The black swan: the impact of the highly improbable. Random House Trade Paperbacks, New York
11. Diamond JM (1999) Guns, germs, and steel: the fates of human societies. W. W. Norton & Company, New York
12. Kurzweil R (2006) The singularity is near: when humans transcend biology. Penguin Books, New York
13. Barnosky AD et al (2012) Approaching a state shift in Earth's biosphere. Nature 486:52–58
14. Porta M (ed) (2008) Dictionary of epidemiology. Oxford University Press, Oxford
15. Ben-Shlomo Y, Kuh D (2002) A life course approach to chronic disease epidemiology: conceptual models, empirical challenges and interdisciplinary perspectives. Int J Epidemiol 31(2):285–293
16. Tokoro M (2015) Open systems dependability: dependability engineering for ever-changing systems. CRC Press, Boca Raton
17. Petherick A (2012) The ten countries most vulnerable to climate change and their corruption rating. Nat Clim Chang 2:144–145
18. Strogatz SH (2014) Nonlinear dynamics and chaos: with applications to physics, biology, chemistry, and engineering. Westview Press, Cambridge
19. Murphy KP (2012) Machine learning: a probabilistic perspective. The MIT Press, Cambridge

20. Funabashi M (2014) Network decomposition and complexity measures: an information geometrical approach. Entropy 16(7):4132–4167
21. Hayes PJ (1971) The Frame problem and related problems in artificial intelligence. Technical Report, Stanford University, Stanford, CA, USA
22. Millennium Ecosystem Assessment (2005) Ecosystems and human well-being: biodiversity synthesis. World Resources Institute, Washington, DC. http://www.unep.org/maweb/documents/document.354.aspx.pdf
23. Yeo-Chang Y, Misum P (2014) Value of traditional knowledge in forest policy process. In: Proceedings of international conference, sustainable management including the use of traditional knowledge in Satoyama and other SELPS, pp 73–78
24. Ministry of Agriculture, Forestry and Fisheries of Japan (2012) Index species assessment manual of biodiversity for agriculture (in Japanese), http://www.niaes.affrc.go.jp/techdoc/shihyo/index.html#mokuji1 (last viewed on 30 Jun 2015)
25. The International Partnership for the Satoyama Initiative (IPSI) (2014) Toolkit for the indicators of resilience in socio-ecological production landscapes and seascapes (SEPLS), http://satoyama-initiative.org/wp/wp-content/uploads/2014/11/TOOLKIT-X-WEB.pdf (last viewed on 30 Jun 2015)
26. Jensen ME, Bourgeron PS (2000) A guidebook for integrated ecological assessment Springer, New York
27. Biggs R et al (2009) Turning back from the brink: detecting an impending regime shift in time to avert it. Proc Natl Acad Sci U S A 106:826–831
28. ISC-PIF (Institut des Systémes Complexes, Paris Île-de-France) (2009) French roadmap for complex systems 2008–2009. ISC-PIF, Paris. https://hal-polytechnique.archives-ouvertes.fr/hal-00392486
29. Funabashi M (2013) IT-mediated development of sustainable agriculture systems–toward a data-driven citizen science. J Inf Technol Appl Educ 2(4):179–182
30. Pool R (1989) Ecologists flirt with chaos. Science 243(4889):310–313
31. Hoffman A et al (2014) Improved access to integrated biodiversity data for science, practice, and policy - the European biodiversity observation network (EU BON). Nat Conserv 6:49–65
32. Pettorelli N, Kamran Safi K, Turner W (2014) Satellite remote sensing, biodiversity research and conservation of the future. Philos Trans R Soc Lond B Biol Sci 369(1643):20130190
33. Gebbers R, Adamchuk VI (2010) Precision agriculture and food security. Science 327:828–831
34. Xu H-L (2001) Nature farming history, principles and perspectives. J Crop Prod 3(1):1–10
35. Fujita M (2001) Nature farming practices for apple production in Japan. J Crop Prod 3(1):119–125
36. Seifert V, Ramankutty N, Foley JA (2012) Comparing the yields of organic and conventional agriculture. Nature 485:229–232
37. Tuck SL et al (2014) Land-use intensity and the effects of organic farming on biodiversity: a hierarchical meta-analysis. J Appl Ecol 51:746–755
38. Bengtsson J, Ahnstrom J, Weibull A-C (2005) The effects of organic agriculture on biodiversity and abundance: a meta-analysis. J Appl Ecol 42:261–269
39. Douglas L, Brian M (1995) An introduction to symbolic dynamics and coding. Cambridge University Press, Cambridge
40. Atsuyuki Okabe A, Boots B, Sugihara K, Chiu S-N (2000) Spatial tessellations - concepts and applications of voronoi diagrams, 2nd edn. Wiley, Chichester, p 671
41. Automated Meteorological Data Acquisition System (AMeDAS). http://www.jma.go.jp/jp/amedas/ (last viewed on 30 Jun 2015)
42. Funabashi M, Chavalarias D, Cointet J-P (2009) Order-wise correlation dynamics in text data. In: Complex networks, studies in computational intelligence, vol 207. Springer, New York, pp 161–171
43. Hirata Y, Judd K, Kilminster D (2004) Estimating a generating partition from observed time series: symbolic shadowing. Phys Rev E 70:016215
44. Baum LE, Ted Petrie T (1966) Statistical inference for probabilistic functions of finite state Markov chains. Ann Math Stat 37(6):1554–1563

Entropy, Earthquakes, and Tumors

Oscar Sotolongo Costa

1 Introduction

Earthquakes and tumors are not closely related as this paper may suggest. Nevertheless, a common feature, i.e., the presence of nonextensivity in both phenomena, deserves some attention. Indeed, long range correlations are present in one way or another in both problems, particularly in the distribution of earthquake energy and the characteristics of the interaction of different types of radiation with living cells.

This feature, that I wish to highlight here, requires more space than the one available here to be treated in detail, but I want to give a bird's eye viewpoint of this theme, exposing the main results obtained by our group. The bibliography may help to which wants to track this problem. References [1–3] refer to the first part of this talk and contain the main results on fractures and earthquakes, while in [4–9] deals with the characteristics of tumors and their interaction with radiation.

2 Genesis

The genesis of our work lies in the attempt to obtain a more efficient combustion of water–oil emulsified fuels. This kind of fuel experiences the process of "microexplosions" of the sprayed droplets when enter a combustion chamber, as illustrated in the photo of Fig. 1.

When trying to reproduce this phenomenon, we wanted to find the droplet size distribution function better to say "fragment size distribution function" (fsdf) to

O.S. Costa (✉)
IICBA, CInC, Universidad Autónoma del Estado de Morelos, Avenida Universidad 1001, Cuernavaca, CP 62209, Morelos, Mexico
e-mail: osotolongo@uaem.mx

P. Bourgine et al. (eds.), *First Complex Systems Digital Campus World E-Conference 2015*, Springer Proceedings in Complexity, DOI 10.1007/978-3-319-45901-1_27

Fig. 1 Record of the process of microexplosion of a drop of water–oil emulsion in a combustion chamber. When entering a region of high temperature, a process of bubble nucleation that leads to the explosion of the drop occurs producing many smaller droplets which present a larger area to combustion, improving the burning process, and reducing waste of unburned fuel and contamination

evaluate how good is the improvement of combustion in presence of microexplosions. The results in detail are in Ref. [1]. One important result is that for the case of microexplosions the fsdf is found as a power law distribution of sizes. Besides, we studied fsdf for the rupture of mercury drops when falling from a given height. In this case the scaling is not always present, but only when the height is large enough, i.e., when the energy of the fragmentation process is large enough (Fig. 2).

If we try to find the reason for this behavior, note that as fragmentation is a long range phenomenon its description in terms of first principles must be made with Tsallis entropy. Boltzmann´s entropy is inadequate for this.

Then, let's use Tsallis entropy:

$$S_q = \frac{1}{q-1}\int (1 - p^q(v)dv),$$

where $p(v)$ is the probability density of finding a drop of volume v. Let us extremize the entropy subjected to the known conditions:

(a) Normalization: $\int p(v)dv = 1$,

(b) Finiteness of the q-mean value: $\int vp^q(v)dv = 1$

Then the obtained fsdf in terms of the droplet radius is

$$p_q(r) = 3Cr^2\left[1 + (q-1)\,\alpha r^3\right]^{\frac{1}{q-1}},$$

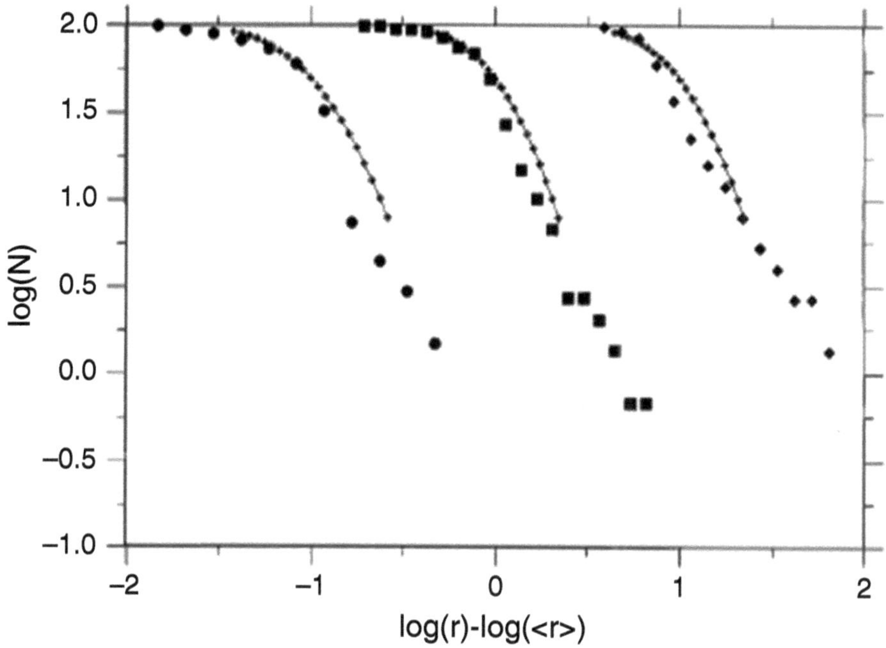

Fig. 2 Experimental results of the cumulative number of droplets (in 10 % and log scale) vs. log(*r*) − log(<*r*>), where <*r*> *l* is the geometrical mean of the droplet radii. The falling heights are 0.5 m (*circles*), 1 m (*squares*), and 5 m (*diamonds*). Note that, to clarify the plot, we have shifted one unit to the left of the results for $h = 0.5$ m and one unit to the right of the results for $h = 5$ m. *Solid lines* represent the theoretical prediction (related to the error function) assuming log-normal distributions. Transition to a scaling law when falling height is increased is evident

where C, α are fitting constants. The asymptotic result when $q \to 1$ gives the well-known (for engineers) Rossin–Ramler distribution function. For larger fragment sizes scaling is present. These are good results, since for the first time this transition is obtained as a result of first principles of physics.

3 Earthquakes

Once obtained these results, we found an inspiring paper from De Rubeis et al. [10] that leaded us to think about the influence of fragments in the occurrence of earthquakes. The model [10], as illustrated in Fig. 3, includes the simulation of the interaction of two profiles representing tectonic plates, with asperities with sizes determined by a Gaussian law.

When one of the asperities casually breaks, the corresponding energy of the earthquake is proportional to the size of the "broken tooth." The simulation gave

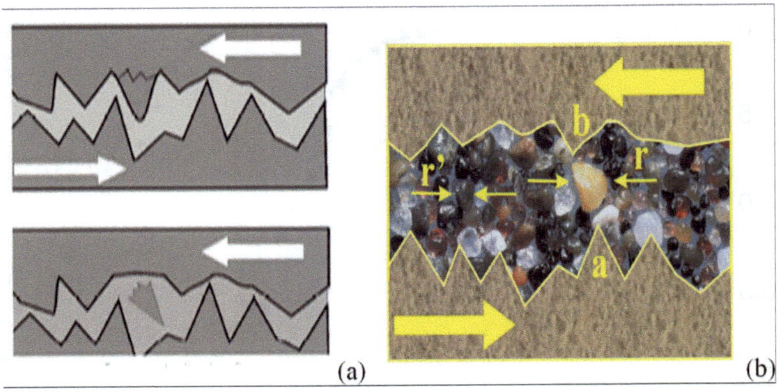

Fig. 3 (**a**) Model of De Rubeis et al. interaction of two profiles (Tectonic plates) with asperities (teeth) gaussianly distributed. When a teeth is broken an earthquake is triggered whose energy is proportional to the area of the fragment. (**b**) Our model. Interaction mediated by hindering fragments. When a fragment is released the energy of the earthquake is, as before, proportional to the area of the fragment, but now the fsdf is the one given by the extremization of the Tsallis entropy by the method previously explained

a power law distribution of the earthquake energy, i.e., the Gutenberg Richter law, which has a limited range of application, as will be seen.

Our variation was to introduce the fragments between the two plates. The fragments are able to not only facilitate the slip of the plates but also to hinder their displacement. Then, the stresses accumulate until the hindering fragment is somehow released (we call it fragment–asperity interaction). Then applying Tsallis entropy extremization [3] in the same way as previously explained to find fsdf, it was possible to find the area distribution of fragments. Besides, assuming proportionality of energy and area as in Ref. [10], we found the energy distribution of earthquakes. Expressed in terms of the magnitude "m" of the seism is

$$G(>m) = \frac{N(>m)}{N} = \left(1 + a\ (q-1)\ (2-q)^{\frac{1-q}{q-2}}\ 10^{2m}\right)^{\frac{2-q}{1-q}}. \qquad (1)$$

It describes the dependence of the exceedance (fraction of earthquakes of magnitude larger than m) with m. Figure 4 shows a fitting in the whole range of magnitudes with catalogs of earthquakes from Cuba and Almeria.

The good agreement shown has also been found with more catalogs like California, Iberian Peninsula [3], and others. This formula has a clear advantage over the Gutenberg–Richter (GR) law, since GR is an empirical law and fits well only in the region of "intermediate" magnitudes but not in the whole range. This highlights the active role of fragments in the very nature of earthquakes. So, our model improves the hypothesis of slipping of tectonic plates without intermediate agents.

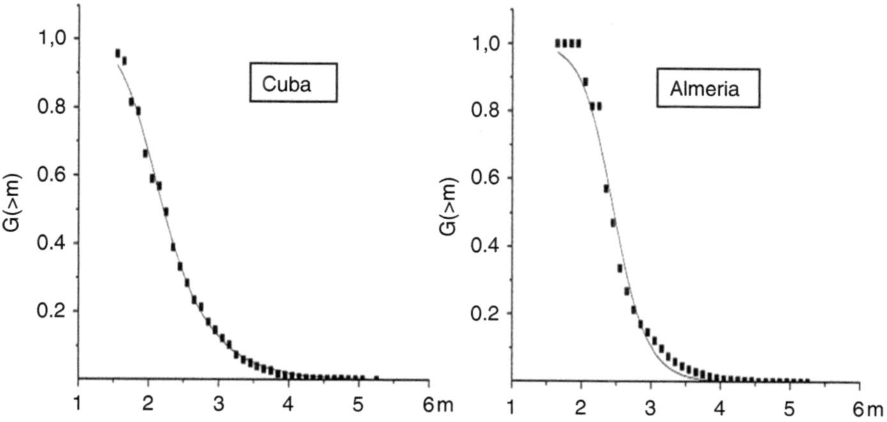

Fig. 4 Fitting of formula (1) with catalogs from Cuba and Almeria. The ordinate is in linear scale to highlight the good fitting

So, if fragments play such an important role in seismic activity, we could judge about a possible relationship between fracking in shale gas exploitation and the triggering of the seismic activity in those areas where fracking wells have grown rapidly.

As is known, fracking consists in the perforation of a well of several kilometers deep and, once the shale layer is reached, the perforation continues some more kilometers. Then underground layers are subjected to a forced breaking which produces cracks through which shale gas or oil can be extracted. Of course, the propagation of these cracks generates fragments.

As far as we know, there is no official recognition of any relation between fracking and earthquakes despite the amazing coincidence of the rhythm of growth of seismic activity and that of fracking wells.

As shall be seen, the energy distribution of the earthquakes in the lands characterized by fracking activity coincides with that produced by the fracking–asperity interaction. So, a link can be stablished between both processes what leads to suspect about a close relationship fracking-earthquakes.

Figure 5 shows the result of fitting Eq. (1) with a catalog of earthquakes of Oklahoma in 2013. This gives a good fitting. Fracking, then, is likely a cause of seismicity in fracking regions.

The earthquakes entering the catalogs we have found are generally of magnitude smaller than 4, i.e., those are earthquakes of relatively small magnitude. But cracks propagate and the formation of larger fragments is to be expected. Consequently, an increase in the magnitude of earthquakes is also to be expected if fracking continues.

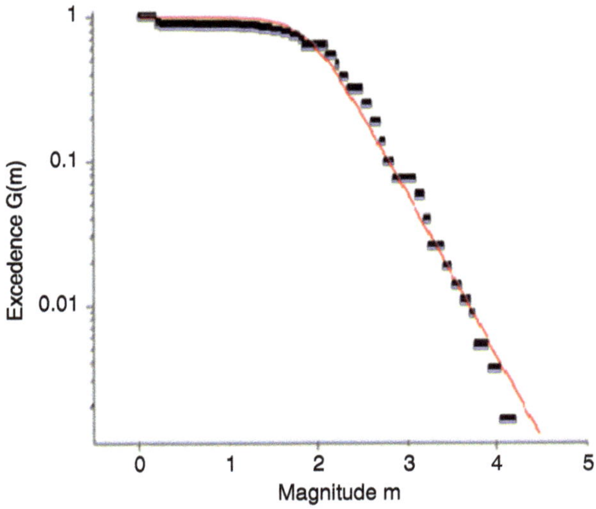

Fig. 5 Fitting of formula (1) with the catalog of earthquakes of Oklahoma in 2013

4 Tumors and Radiobiology

Another important application of this viewpoint is the interaction of radiation with living tissues, what is one of the more used ways to fight tumors, specially cancer tumors.

The interaction of radiation with living tissues is a very open, difficult and, in spite of many efforts, an almost unexplored theoretical problem. This is understandable considering the huge difficulties involved in the description of the system and of the interactions.

Then, a description in terms of general laws of physics, applicable in many different systems, is desirable. Here is, where, similar to the problem of fragmentation, Tsallis entropy is a tool to be applied.

Empirically, it has been found that the survival probability F depends exponentially with a variable called "tissue effect," E, which in the linear-quadratic (LQ) model depends of the applied dose D in a quadratic way, $E = \alpha D + \beta D^2$, being α and β adjustment constants. For a description of all magnitudes involved in radiobiology see Ref. [11].

Let us build the entropy with the probability density $p(E)$ of kill a cell (tissue) with a radiation of tissue effect E. Then:

$$S_q = \frac{1}{q-1}\left(1 - \int_{\Omega} p^q(E)dE\right),$$

and extremize it with the constraints

$$\int_{\Omega} p(E)dE = 1,$$

$$\int_{\Omega} Ep^{q}(E)dE = \langle E \rangle_q < \infty.$$

The last condition demands the existence of a finite value for the q-mean value.

Let us add the fact that beyond a given level of radiation dose D_0 traduced in a value of the tissue effect E_0 beyond which no cell survives. The survival probability, in terms of the dose, is

$$F_s(D) = \begin{cases} \left(1 - \frac{D}{D_0}\right)^{\gamma} & \forall D < D_0 \\ 0 & \forall D \times D_0 \end{cases}, \tag{2}$$

where

$$\gamma = \frac{2-q}{1-q} \quad \text{and} \quad D_0 = E_0/\alpha$$

$$E_0 = \frac{2-q}{1-q}\left(\frac{\langle E \rangle_q}{2-q}\right)^{\frac{1}{2-q}}.$$

The formula (2) involves two factors: γ, which in the analysis of experimental data results to depend only of the tissue, and D_0 determined mainly by the radiation.

To compare our model with experiments a collection of diverse data was taken. The fitting of data using (2) is in Fig. 6 (see Ref. [4]).

A surprising collapse of all the data irrespective of the tissue and the radiation is seen, as predicted by (2). In other words, a universal behavior is revealed.

we also obtained that, if the parameters are determined both for healthy and cancerous cells, as both are bombarded by the same radiation, it can be derived the most efficient and less harmful treatment van be selected so that the treatment kills the most of the cancerous cells and the less of healthy cells. In other words this permits in principle to choose the radiation, dose, and periodicity of treatment that minimize the damage on the patient, or even to decide an alternate treatment. This is of crucial importance in radiotherapy.

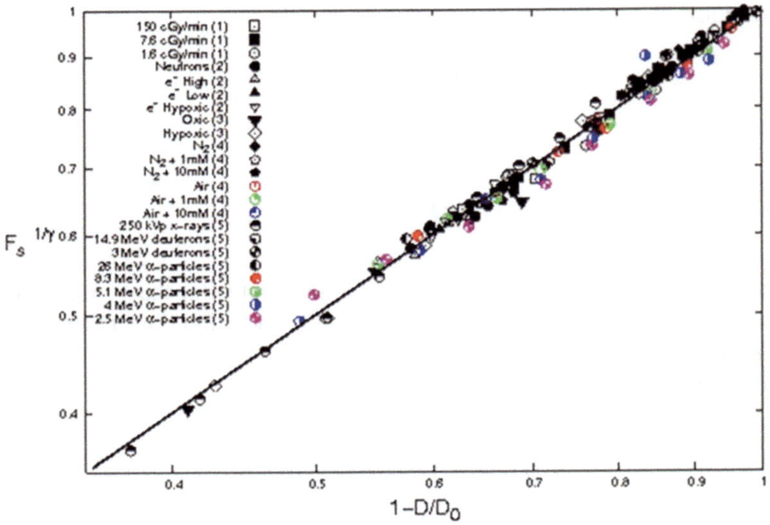

Fig. 6 Fitting of the survival probability $F^{1/\gamma}$ vs. $1-\frac{D}{D_0}$. The collapse of data for all radiations and all tissues is evident. The straight line represents formula (2) conveniently written as $F^{\frac{1}{\gamma}} = 1-\frac{D}{D_0}$

5 Conclusions

Not all the results have been exposed. Of course, this viewpoint has opened many new perspectives in the fields we exposed and there are open questions on which work is in progress. But some results are mandatory to be mentioned to sum up.

Particularly, we have seen that fragment–asperity model describes earthquake magnitude distribution in fracking zones, not only in well established seismic zones, so that the relation fracking-seisms cannot be excluded. Besides, if the model is correct, an increase of seismic magnitudes is to be expected.

The Tsallis formalism gives a universal way to find the survival fraction after radiation sessions. Also, if the characteristic coefficients are known for the tumor and the surrounding tissue (this demands a lot of experimental work), then some hints can be given to choose the less harmful treatment to apply.

References

1. Sotolongo O, Moreno Y, Lloveras J, Antoranz JC (1994) Criticallity in droplet fragmentation. Phys Rev Lett 76(1):42–46. See also Sotolongo O, Lopez E, Barreras F, Marin J (1994) Scaling in drop distributions. Application in combustion. Phys Rev E 49:4027
2. Sotolongo-Costa O, Rodriguez AH, Rodgers GJ (2000) Tsallis entropy and the transition to scaling in fragmentation. Entropy 2(4):172–177

3. Sotolongo-Costa O, Posadas A (2004) A fragment-asperity interaction model for earthquakes. Phys Rev Lett 92:048501
4. Sotolongo-Costa O, Morales L, Rodriguez Perez D, Antoranz JC, Chacon Reyes M (2003) Behavior of tumors under non stationary therapy. Physica D 178(3–4):242–253
5. Rodriguez-Perez D, Sotolongo-Grau O, Riquelme RE, Sotolongo-Costa O, Antonio Santos Miranda J, Antoranz JC (2007) Assessment of cancer immunotherapy outcome in terms of the immune response time features. Math Med Biol 24:287–300. doi:10.1093/imammb/dqm003
6. Sotolongo-Grau O, Rodriguez-Perez D, Santos-Miranda JA, Sotolongo-Costa O, Antoranz JC (2009) Immune system-tumour efficiency ratio as a new oncological index for radiotherapy treatment optimization. Math Med Biol 26:297–307. doi:10.1093/imammb/dqp005
7. Sotolongo-Grau O, Rodriguez-Perez D, Santos-Miranda JA, Sotolongo-Costa O, Antoranz JC (2009) Factor de supervivencia celular bajo la radiación. In: Proceedings of the XXV Congreso Anual de la Sociedad Española de Ingeniería Biomédica, CASEIB, Cádiz, pp 10–13
8. Sotolongo-Grau O, Rodriguez-Perez D, Antoranz JC, Sotolongo-Costa O (2010) Non-extensive radiobiology Bayesian inference and maximum entropy methods in science and engineering. In: Proceedings of the 30th international workshop on Bayesian inference and maximum entropy methods in science and engineering, 4–9, July 2010, Chamonix, France, edited by Mohammad-Djafari A, Bercher J-F, Bessiere P (AIP, 2010), vol 1305 of AIP Conference Proceedings, pp 219–226. ISSN: 0094-243X E-ISSN: 1551-7616 http://arxiv.org/abs/1006.3410
9. Sotolongo-Grau O, Rodriguez-Perez D, Antoranz JC, Sotolongo-Costa O (2010) Tissue radiation response with maximum Tsallis entropy. Phys Rev Lett 105:158105
10. De Rubeis V, Hallgas R, Loreto V, Paladin G, Pietronero L, Tosi P (1996) Self-affine asperity model for earthquakes. Phys Rev Lett 76:2599
11. Tubiana M (1990) Introduction to radiobiology. Taylor & Francis, London

Coarse-Grained Molecular Dynamics for Copolymer-Vesicle Self-Assembly. Case Study: Sterically Stabilized Liposomes

Alexander Kantardjiev and Pavletta Shestakova

1 Introduction

Our study aims to understand the principles governing polymer–lipid vesicle liposome assembly and ensuing hybrid liposome stability by combination of coarse-grained molecular dynamics simulations and complementary NMR-DOSY experiments. Besides fundamental interest (principles of vesicle formation and evolution) the problem has a strong application side—self-assembly of synthetic vesicles for nanotechnology [1]. A prominent example for the nanotechnological interest in liposome systems is the intelligent design of drug delivery systems that require polymer stabilized vesicles. Such sterically stabilized liposomes fall within mesoscopic ordering of scales and thus the problem falls within spatial and temporal scales extending far beyond current computational capability for atomistic dynamics.

1.1 Some Musings About Simulation of Hybrid Vesicles

A first guess approach is to apply coarse-graining which allows to tackle consistently the problem with adequate system size and time scale. Coarse graining implies fewer degrees of freedom, cheaper potential calculation, and larger time step. In this way reliability and accuracy of fine grained (FG), i.e., atomistic models, is combined with the efficiency of a reduced representation. Thus our million-atoms simulations yield trajectories starting from a randomized distribution

A. Kantardjiev (✉) • P. Shestakova
Institute of Organic Chemistry with Centre of Phytochemistry, NMR Lab,
Bulgarian Academy of Sciences, Sofia, Bulgaria
e-mail: alexkant@orgchm.bas.bg; psd@orgchm.bas.bg

P. Bourgine et al. (eds.), *First Complex Systems Digital Campus World E-Conference 2015*, Springer Proceedings in Complexity, DOI 10.1007/978-3-319-45901-1_28

of lipid molecules (1,2-Dipalmitoyl-sn-glycero-3-phosphocholine—DPPC and Cholesterol) and varying concentration of copolymers with lipid mimetic units in explicit solvent (order of million water molecules). Polyethylene glycol (PEG) grafted on vesicle surface has a stabilizing effect and leads to prolonged circulation times of the liposome in the blood stream. PEG covalently coupled to a phospholipid moiety, known as PEG-lipid, possesses the status of the current most effective steric stabilizer for liposomes. However conventional PEG-lipids suffer severe limitations due to the addition of a negative net charge stemming from the carbamate linkage (the covalent link of the PEG chain). On the other hand, an ether bond (PEG moiety linked to the glycerol skeleton via ether linkages) leads to changes in the dipole potential. Such effects can be detrimental to some properties of the sterically stabilized liposomes (e.g., their pH-dependence). Here we focus on a new class promising stabilized copolymers with ethylene oxide (EO). Short blocks of aliphatic double chains mimic lipid tails. The C12 aliphatic chains are attached to a glycerol skeleton via ether linkages. The PEG chain can be of variable length and this fact is to be exploited for search of a better stabilizing agent. Again, ether bonds are used to attach PEG chains to the lipid mimetic units.

1.2 Thesis Proposal

Our work is focused at computational and DOSY-NMR study of these novel copolymers with the following basic premise in mind—the number of the lipid mimetic anchors and the size of the PEG chain control the stabilizing potential of the copolymer. The number of lipid mimetic units is a prerequisite for pushing the saturation limit and thus enhancing the formation of a denser protective PEG layer around the vesicle. We used the following copolymers based on 1,3-didodecyloxy-propane-2-ol (DDP) and 1,3-didodecyloxy-2-glycidylglycerol (DDGG) lipid mimetic anchors. Next we deal with methodological issues regarding MD membrane simulations—MD setup, coarse-grained representations of lipids, and observing properties computationally (Fig. 1).

2 Methods

2.1 Molecular Dynamics Setup

Molecular dynamics simulations were carried out under the isothermal–isobaric conditions in order to emulate NPT ensemble. Pressure was fixed at 1.01 bars and coupled isotropically with a compressibility of 3×10^{-3} bar^{-1}. We applied Berendsen barostat with a coupling constant of 5 ps. Velocity Verlet algorithm was used to propagate the equations of motion. A time step of 20 fs was used. Actually

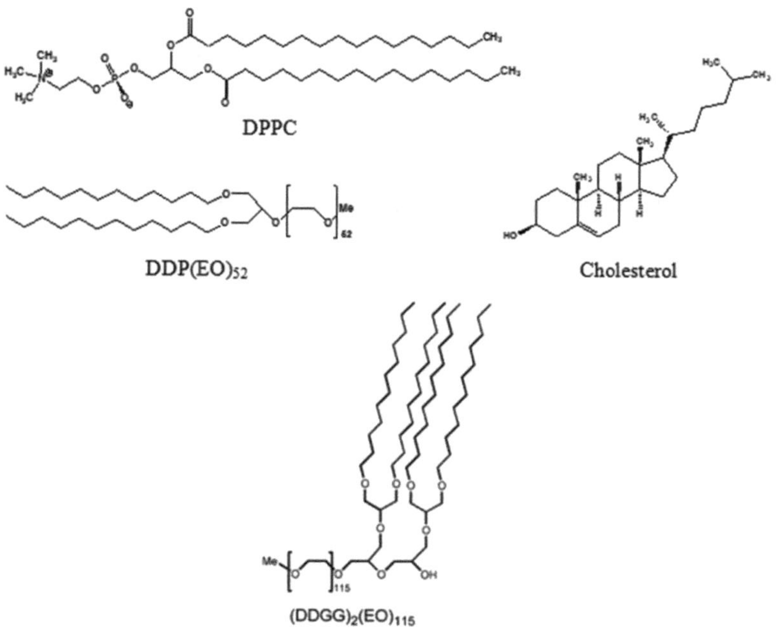

Fig. 1 Chemical structures of copolymer-vesicle constituents

a time stem of 40 fs was initially set (as recommended) but the stability of the simulation was compromised (due to heating that goes beyond the capability of the thermostat). Equilibration of the systems was monitored by looking at the system energy and some structural parameters as, for example, area per lipid. A cutoff of 12 Å was used for the Lennard-Jones potential and electrostatic interactions.

2.2 *Coarse-Grained Models of Major Player Molecules*

All systems were modeled with the MARTINI force field [2, 3] which is reasonably close to the atomic details. Its scaling is nearly fourfold (with some exceptions mentioned below). On average four heavy atoms are mapped to a single coarse-grain bead—interaction center. The philosophy is based on a modular principle by definition of an array of particle-bead types with specific chemical properties. As is the case with fine-grained atomistic case—the strength of the interaction is a function of the interacting particle types and the distance. Bonds and bond angles are described with a set of standard bonded potential energy functions-harmonic potentials for bonds/cosine type potential for angles. In our case MARTINI force field is used to model liposome assembly of DPPC/Cholesterol/Copolymer (DDP/DDGG) in explicit water molecules. DDP and DDGG polymers were modeled by combining the available parametrization for the PEG chain and the structure of the glycerol skeleton and corresponding aliphatic chains.

2.3 *Molecular Dynamics Simulation Software*

Molecular dynamics simulations were performed using the Gromacs software simulation package [4]. The software was installed in a GPU CUDA computer environment running Scientific Linux 6.3 as operating system. Each simulation was run using 4 GPU nodes. Each node was equipped with Intel processors at 2.66 GHz and 16 GB local memory (2 GB per processor core).

3 Results and Discussion

3.1 *System Setup Organization*

A random ternary mixture of DPPC, Cholesterol, and DDP/DDGG molecules in ratio corresponding to the experimental mol % was generated by the *genbox* utility of Gromacs, i.e., without a biased coarse-grained lipid template to construct the vesicle and therefore closer to the initial experimental setup. Phospholipids (DPPC) and Cholesterol molecules were mixed in 2:1 molar ratio. We generated a concentration ladder of the selected promising copolymers—in MD search for the saturation limit and detailing the dependence of liposome assembly on concentration and structure of copolymers—DDP $(EO)_{52}$ or DDGG $(EO)_{115}$ or $(DDGG)_2$ $(EO)_{115}$. A rectangular (cubic) box with a 20 nm size was used. Periodic boundary conditions were applied in all directions. The mix was solvated via the available Gromacs package utility (genbox). The initial configuration of the system was subjected to a round of energy minimization using the steepest descent algorithm until reaching a predefined cutoff (in effect—several hundred minimization steps). For such a system the corresponding atomistic representation would require on the order of million particles (atoms) and the prohibitive time step of 2 fs thus obliterating achieving adequate results with the available contemporary computational resources. Membranes with several species (components) with fixed proportions are tricky for the equilibration and in our setup it required hundreds of nanoseconds. We have calculated the average of the order parameter through the averaging of the second Legendre polynomial over all lipids and time for each of the described vectors.

3.2 Spontaneous Self-Assemblies Within Hundreds of Nanoseconds

The process of self-assembly to vesicle closures and concurrent bicelles/micelle structures is followed along several hundreds of nanoseconds long MD trajectories. The concise format of this contribution prohibits presentation of the overwhelming amount of data—thus we reveal several informative snapshots [5] which directly support the hypothesis of the "PEG Length to Lipid Mimetics" ratio as a major structural determinant controlling the shielding property of the copolymer in vesicle assembly (Fig. 2).

3.3 Credulity Addendum to MD Simulations: Coincidence with DOSY-NMR Data

For a specific DDP $(EO)_{52}$ concentration (a 5 mol%) a DOSY-NMR experiment was carried out. Simulation data is successfully verified against the experimental NMR results (Fig. 3).

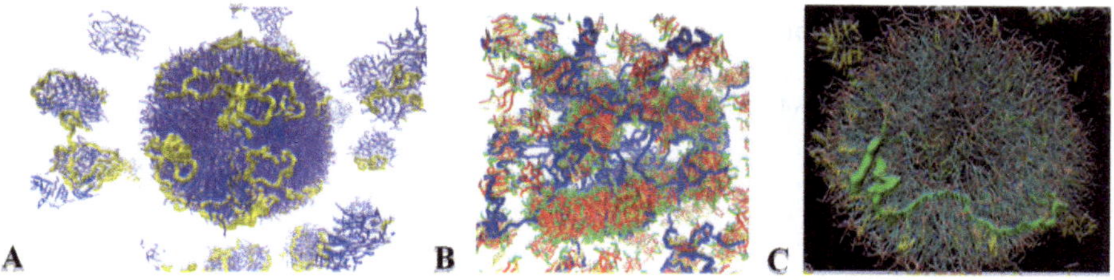

Fig. 2 Unequivocal direct support for the "PEG length vs Lipid Mimetics" ratio hypothesis. Major players at around 100 ns for a 5 mol% copolymer concentration with respect to DPPC. (**a**) DDP $(EO)_{52}$ "PEG-**52** vs **single** DDP lipid-mimetics." DDP $(EO)_{52}$—*yellow*. (**b**) DDGG $(EO)_{115}$ "PEG-**115** vs **single** DDGG lipid-mimetics." DDGG $(EO)_{115}$ is in *blue*; DPPC heads are *green*; DPPC tails are *red*. Bicelle refused closure to a vesicle. (**c**) $(DDGG)_2$ $(EO)_{115}$ "PEG-**115** vs **double** DDGG lipid-mimetics." $(DDGG)_2$ $(EO)_{115}$ is in *green*. Restored potency for vesicle formation due to the additional lipid mimetic

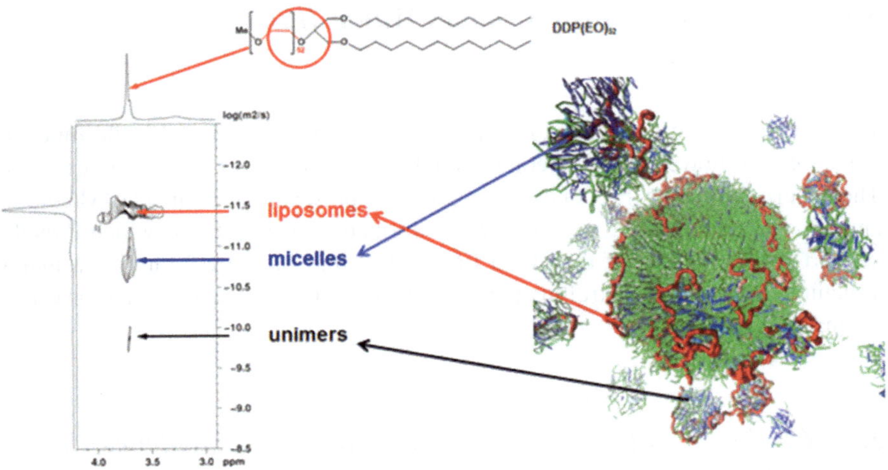

Fig. 3 DOSY-NMR spectra of a DDP(EO)$_{52}$ Copolymer-Liposome against a simulation snapshot at around 100 ns with the same components and concentrations. Nice fit of both the liposome itself and the concurrent satellite structures

4 Conclusions

1. "PEG length vs Lipid Mimetics" ratio might be a determinant for copolymer "stealth" potential.
2. Consistency with NMR-DOSY findings.

References

1. Lasic DD, Needham D (1995) Stealth liposomes: a prototypical biomaterial. Chem Rev 95:2601–2628
2. Marrink SJ, De Vries AH, Mark AE (2004) Coarse grained model for semiquantitative lipid simulations. J Phys Chem B 108:750–760
3. Marrink SJ, Risselada HJ, Yefimov S, Tieleman DP, De Vries AH (2007) The MARTINI force field: coarse grained model for biomolecular simulations. J Phys Chem B 111:7812–7824
4. van der Spoel D, Lindahl E, Hess B, Groenhof G, Mark AE, Berendsen HJC (2005) GROMACS: fast, flexible, and free. J Comput Chem 26:1701
5. Humphrey W, Dalke A, Schulten K (1996) VMD - Visual molecular dynamics. J Mol Graph 14:33–38

Understanding and Modeling the Complexity of the Immune System

Véronique Thomas-Vaslin

1 Introduction

The immune system is a complex biological micro-ecosystem, adaptive, highly diversified, self-organized cognitive network of cells, and molecular entities with degeneracy properties, allowing in healthy individuals for a robust and resilient system with emergent properties such as anamnestic responses and regulations. The adaptive immune system has evolved into a complex system of billions of highly diversified lymphocytes all interacting as a connective dynamic multi-scale organized and distributed system, in order to collectively insure body identity and integrity, and species preservation of symbiotic organisms exposing polygenomic antigens. The immune system is characterized by complexity at different levels: network organization through fluid cell populations with inter- and intra-cell signalling, an extraordinary lymphocyte receptor diversity, cell clonotype selection and competition at cell level, migration and interaction inside the immunological tissues and fluid dissemination through the organism, homeostatic regulation while rapid adaptation to a changing environment. Lymphocytes are the key actors of the immune system of vertebrates, in the middle of a multi-scale biological organization "from molecule to organism," and at the confluence with other different biological systems and the environment. The perception of antigens induces a network of immuno-receptors that could be viewed as an internal representation of antigens. Fluctuations and variability are key factor for the immune system to adapt perturbations and aging.

V. Thomas-Vaslin (✉)
Sorbonne Universités, UPMC Univ Paris 06, INSERM, UMRS959, CNRS, FRE3632, Immunology-Immunopathology-Immunotherapy (I3), 83 Boulevard de l'Hôpital, Paris 75013, France
e-mail: veronique.thomas-vaslin@upmc.fr

P. Bourgine et al. (eds.), *First Complex Systems Digital Campus World E-Conference 2015*, Springer Proceedings in Complexity, DOI 10.1007/978-3-319-45901-1_29

A deeper understanding of T-cell differentiation, diversity, dynamics, repertoires selection, and regulation processes is key for fundamental research, medical advancement, and drug discovery. Moreover the immune system represents a complex system sharing some transversal properties with other complex systems (organization as dynamical properties, resilience to perturbations, etc.).

2 Crossing Interdisciplinary Transversal Questions from Other Complex Systems to Study the Complexity of the Immune System

Understanding the organization and the regulation of the complex immune system refers to transversal questions already mentioned for other "live" complex systems such as other biological, social, or ecological macrosystems, with some peculiarities specific to the immune system. Thus, the immune system can be observed and modeled from different point of view with the theoretical and methodological approaches shared by other interdisciplinary domains using biological, physical, philosophical, mathematical, statistical, and computer approaches. Some examples of the cross-fertilization required to solve open questions concerning specifically the immune system are proposed in Table 1.

3 Open Question and Challenges Related to the Complexity of the Immune System

More than transversal questions common to other complex systems, the immune system presents some peculiarities that require particular investigations and modeling and this represents new challenges to overcome.

3.1 Objective Identification of Immune System Cell Populations

Innate and adaptive immune system subpopulations are currently defined on the basis of the revelation of a combination of cell surface or intra-cellular/nuclear molecules that the researcher has to define to explore the phenotype and functionality of cells. Thus, the cell populations revealed by cell staining are largely dependant on the mixture and the number of chosen parameters (n) that will drive the number of subpopulations (2n). Techniques like flow cytometry analysis allow quantification of several parameters from individual cells allowing characterizing cell size, structure, specific phenotype, and function of millions of cells in a multi-dimensional way.

Table 1 Crossing transversal questions identified for the investigation of the complex systems[a] with the questions specific to the immune system

Transversal questions commons to other complex systems	Questions specific to the immune system
Study and modeling of adaptive multi-scale system	The adaptive immune system
Integration of high-throughput multi-scale and multi-parametric data and metadata sharing	Biological generic data from transcriptome, proteome, but also cytome, repertome, etc.
Computer or mathematical tools for exploration and formalization	Supervised and non-supervised statistical modeling, mechanistic reconstruction of immune system, and lymphocyte behavior
Theoretical reconstruction	Multi-scale analysis, representation of heterogeneous data, organization of knowledge's database describing the immune system around the lymphocyte level, from molecule to organism, for data-driven and hypothesis-driven reconstruction
Metamodels, multi-formalism for reconstruction and visualization of dynamics, differentiation, and behavior	Mathematical and computer modeling for lymphocyte cell population dynamics, activation, regulation, and selection processes; use of oriented object, graphical languages, SMA, ontologies, for modeling the multi-scale entities of the immune system, etc.
Fluctuations, stability, variability, regulations at multi-scale levels	Multi-level/Multi-scale: organism, lymphoid tissues, lymphocyte populations, cellular and molecular lymphoid repertoires
Robustness/resilience and relation to organization	Behavior of immune system from development to aging; resilience to perturbations, transition to immunopathologies (infection, autoimmunity, cancer, etc.); immunotherapy/vaccination
Model the relationships between biodiversity, functioning and dynamics of the (eco)systems	Diversity, stability/perturbation of immune repertoires, and lymphocyte populations
Self-organization, simulation of virtual landscapes	Auto-organization of cells in lymphoid organs and development, cell network, and immune repertoire
Data mining, extraction, visualization of data, and semantic and syntactic analysis of scientific literature requires artificial intelligence and automatic learning approaches	Information extraction and visualization of immune literature with concepts specific to immune system

[a] In the context of the Complex Systems Digital Campus '15 - World e-Conference: http://cs-dc-15.org/papers/organisms/cognitive-immune-sys/understanding-and-modelling-the-complexity-of-the-immune-system/

However, current analyses performed by manual gating inspect parameters 2 by 2 and do not reveal the complexity of the lymphocyte subpopulations that coexpress several parameters.

The challenge is developing methods and software tools for current immunological analysis and automatic identification of cell subsets. This allows objective investigation and identification of cells subpopulations that have certainly been ignored by immunologists, to identify variability/stability, resilience, or perturbations among development/aging, through genetic backgrounds, and during the course of perturbations as immunopathologies or immunotherapies.

3.2 Lymphocyte Population Dynamics and Repertoire Selection: Integration of Multi-Level/Multi-Scale Data and Reconstruction of Dynamic Interactions

The most important feature of the immune system is the availability of a diverse cell repertoire and its selection constraints. Lymphocytes are produced in primary lymphoid organs from precursors that differentiate and somatically rearrange DNA variable genes independently, leading to the expression of unique type of immuno-receptor in each lymphocyte. T or B cell repertoires are thus collections of lymphocytes, each characterized by its antigen-specific receptor produced by random somatic rearrangements of V(D)J gene segments during lymphocyte differentiation. The potential repertoire of 10^{15}TCR/Ig receptors is far beyond the lymphocyte count in a single individual. Then, process of lymphocyte selection with high cell death or amplification of particular antigen-specific clones represents a network of dynamical interactions conferring tolerance to avoid autoimmunity though retaining the potential to respond to a very large collection of antigens. Thus, the dynamics of cell fluxes and turnover, cell selection through division, and cell death is able to adapt a dynamic equilibrium according to the internal genetic variability but also external antigen challenges in various lymphoid populations. The rules governing the clonal selection processes and cell population dynamics stability or disturbance in immune pathologies and aging are far to be fully understood. Integration of high-throughput data describing qualitatively and quantitatively cell populations, repertoire diversity, and gene expression should allow data mining, signature discovery, and reconstruction of dynamics behaviors.

The challenge is the integration of multi-scale data and metadata describing cell populations, cellular and molecular lymphocyte repertoires, gene expression and proteome across time, lymphoid organs, and genetic background, in various conditions describing physiological or pathological states or treatments. Organization of knowledge's using standardized database with ontologies and state transition diagrams should improve organization of data for data mining and dynamic computational modeling. Object-oriented computer modeling taking into account the levels of the "organism," the "organs," the "cell," and the "molecule" through various time scales should improve the interoperability of mathematical and computer models already developed in the field, allowing also the direct intervention of the biologist to implement the models and suggest new experiments or treatments.

3.3 *Understanding Resilience or Instabilities to Perturbations, Immune Dysfunction in Order to Improve Immuno-Intervention Strategies*

Internal cell instability or external cell perturbations can impact the stability and reactivity of the immune system at various biological levels (from molecules to organism). Thus not only physiological aging and immunopathologies like infectious, autoimmune or inflammatory diseases, and cancer but also immunotherapies or preventive immuno-intervention like vaccination can perturb the system. Genetic or environmental component alterations (antigens, infections, chemicals, nutriments, etc.) or other biological instabilities (as in nervous, hormonal, metabolic systems, etc.) can affect the organization of immune system, its dynamics and lymphocyte repertoires and turn the physiologic equilibrium to immunopathologies. The identification and quantification of variability and perturbations at these different levels and through time should allow understanding the resilience or instability of the system. Conversely, improving knowledge on the physiological or pathological dynamic behavior of the immune system should also reveal keys for immuno-interventions.

The challenge is to connect and better integrate knowledge, as a result of data mining, with dynamic computer modeling and simulations to be able to understand the system behavior under such perturbations. The resilience and homeostatic regulation of the system (steady state dynamic equilibrium) but also variability/fluctuation (according to genetic background, physiological development and aging) to pathological perturbations of the immune system should thus be investigated by multi-disciplinary approaches. This requires the development of original biological, mathematical, and computer modeling tools. This might allow assessing the quantity/quality of small perturbations, at various scale levels, that can impact and/or dys-balance the whole immune system equilibrium with the search of threshold effects. On the opposite it might allow estimating the maximal variability that the system can endure without global perturbations at the organism level.

3.4 *Extract, Visualize, and Organize Immunological Knowledge from Scientific Immune Literature*

The complexity of the immune system, related to biological multi-scale levels, but also of the data generated by multiple technologies, published in the form of unstructured text requires the development of innovative techniques of data and literature mining to enhance information retrieval, visualization of enormous quantities of data, and organization of knowledge.

The challenge is to develop data mining and machine learning methods to have a better understanding of the complexity of the immuno-physiome. Innovative data mining, semantic and syntactic analytical approaches should help define the concepts that have to be extracted and algorithms to automatically extract the data with a maximum of accuracy.

3.5 *Contribute to Global Evaluation of Complex Systems and Risk Issue*

A global evaluation of the behavior of complex systems should be undertaken under philosophical and scientific aspects. The behavior of complex systems is related to their multi-scale organization. While the immune system is related to micro-levels from molecule to organism, the biosphere is related to macro-levels from organisms to global environment, through social interactions, migration, ecosystems, climate, and biosphere. Indeed data, simulations, and predictions are difficult to establish for some systems. Organization of systems results from the selection among diversities of only a small fraction of all potential possibilities, with an infinite combination of parameters. This contributes to selected dynamic equilibrium allowing the systems to resist time and perturbations unless the resilience is disrupted.

The challenge is to provide a global analysis of common properties of complex multi-scale systems in order to understand the robustness and the degree of resilience of systems selected on the basis of their organization and risk of changes in the dynamic equilibrium under various perturbations. The notion of emergence and immergence have to be analyzed in order to understand whether or not the aging and evolution of a system and the threshold effects are involved in the resilience of systems or could induce their disorganization and fragility.

3.6 *The Human Perspective: Revisiting the "Immune System," Limits, Definition, Characteristics, Functions, and Stability*

Complex systems are often viewed as multi-scale, self-assembly, adaptive dynamic and cognitive networks of diverse interacting agents capable of sensing patterns with degenerative properties. In biology, links between entities are processes that occur at various scales. At the macroscopic level, the nervous system insures the cognition, perception, and memory of "self," through mental links and related relations to macroscopic external environment, allowing to define ego as well as physical/somatic identity based on consciousness, memory, and social interactions. Similarly, at the microscopic level, the immune system as a cognitive, diverse, dynamic, fluid, and anamnestic system senses the quality and quantity of microscopic patterns, either from body or environment. The immune system is thus at the interface of the dynamic symbiotic organization of the individual (composed in adults of ten times more prokaryotic than eukaryotic cells) and constantly senses its environment and its own components (idiotopes). This leads to individual cell signalling up to collective decision-making allowing for discrimination and memory of microscopic entities in a systemic way.

Biological and philosophical conceptual questions remain about the notion of organization, organism, immune system, the perception of self, identity, memory, tolerance, and resilience.

- Are there limits between the immune system, the organism, and the environment?
- Has the immune system a role to define the identity of an organism?
- What are the major characteristics of an organization and an organism?
- How do cognition, diversity, selection, memory, and dominant tolerance contribute to the dynamic stability of the system and the resilience of the organism?
- How can we define the immune system? Is the term "immune," with its etymologic origin meaning "exempt," adequate according to current knowledge?

3.7 *Bio-Inspired and Artificial Immune Systems*

Knowledge's assembled to understand, reconstruct, simulate, and predict the complex multi-scale behavior and resilience of the immune system dynamics in health, aging, diseases, or treatments could be useful for the design of innovative artificial immune systems reproducing or inspired from the behavior of the natural immune system. Thus, understanding this natural organization, selected during the evolution of species and cells in organism, can help to design innovative evolutive artificial immune systems, and referring to the "war" metaphor to design the best "immunologistic" optimization. This could help understanding the characteristics of an organization as a process and as a result, from organisms to societies and for the design of regulatory processes and security purposes.

The challenge is to overcome the conceptual and technical limitations to design self-organized artificial immune systems resilient to perturbations and able to preserve the identity and integrity of the organism or societies.

4 Conclusions

Identifying theoretical and methodological questions related to the complexity of the natural and artificial immune systems will help to structure new ideas and collaborative work across complex systems. Responding to these challenges will improve the global data exploration, the emergence of new concepts, and understanding of the immune system linked to other biological systems or to macroecological/biosphere systems.

5 Biography of the Author

PhD in Immunology at University Pierre et Marie Curie, researcher at CNRS, Véronique Thomas-Vaslin has founded and directed the Integrative Immunology: Differentiation, Diversity, Dynamics team (https://www.i3-immuno.fr/en/#People/VTV). As a member of "Réseau National des Systèmes Complexes" she has founded the ImmunoComplexiT network (http://www.immunocomplexit.net/) and is in the steering committee of Institut des Systèmes Complexes-Paris Ile-de-France (http://iscpif.fr/).

Her research focuses on lymphocyte population dynamics, diversity, selection of repertoires, studying the physiology of the immune system, tolerance, and the regulation of immune responses in health and diseases. Her current aim is the integration of the complexity of the immune system as an ecosystem, responding to perturbations and aging. Computational model is designed to reconstruct the multi-scale T-cell dynamics, from their generation in thymus to the control of immune responses.

5.1 Some Publications

Thomas-Vaslin, V. and A.A. Freitas, Lymphocyte population kinetics during the development of the immune system. B cell persistence and life span can be determined by the host environment. Int. Immunol., 1989. 1: pp. 237–246.

Thomas-Vaslin, V., L. Andrade, A. Freitas, and A. Coutinho, Clonal persistence of B lymphocytes in normal mice is determined by variable region-dependent selection. Eur J Immunol, 1991. 21(9): pp. 2239–2246.

V. Thomas-Vaslin.*, Modigliani, Y*, A. Bandeira, M. Coltey, N.M. Le Douarin, A. Coutinho, and J. Salaun, *contributed equally Lymphocytes selected in allogeneic thymic epithelium mediate dominant tolerance toward tissue grafts of the thymic epithelium haplotype. Proc Natl Acad Sci U S A, 1995. 92(16): pp. 7555–7559.

Thomas-Vaslin, V., J. Salaun, B. Gajdos, N. Le Douarin, A. Coutinho, and A. Bandeira, Thymic epithelium induces full tolerance to skin and heart but not to B lymphocyte grafts. Eur J Immunol, 1995. 25(2): pp. 438–445.

Le Douarin, N., C. Corbel, A. Bandeira, V. Thomas-Vaslin, Y. Modigliani, A. Coutinho, and J. Salaun, Evidence for a thymus-dependent form of tolerance that is not based on elimination or anergy of reactive T cells. Immunol Rev, 1996. 149: pp. 35–53.

Thomas-Vaslin, V., D. Damotte, M. Coltey, N.M. Le Douarin, A. Coutinho, and J. Salaun, Abnormal T cell selection on nod thymic epithelium is sufficient to induce autoimmune manifestations in C57BL/6 athymic nude mice. Proc Natl Acad Sci U S A, 1997. 94(9): pp. 4598–4603.

Thomas-Vaslin, V., B. Bellier, J.L. Cohen, O. Boyer, N. Raynal-Raschilas, D. Glotz, and D. Klatzmann, Prolonged allograft survival through conditional and specific ablation of alloreactive T cells expressing a suicide gene [see comments]. Transplantation, 2000. 69(10): pp. 2154–2161.

Bellier, B., V. Thomas-Vaslin, M.F. Saron, and D. Klatzmann, Turning immunological memory into amnesia by depletion of dividing T cells. Proc Natl Acad Sci U S A, 2003. 100(25): pp. 15017–15022.

Mesel-Lemoine, M., M. Cherai, S. Le Gouvello, M. Guillot, V. Leclercq, D. Klatzmann, and F. Lemoine*, V. Thomas-Vaslin* *contributed equally Initial depletion of regulatory T-cells: the missing solution to preserve the immune functions of T lymphocytes designed for cell-therapy. Blood, 2006. 107: 381–388

Giraud, S., Barrou, B., Sebillaud, S., Debre, P., Klatzmann, D., and Thomas-Vaslin, V. (2008). Transient Depletion of Dividing T Lymphocytes in Mice Induces the Emergence of Regulatory T Cells and Dominant Tolerance to Islet Allografts. Am J Transplant 8, 1–12.

Thomas-Vaslin, V., Altes, H. K., de Boer, R. J., and Klatzmann, D. (2008). Comprehensive assessment and mathematical modeling of T cell population dynamics and homeostasis. J Immunol 180, 2240–2250.

Bersini, H., D. Klatzmann, A. Six and V. Thomas-Vaslin (2012). “State-Transition Diagrams for Biologists.” PloS one 7 (7), e41165.

Thomas-Vaslin, V., A. Six, H. P. Pham, C. Dansokho, W. Chaara, B. Gouritin, B. Bellier and D. Klatzmann (2012). Immunodepression & Immunosuppression in aging mice “Immunosuppression” InTech - Open Access Publisher, ISBN 978-953-308-19-8 – doi:10.5772/29549

Thomas-Vaslin, V., A. Six, J. G. Ganascia and H. Bersini (2013). “Dynamical and mechanistic reconstructive approaches of T lymphocyte dynamics: Using visual modelling languages to bridge the gap between immunologists, theoreticians and programmers.” Frontiers in Immunology 01 October 2013 | doi:10.3389/fimmu.2013.00300

Abi Haidar, A., A. Six, J.-G. Ganascia and V. Thomas-Vaslin (2013). The Artificial Immune Systems Domain: Identifying Progress and Main Contributors Using Publication and Co-Authorship Analyses. Advances in Artificial Life, ECAL. 12: 1206–1217 http://mitpress.mit.edu/sites/default/files/titles/content/ecal13/ch185.html

Six, Mariotti-Ferrandiz ME, Chaara W, Magadan S, Pham H-P, Lefranc M-P, Mora T, Thomas-Vaslin V, Walczak AM and Boudinot P (2013) The past, present, and future of immune repertoire biology – the rise of next-generation repertoire analysis. Front. Immunol. 4:413. doi:10.3389/fimmu.2013.00413

Thomas-Vaslin, V. (2014). “A complex immunological idiotypic network for maintenance of tolerance.” Front Immunol 5: 369. http://www.ncbi.nlm.nih.gov/pubmed/25132837

Thomas-Vaslin, V. (2015). Complexité multi-échelle du système immunitaire: Evolution, du chaos aux fractales. Le vivant critique et chaotique. E. Matériologiques. Paris: 333.http://www.materiologiques.com/Le-vivant-critique-et-chaotique

Acknowledgments This material is based upon work supported by the Complex Systems Institute of Paris Île-de-France (ISC-PIF) and DIM Ile de France; RNSC and support to the ImmunoComplexiT network; CNRS Programme Interdisciplinaire Longévité & Vieillissement, ATC Vieillissement INSERM; Emergence-UPMC program; PHC Tournesol EGIDE, CNRS PEPS BMI; and Investissements d'Avenir program (ANR-11-IDEX-0004-02), LabEx Transimmunom.

A Simple Paradigm for Nooconomics, the Economy of Knowledge

Idriss J. Aberkane

The purpose of this article is to outline a simple and improvable paradigm for noodynamics, the study of knowledge flows. If we endow this paradigm with an economic perspective, and consider its agents buyers and sellers, it can then also be considered an early working paradigm, or more precisely, an early building block towards a working paradigm, for nooconomics, the economy of knowledge. I thus attempt to demonstrate that noodynamics can be codified extremely simply, although not entirely, and will leave the many exceptions to the present paradigm to discussion.

From an economic and political perspective, the reason to practise nooconomics is much-too-obvious, and it is that knowledge, just like human stupidity one may add, is infinite. The argument of knowledge's intrinsic infinity may be constructed as follows, from simple though not easily demonstrated premises: let us at least consider that any knowledge of knowledge is another knowledge (ie. that knowing has no fixed point), then one can establish that the universe of all possible knowledge has a non-surjective injection within itself, and that it is thus infinite. This of course leaves the question of its boundedness open. A more elegant demonstration would be that the powerset of any set of knowledge is making for distinct knowledge and thus, that for any set of knowledge, a distinct powerset, which is mathematically

I.J. Aberkane (✉)
Unesco-Unitwin Complex Systems Digital Campus, 7 rue René Descartes, F-67000 Strasbourg, France

Centre de Recherche en Gestion, Ecole Polytechnique, Bâtiment Enta, 828 Boulevard des Maréchaux, 91762 Palaiseau Cedex, France

Kozmetsky Global Collaboratory, Stanford University, Kozmetsky Family Center, 224 Panama Street (Nora Suppes Hall), Suite 104, Stanford, CA 94305-4110, USA
e-mail: idriss.aberkane@polytechnique.edu

P. Bourgine et al. (eds.), *First Complex Systems Digital Campus World E-Conference 2015*, Springer Proceedings in Complexity, DOI 10.1007/978-3-319-45901-1_30

larger, can be constructed. The notion of serendipity, which I will use later as one of the fundamental principles of noodynamics, can also allow the construction of a sound argument that human knowledge is potentially infinite, as the possible serendipities of a knowledge set are calculated over elements of its powerset.

Thus still, from an economic and political point of view, knowledge may very well be the only potentially infinite resource that is offered to Humanity. We know very well that infinite growth is impossible if it is purely based on raw materials, which are finite, even when they are renewable, granted that we still consider time finite. But infinite growth is not impossible if it is knowledge-based. The question of global economic growth's current confrontation with the ecosphere's material finiteness can indeed be transcended by the question of better allocating our growth between material and immaterial origins. Besides, the ecosphere, much more than a source of raw materials, is also a source of knowledge in itself.

1 Basic Noodynamics

Probably the most basic property of knowledge exchanges is what we may call "Soudoplatoff's law", after its many formulations by Internet entrepreneur Serge Soudoplatoff when he worked for IBM in 1984. Although many scholars and philosophers must have been aware of it in the past, especially since it was well stated by Sufi master Amadou Hampâté Bâ and understated in one of the famous Mollah Nasruddin stories—and indeed there are several laws that were named after not their first, but their clearer presenters—Soudoplatoff has made, in my sense, the clearest formulation of it in recent years, and simpler for the layman than the economically consecrated "non-rivalry of non-material goods" of Lawrence Lessig.

Soudoplatoff's Law

> *when one shares a material good, one divides it. When one shares an immaterial good, one multiplies it.*

As we will see, this law may also be captured in one of the three laws of noodynamics that I outline here. Since knowledge is an immaterial good indeed, sharing it does not divide it, but rather multiplies it and also makes it evolve. Knowledge therefore may not be considered a thing but rather a process, in a manner comparable to the process of life. Its economics, also, is one of fluxes rather than one of stocks, in which ownership must clearly be redefined.

Soudoplatoff's law has many non-trivial implications. From a geopolitical perspective, one must admit that most of the conflicts that Humanity has been knowing have been a consequence of that sharing a material good implies its division. Be the good a piece of meat, a reserve of freshwater, an oil deposit or mere territory, the fact that sharing something material divides it has been an endless, and certainly the very most prolific source of conflicts. From the perspective of sustainable development also, Soudoplatoff's law very well underlines the economic and political interest of biomimicry: if one considers the ecosphere a source of raw materials, then indeed, anything that is taken from it is divided. If it is otherwise considered a source of knowledge, then anything that is taken from it is multiplied. Thus, the very political interest of biomimicry is indeed that it is a form of economic development, knowledge-based in nature that does not imply the destruction of nature but rather its protection.

Since nooconomics is an economy of fluxes rather than of stock, it is interesting to remark that the legal ownership of knowledge is very different from its practical ownership. One may own a library, even as the depositor or heir of its intellectual property, but one may not have read its books. Knowledge also could be defined in many ways: the tongue-in-cheek one of Idries Shah "knowledge is something you can use, belief is something that uses you", the classical one of Socrates and Plato, namely the intersection between belief and reality, the "KID" paradigm of Peter Drucker, that Information (I) is Data (D) endowed with relevance and purpose, and that converting D into I requires Knowledge (K), or finally we may place all these categories of the knowledge economy along a continuous differential spectrum beginning with data and ending with wisdom. The perspective on data can generate information, the perspective on information can generate knowledge and the perspective on knowledge can generate wisdom, which is not time-dependent, interestingly. Another simple way to separate between information and knowledge is that knowledge is reproducible while information is not.

In any case, just as it can be extremely difficult to define numbers *per se*, yet easy to establish a simple paradigm for arithmetics, we may not attempt to define knowledge, but merely the relations that its study encompasses. The study of these relations establishes noodynamics. Of course that noodynamics and nooconomics be so much inter-related makes them a very exciting paradigm, as if economics and thermodynamics had been plainly entangled from the very beginning of their conceptual inception.

It remains interesting to remark that information is essentially more perishable than knowledge, which regards the laws of the universe. Yet although knowledge does not really perish, it can be made obsolete by better, transcendent knowledge. Wisdom in turn, the perspective on knowledge, or in the classical definition, self-knowledge, is not even dependent upon space and time.

Another implication of Soudoplatoff's law is to be considered in models of bargaining power, such as the classical forces of Michael Porter, and in the behaviour of maximising agents in general. Indeed, maximising agents are to be expected to behave differently when bargaining over something material and something immaterial. What is to be expected of maximising agents in nooconomics? This is an interesting open scholarly question.

2 A Simple Paradigm for Nooconomics

Let us consider first at least two fundamental social properties of knowledge.

Property 1 Prolificity

Knowledge is prolific; it is growing exponentially

There is somehow a kind of biology to knowledge, and it surely has an exponential growth, which again from a Malthusian point of view is a very good thing. As a rule of thumb, we may consider the list of newly solved problems a reliable correlate of the "quantity" of knowledge (not its "quality" indeed, which may only be assessed *in futurum*, as one rarely knows in advance which knowledge will turn out revolutionary or crucial), and this list in itself has a doubling time of about seven to nine years. In The Technopolis Phenomenon, venture capitalist Regis McKenna already considered in 1991 that this doubling time, which is clearly dynamic in itself, was of 10 years. The doubling time of the quantity of scientific publications coming from the People's Republic of China is of five years, for example, although the list of truly solved problems is a much more accurate correlate of the knowledge mass than the latter.

Property 2 Collegiality

Humanity makes knowledge collegial, namely "truth is a shattered mirror" and everybody owns a bit. Since human beings—and especially more so regarding academics one must admit—have an ego, they display two very counter-productive tendencies with respect to optimal noodynamics. First human beings tend to consider "their little bit the whole to own", and second, they tend to resist the collectivisation of knowledge, which projects such as Wikipedia have still demonstrated to be of tremendous benefit to Humanity. Academic peer-review in this respect is also proving extremely immature in process and mindset, in that it should evolve into peer-improvement, rather than pass-or-fail admission/rejection, a process that belongs to a time when the limitation on publications was merely coming from that on available printing space and also from an era where hypertext and modification mark-up did not exist. An economic model properly rewarding peer-improvers remains to be developed, however.

Knowledge is a shattered mirror, of which everybody owns a bit, and it is in the most fundamental interests of nooconomists to favour the dynamic collectivisation of it. Noocollectivism, interestingly enough, is of course profoundly different from the collectivisation of material goods, once again owing to Soudoplatoff's law, and dogmatically speaking, me be more considered the result of *laissez-faire* than dirigisme.

Let us now establish at least three fundamental laws of noodynamics. From a social perspective two of them are excellent news, and one of them is bad news.

Law 1 Positive Sum

Knowledge exchanges are positive sum. This is a re-expression of Soudoplatoff's law: when I give away say 20 euros, I lose them. When I give away knowledge, I do not. Material exchanges are null sum, immaterial exchanges, such as knowledge exchanges, are positive sum. This, of course, is excellent news.

Law 2 Not instantaneous

Property exchanges may be considered instantaneous and, of course, scalable. It takes virtually the same time to transfer the ownership of 20 euros or 20 million euros, e.g. a signature, whether physical or electronic. Exchanges of legal ownership are thus virtually instantaneous so much so that the current legal limitation to their frequency, that of high-frequency trading, is the nanosecond. However, there is no high-frequency trading in nooconomics, because knowledge exchanges take time. It takes time to read this article, it cannot be acquired at high-frequency yet. Thus knowledge exchanges, unlike properties exchanges, are flows. This may be considered either a bad news (so far) or a fertile opportunity, that of the technology-driven increasing of the micro- and macro-knowledge flows, in which neuroergonomics is destined to play a major role.

Law 3 Superlinear

Property compositions may be linear in general. This is of course not true from a venture capitalist's point of view, in that an entrepreneurial project is more that the sum of its parts. But from a saver's point of view adding one kilograms of rice to another kilograms of rice is something linear, and makes two kilograms of rice. As long as processes are not involved thus, property composition is linear. Adding one thousand euros to an account already credited with one thousand euros makes two thousand euros. This does not apply to knowledge compositions, however, precisely in that knowledge is a process. The composition of two bits of knowledge systematically generates a third one, which is anywhere between trivial (but non-null) and revolutionary.

We may capture this property with the simple inequality

$$\mathrm{K}(\mathrm{A} \wedge \mathrm{B}) > \mathrm{K}(\mathrm{A}) \wedge \mathrm{K}(\mathrm{B}) \tag{1}$$

"knowing A and B together is more than knowing A and knowing B separately", which does not apply to "owning" from a saver's point of view.

Knowledge may thus be considered to "reproduce" in a way, and the difference between knowing two things separately and knowing them together may be called the "fertility" of knowledge, which is somehow comparable to the notion of entrepreneurial added value. This fertility accounts for the intrinsic prolificity of knowledge. Since knowledge is always action-oriented indeed, and not something to save in nature, its composition shares some close similarities with that of entrepreneurial means. We may also oversimplify our paradigm by calling the fertility of knowledge "serendipity", that is, the fortuitous creation of novel knowledge from the collision of several distinct bits of knowledge, although it remains unsure whether serendipity covers all of the fertility of knowledge. Is all the fertility of knowledge the result of serendipity? This question may prove very scientifically fertile in itself.

I know introduce what I consider to be the simplest possible non-trivial knowledge flow equation. It will at least apply to the knowledge flow of any reader of this article, or of any audio-visual or haptic content (such as the Braille writing), or of any Wikipedia article, for which one may also call this knowledge flow equation the "Wikipedia equation". Its construction is very intuitive and straightforward: what is it that we spend indeed in acquiring conceptual knowledge from a book, a video game or a course? Attention and time. Since these two currencies of nooconomics are crucial to each other and must be spent together to achieve any result, the amount of flowed knowledge is surely not proportional to their sum, but rather to their product, in that spending one hour and zero attention surely implies a null transfer, and spending all one's attention for less than even the subliminal perceptual threshold also implies a null flow. If one admits, along with Dehaene and others, that there are at least three qualitatively distinct levels of arousal, namely subconscious, preconscious and conscious (the possibility of more advanced states of consciousness is not at all negated), then one could formulate the following equation:

Equation 1
Simplest knowledge flow—from an economic point of view

$$\varphi\,(\mathrm{k}) \propto \mathrm{At} \tag{2}$$

"the amount of flowed knowledge follows the product of attention and time"

Some constants c1, c2 and c3 would then be left each to account for a different level of cortical arousal, so as to distinguish clearly subliminal learning from conscious learning. And since attention, just as numbers and knowledge, is hardly well-defined *per se* today but rather in its relation to other scientific objects such as consciousness (with which it may one day be fully united scientifically), one may, for the moment, simply scale the attention factor (A) from zero to one. The product of attention and time becomes the most essential currency of nooconomics. The unity of attention in the knowledge flow equation could be called "deciPosner" or "deciDehaene" in the manner of the decibel, but this choice will be left to the community.

The unity of the product At, as the fundamental currency of knowledge transfers, should also be given a simple name. In this article I will simply call them Ats, or @.[1] 1 @ is equal to one hour at full attention (whatever its range, subliminal, preconscious or conscious, them being defined by the constants used in the equation and not by the A variable), namely an attention of 1. How could we measure

[1]Two horizontal strokes may be added so as to indicate one @ is a currency.

attention from an economic point of view? Maximal attention, intuitively, is achieved when one is so much taken by what he is doing that he misses any external stimulus (eg. you are so absorbed by a book that you miss your subway station). Otherwise a correlated measure of foveation (eye-tracking) and cortical arousal could be an introduction to such a problem in cognitive neurosciences, but there are surely much simpler ways to extract empirical measures of attention. The key to their finding will surely be to distinguish between collective (macro), where the law of large numbers may apply, and individual (micro) attention, with the study of the mesoscopic scale a very exciting problem of nooconomics.

The knowledge flow equation has many interesting economic and political implications. One of them is bad news, and mostly all of the others are excellent news. We may understand them with the thought experiment of a "knowledge marketplace" (which the Internet has clearly become) as opposed to a "material marketplace". The difference between regular capital and knowledge capital (@) is that, with regular capital, if one visits a marketplace and buys nothing, one's purchasing power has remained essentially the same, which is not the case with knowledge capital, since one, visiting a knowledge marketplace yet buying nothing, would still have spent time and thus reduced one's purchasing power. Knowledge capital flies whether it is used or not, unlike regular capital, and this is the bad news. The economy of knowledge is not one of savings, but one of revenues.

One behavioural difference between a regular marketplace and a knowledge one is the counter-intuitive relation between envy and purchasing power as limiting agents in the purchasing act. In purchasing a commercial good, purchasing power is the main limitation, not envy, which marketing attempts to maximise for any good. In a knowledge marketplace, however, purchasing power is not quite the practical behavioural limitation, but envy is. When having to choose between say, a Wikipedia page on photonic molecules and something much more attention-enticing online, the average spender of @ may not easily choose the former, unless he or she is passionate about photonics molecules, and this is an interesting point of nooconomics.

The good news of course is that anybody is born with a certain, non-null quantity of @ in his or her lifetime. The knowledge economy is the only one in which everybody is naturally born with a non-null purchasing power, assuming an equal access to the knowledge marketplace, a simplification made all the more realistic by the advent of the Internet. Nooconomics is also, thus, a paradigm in which the prime limitation to the act of buying is not purchasing power but desire.

If @ is the unit of purchasing power in nooconomics, it becomes possible to establish a price tag for any symbolic knowledge online, from learning a new language to knowing how to cook traditional French food to handling algebraic topology. One could consider the global knowledge marketplace a global "app store" in which anybody can virtually download any app into their brain, granted they pay enough @. Interestingly, knowing in advance how many @ a certain knowledge costs can also be a tremendous incentive or motivator. Let us say that, before @s were consider, anybody wanting to learn something was like someone ordering a dish from a menu with no price tags. Also, one must note that in nooconomics, the unemployed intrinsically enjoys more purchasing power than those in work, having more @ to spend. This should be closely studied in the field of public policy, as it provides the basis for a straightforward conversion between poverty and wealth under certain conditions.

Since Beck and Davenport have elsewhere defined that attention is the "new currency of business", their is also a two-sided dimension to the spending of @ in a knowledge market, since receiving @ is convertible into money, and spending @ is convertible into knowledge. Buying knowledge, to a certain extent, is also creating regular capital, though a decaying one; this forms a fascinating bridge between classical economics and nooconomics. All in all, though, nooconomics seems just as different from classical economics then quantum mechanics was from classical physics.

If we consider maximising agents in nooconomics, we can also observe the interesting conditions under which individuals will maximise their knowledge flow. Under which circumstances do we maximise our expenditures of attention and time indeed? Undoubtedly, the supreme circumstance is love, and the one right under it is addiction (all love is addictive but not any addiction is love). Thus, nooconomics is the only economic paradigm maximising the purchasing power of those who are in love. As da Vinci clearly reminded, the origin of knowledge is love. And since addiction, right below it, is also a maximiser of knowledge flows, (video) games end up being remarkable "nooducts", namely aqueducts of knowledge, with a very high bandwidth. If knowledge is the new oil, surely (video) games are the new pipeline, and neuroergonomics a fascinating new fluid dynamics.

3 Conclusion

The fertility of nooconomics, as a paradigm, lies in the many original scientific problems it poses. One could already consider that the knowledge flow equation I have introduced does not capture any kind of flows. Indeed, it lacks a notion of synergy: what if the entrant knowledge resonates with already existing knowledge? Surely one should consider this case, in which the knowledge flow could exhibit a form of positive or negative feedback. Attention could either increase or decrease through time, but surely attention is in itself a function of time, which this initial equation does not consider. Thus, one could probably suggest the following equations instead:

Equation 2
Simplest knowledge flow with dynamic attention[2]

$$\varphi\,(\mathrm{k}) \propto \mathrm{A}\,(\mathrm{t}) \tag{3}$$

[2]In this case φ(k) represents a flow in the physical meaning of "instant quantity being transferred", not in the economic meaning of "total transferred quantity".

Equation 3
Simplest knowledge flow with synergy

$$\varphi\,(\mathrm{k}) \propto \mathrm{A}\,(\mathrm{t}) + \mathrm{Syn}\,(\mathrm{k}, \mathrm{t})\,. \tag{4}$$

where Syn (k,t) would represent the interference between entrant and preexisting knowledge at a given time (whether positive or negative). Also just as some enzymes may be "michaelian" or "non-michaelian", some knowledge flows may be simply captured by the non-synergic regime (Eq. 1), and others not. I suppose a more general theory of noodynamics could be eventually captured by equations in the form of:

Equation 4
Outline for a general theory of knowledge flows (general noodynamics)

$$\varphi\,(\mathrm{k}) \propto \mathrm{Res}\ \ (\mathrm{Sp}, \mathrm{Ev}) \tag{5}$$

where the flow of knowledge is proportional to the resonance (Res) between the spontaneous (Sp) and evoked (Ev) activities of the brain, or the learning system at large (thus, why not, considering the knowledge flow of such other cognitive systems as the immune system, for example, and not only the sentient brain). In such an equation, however, both the Res operator and the variables Sp. and Ev. remain to be defined.

Other interesting questions will regard the macroeconomics and political economy of knowledge flows. For example, if global knowledge has a dynamic doubling time, and that the individual knowledge flow is in the form of Eq. (1), what does it predict in terms of the macroscopic equilibria human groups (such as organisations or states) should achieve? Should states aim to equate their variation of the knowledge flow with the velocity of knowledge or with the size of the noosphere (global knowledge) itself? From a microeconomics perspective, what does the fundamental attention-knowledge transaction of individual agents imply (namely, any knowledge-giver is an attention receiver), especially in terms of cognitive and behavioural psychology? Could we eventually develop a working paradigm for mesoscopic nooconomics?

References

1. Acs ZJ, de Groot HLF, Nijkamp P (2002) The emergence of the knowledge economy: a regional perspective. Springer, Berlin
2. Adler PS (2001) Market, hierarchy, and trust: the knowledge economy and the future of capitalism. Org Sci 12:215–234
3. Amidon DM (2009) Innovation strategy for the knowledge economy. Taylor & Francis, Abingdon
4. Boden M, Miles I (2000) Services and the knowledge-based economy. Continuum, Abingdon

5. Brown P, Hesketh A, Williams S (2004) The mismanagement of talent: employability and jobs in the knowledge economy. Oxford University Press, Oxford
6. Burton-Jones A (2001) Knowledge capitalism: business, work, and learning in the new economy. OUP Catalogue
7. Carayannis EG, Kaloudis A, Mariussen A (2008) Diversity in the knowledge economy and society: heterogeneity, innovation and entrepreneurship. Edward Elgar Publishing, Inc., Cheltenham
8. Clarke T (2001) The knowledge economy. Edu Train 43:189–196
9. Cooke P (2001) Regional innovation systems, clusters, and the knowledge economy. Ind Corp Change 10:945–974
10. Cooke P, Piccaluga A (2006) Regional development in the knowledge economy. Taylor & Francis, Abingdon
11. Cunningham M (2006) Exploiting the knowledge economy: issues, applications and case studies. IOS Press, Amsterdam
12. Cunningham PM (2007) Expanding the knowledge economy: issues, applications, case studies. IOS Press, Amsterdam
13. Cunningham P, Cunningham M (2004) Eadoption and the knowledge economy: issues, applications, case studies. IOS Press, Amsterdam
14. Cunningham P, Cunningham M, Fatelnig P (2003) Building the knowledge economy: issues, applications, case studies. IOS Press, Amsterdam
15. Dahlman CJ, Aubert JE (2001) China and the knowledge economy: seizing the 21st century. World Bank, Washington
16. Dahlman CJ, Utz A, Institute WB (2005) India and the knowledge economy: leveraging strengths and opportunities. World Bank
17. Davenport TH, Beck JC (2013) The attention economy: understanding the new currency of business. Harvard Business Review Press, Watertown
18. Dehaene S, Changeux J-P, Naccache L, Sackur J, Sergent C (2006) Conscious, preconscious, and subliminal processing: a testable taxonomy. Trends Cogn Sci 10:204–211
19. Dehaene S, Izard V, Pica P, Spelke E (2006) Core knowledge of geometry in an Amazonian indigene group. Science 311:381–384
20. Dolfsma W, Soete L (2006) Understanding the dynamics of a knowledge economy. Edward Elgar Pûblishing, Inc., Cheltenham
21. Doz YL, Santos J, Williamson P (2001) From global to metanational: how companies win in the knowledge economy. Harvard Business School Press, Brighton
22. Drahos P, Braithwaite J (2002) Information feudalism: who owns the knowledge economy? Earthscan, London
23. Dunning JH (2002) Regions, globalization, and the knowledge-based economy. Oxford University Press, Oxford
24. Etzkowitz H, Leydesdorff LA (1997) Universities and the global knowledge economy
25. Formica P (2004) Strengthening the knowledge economy: essays on knowledge policy and international entrepreneurship. EffeElle Edition, Ferrara
26. Jaffe AB, Trajtenberg M (2002) Patents, citations, and innovations: a window on the knowledge economy. MIT Press, Cambridge
27. Jägers PH, Jansen W, Steenbakkers W (2012) New business models for the knowledge economy. Ashgate Publishing, Farnham
28. Julien PA (2008) A theory of local entrepreneurship in the knowledge economy. Edward Elgar Publishing, Inc., Cheltenham
29. Kenway J, Bullen E, Fahey J, Robb S (2006) Haunting the knowledge economy. Taylor & Francis, Oxford
30. Lengnick-Hall ML, Lengnick-Hall CA (2003) Human resource management in the knowledge economy: new challenges, new roles, new capabilities. Berrett-Koehler, Oakland
31. Leydesdorff L (2006) The knowledge-based economy: modeled, measured, simulated. Universal Publishers, Boca Raton
32. Luque E (2001) Whose knowledge (economy)? Soc Epistemol 15:187–200

33. Meier A (2012) eDemocracy & eGovernment: stages of a democratic knowledge society. Springer, Berlin
34. Neef D (1998) The knowledge economy. Butterworth-Heinemann, Oxford
35. Powell WW, Snellman K (2004) The knowledge economy. Annu Rev Sociol 30:199–220
36. Romer PM, Kurtzman J (2004) The knowledge economy. In: Handbook on knowledge management, vol 1. Springer, pp 73–87
37. Rooney D, Hearn G, Kastelle T (2012) Handbook on the knowledge economy. Edward Elgar Publishing, Inc., Cheltenham
38. Schienstock G (2004) Embracing the knowledge economy: the dynamic transformation of the Finnish innovation system. Edward Elgar Publishing, Inc., Cheltenham
39. Shattock M, International Institute for Educational Planning (2009) Entrepreneurialism in universities and the knowledge economy. McGraw-Hill Education, New York
40. Spelke E, Lee SA, Izard V (2010) Beyond core knowledge: natural geometry. Cogn Sci 34:863–884
41. Stewart J, Tansley C (2002) Training in the knowledge economy. Chartered Institute of Personnel and Development, London
42. Cross R, Parker A, Sasson L (2003) Networks in the knowledge economy. Oxford University Press, Oxford
43. Westeren KI (2012) Foundations of the knowledge economy: innovation, learning and clusters. Edward Elgar, Cheltenham
44. Westlund H (2006) Social capital in the knowledge economy: theory and empirics. Springer, Berlin

From Waste to Kwaste: On the Blue Economy in Terms of Knowledge Flow

Idriss J. Aberkane

1 Introduction

Currently, the Blue Economy may be the most sophisticated biologically inspired economic theory, and at the same time the most practical. It is a paradigm considering that waste-free production can be more profitable than current industrial production. Contrary to the so-called Green Economy, it does not consider that the solving of pollution will inevitably impede growth, but rather that there is a paradigm shift to transcend the conflict of interest between growth and sustainable development. This paradigm shift is fundamentally bio-inspired and synergistic, and may be summed up in the sentence

Axiom of the Blue Economy (1)

We should not ask nature to produce like our economy, we should ask our economy to produce like nature.

I.J. Aberkane (✉)
Unesco-Unitwin Complex Systems Digital Campus, 7 rue René Descartes, F-67000 Strasbourg, France

Centre de Recherche en Gestion, Ecole Polytechnique Bâtiment Enta, 828 Boulevard des Maréchaux, 91762 Palaiseau Cedex, France

Kozmetsky Global Collaboratory, Stanford University, Kozmetsky Family Center, 224 Panama Street (Nora Suppes Hall), Suite 104, Stanford, CA 94305-4110, USA
e-mail: idriss.aberkane@polytechnique.edu

P. Bourgine et al. (eds.), *First Complex Systems Digital Campus World E-Conference 2015*, Springer Proceedings in Complexity, DOI 10.1007/978-3-319-45901-1_31

2 Bio-Inspired Economics

Nature has poorly been studied as an economic system in its totality, rather than in the limited consideration of its tradable outputs (eg. fisheries), but were it investigated as such, its ingenuity and multiscale optimality would surely deserve several Nobel prizes, and expand the field of bioeconomics. In Nature, wastes are an exception, and they exist only at the micro-ecologic scale, for the ecosphere, globally, is in homeostasis, and does not accumulate any particular chemical. On the other hand, our economic systems are not in homeostasis, and do accumulate various products. The point is that for anything that Nature produces, there always ends up to be a demand.

Macroeconomic Optimality of Nature (2)

For anything that the ecosphere currently produces, there is a demand from the ecosphere itself.

Let us note, however, that this optimality is global (macroeconomic) for it does not always apply at the ecosystem level. It is also time-dependent, at equilibrium precisely; as we will see, the ecosphere may have been seen to have produced wastes in the past. In human systems, on the other hand, most value chains end up producing wastes, which are not an exception but the rule of our industrial production, and which accumulate globally. We are probably the only species to produce things that nobody wants. Wastes can indeed be defined as products for which there is a negative demand, products that one would accept only with a payment. Nuclear wastes clearly fall under this category, so do persistent organic pollutants (pop) or even, intriguingly enough, most of the world's output of used ground coffee today, although this is rapidly changing.

Macroeconomic suboptimality of Humanity (3)

Humanity produces things for which their is a negative demand, at both the microeconomic (local) and macroeconomic (global) levels.

If we were producing any of our goods and services in the circular way of nature indeed, there would be no waste. This observation has been made by many such as Ellen MacArthur, Janine Benyus, Fritjof Capra and of course Gunter Pauli.

Comparative Optimality of Nature and Humanity (4)

In Nature, macroeconomic optimality is the rule; in Humanity, it is the exception. Nature achieves economic optimality faster than Humanity, at any scale. Nature is the best representation of the Invisible Hand, before human-made economic systems.

Trial and Error v. Planification (5)

Man is more risk-advert than Nature. Nature's route to market optimisation is multiscale trial-and-error. Man's route to market optimisation is planification at the micro, meso and early macroscopic levels: individuals plan their decisions, enterprises plan their decisions, states plan their decisions, and so do central banks. Nature does not plan, it tries, regrets not the past, fears not the future. Man's fear of the future and regret of the past is the essence of Humanity's multi-level suboptimalities.

Sustainable Development is essentially a Man-made answer to Man-made problems. Without Humanity, there would be no sustainable development, for ecosystems are fundamentally converging to sustainability. Better, any ecosystem suboptimality is transformative in Nature. It could be so of Man-made markets, and this scientific ambition founds biologically inspired economics.

How about the kinetic and the thermodynamic of a transition between a linear and a circular (bio-inspired) economy, however? The contribution of Gunter Pauli regards both the equilibrium state, for which MacArthur has already well argued that it is in favour of the circular economy, and the activation energy or barrier to entry for a Blue Economy retrofitting. This article studies the latter: the micro-, meso- and macroeconomic interests of a conversion to the Blue Economy, and their cognitive barriers among others. It also studies both its possible stygmergies from the perspective of a "Technopolis Phenomenon": Silicon Valley indeed, the best known contemporary technopolis, self-organised through entrepreneurial stygmergies, that is, the phenomenon of constructive peer-inspiration. Can't a global conversion from linear to self-profitable circular economic go the same way, especially more so with the network effects allowed by both crowdsourcing and the Internet in general?

3 Why Blue Economy?

Pauli named the "Blue Economy" not only in reference to the oceans, but primarily in the idea that "blue is the new green", in the sense that the blue economy, unlike the green one, must be profitable in itself, because Nature is both a waste-free industrial system and still a cheaper, leaner, more effective one than ours. It is thus not ridiculous to consider that pollution-free production could be more competitive than the current production, without subsidies or even externality taxes.

If Pauli defined "blue" as simply "green 2.0" I may add another metaphor to understand the origin of this "blue" name. Has Nature produced waste in the past, or is its current state of globally waste-free ecosystem production an attribute of even the earliest ecosystems? Nature has indeed produced some wastes in the past, the most dreadful of all being... gaseous dioxygen. Dioxygen did provoke a relative extinction of biodiversity that was greater, by several orders of magnitude indeed, than the one that the very simultaneous release of absolutely all man-made pollutants would ever cause. The Great Oxygenation as it is referred to in geology was, in relative biodiversity, the greatest extinction of all the biosphere. Yet, it is thanks to oxygen that planet Earth is this "pale blue dot" we know of, and is green as well. If the very worst waste of natural history could be turned into a blessing, ours, even the most dreadful, are just trivial in comparison.

Transformative nature of wastes (6)

In Nature, wastes induce forward-leaning transformation. In Humanity, they induce backward-leaning conservation.

If we consider that human societies are antifragile (in the sense of Taleb 2012), then adding a constraint to them does not necessarily imply impeding their development; one could rather verify the opposite throughout history, an observation that was at the heart of Leibniz's optimistic theodicy. One such case of industrial anti-fragility was the abolition of slavery during the US Civil War: far from being a handicap to the Union, it ended up an economic blessing, because it accelerated the industrial adoption of the steam engine on the short term, and the advent of a consumer economy on the long term.

One usually attributes to Schopenhauer the three stages of the adoption of a revolution in Human history, be it scientific, philosophical, technological, political or moral:

"Schopenhauer's" Three Stages of Revolutionary Adoption (7)

First it is considered ridiculous, then it is considered dangerous, and then it is considered self-evident.

Gandhi defined a similar dynamic: "first they ignore you, then they laugh at you, then they fight you, then you win". That any revolutionary idea, such as the founding principle of quantum mechanics, the marginal reproduction of mature neurons or the existence of a cerebral lymphatic system, among many others must first be considered ridiculous and then dangerous also explains the well-known suboptimality of academic peer-review, which is but peer-pressure in essence, and thus fundamentally inimical to paradigm shifts, normative in nature.

The Blue Economy is surely crossing the three stages of revolutionary adoptions. If the abolition of slavery was a moral, industrial, political and intellectual challenge of the nineteenth century, and the abolition of apartheid of the twentieth century, the abolition of pollution should be the one of the twenty-first century. Probably the only way to achieve it is to demonstrate that this Pareto-optimum can also be a Nash equilibrium. What allows for such a dramatic phase transition is knowledge economy: assuming one demonstrates that the game of prosperity has much different rules than the ones we took for granted, one could establish a demonstration that pollution zero is not only Pareto-optimal but also a Nash equilibrium. I believe the Blue Economy is the most promising paradigm in this direction. It shows that prosperity is a game with dynamic rules, and I may formulate it as the simplest possible equation:

Simplest Formulation of the Blue Economy (8)

$$\textit{waste} + \textit{knowledge} = \textit{asset}.$$

If in any economic game, holding waste were equivalent to a loss, holding waste plus the adequate knowledge could be a win. The knowledge flow can dramatically alter the dynamic of economic games regarding wastes, at the microeconomic level of individuals or small groups holding wastes, at the mesoeconomic level of enterprises, for which the knowledge flow could alter their accounting in the non-linear way of turning liabilities into assets, and finally at the macroeconomic level, where states could consider wastes as an opportunity for growth.

4 In Terms of Knowledge Flow

What if sustainable development was a proper subfield of the knowledge economy? One could define sustainable development as the art of not trivially wasting resources today, that could be better used tomorrow. Biomimicry is thus a perfect example of sustainable development, as it consists of considering Nature not a source of raw material but a source of knowledge. The Blue Economy posits, in the same way, a certain "knowledge panacea doctrine". There is surely more than a synergy between nooconomics (the knowledge economy) and the Blue Economy, the latter is rather an industrial and managerial subfield of the former.

One should first consider the Blue Economy in microeconomic terms. I have reminded of the conservative cognitive bias towards the adoption of revolutions, and it has a decisive impact on the individual knowledge flow. If we were to consider industrial decision-makers as purely rational agents of course, we would expect them to convert all the useful knowledge available to them into action, and in particular that all the knowledge relevant to profitable waste transformation be used to that end. Individual behaviour is not that rational, however. First, for a given knowledge to be adopted, it needs to resonate with the existing paradigm of the receiver, otherwise, no matter how well it is demonstrated, it will be rejected.

Of the most significant cognitive biases I should underline in the study of the Blue Economy in terms of knowledge flow is the bias of conformity. When faced with the dilemma of adopting the truth and leaving one's comfort zone or staying within one's comfort zone and rejecting the truth, most human beings will choose the later. An evolutionary reason that is often invoked to explain this phenomenon is that we, contemporary humans, are mostly the descendent of the weak-minded humans who survived by preferring to remain in their tribe and reject the truth every time they were faced with the dilemma. The others have died. For, between leaving the group and adopting the truth and leaving the truth and remaining in the group, be it in the savannah or during the Ice Age, the latter meant survival, and the former quasi-certain death. Although the adoption of disruptive knowledge is not really equivalent to death anymore, our brain has evolved to discourage it as much as possible, as long as it goes against in-group favouritism. Hence the following strong proposition that the Blue Economy should be "won in the middle":

$$\text{Mesoscopic synergy of the Blue Economy} \tag{9}$$

The adoption dynamic of the Blue Economy, in terms of knowledge flow, does not favour the microscopic but rather the mesoscopic scale. The Blue Economy will be all the more successful as it targets neither individuals (micro) nor nations or even cities (macro) but groups and groups of groups (meso).

It is a testable hypothesis that the adoption dynamic of the Blue Economy should give a premium to the mesoscopic scale, for it is the scale in which the all-or-none mechanism for the adoption of disruptive knowledge could prove the least disruptive. Groups and groups of groups—not individuals—should be the targets.

If knowledge holders are to be modelled as biased maximising agents, as in regular economics, then one should study the social conditions under which they will be *willing* to maximise their knowledge flow. For in nooconomics (the economy of knowledge), the willingness to acquire knowledge is the very first bottleneck of knowledge logistics (applied noodynamics). From a macroeconomic perspective, one should observe that statement (8) implies that economies producing a lot of waste and at the same time a lot of knowledge could most benefit from the Blue Economy in terms of GDP. This alone, of course, could explain the political interest of China for the field. The general statement is that the best way to manage a waste is not to bury it deep in either dirt or water, but in knowledge.

At the transition between the meso- and macroeconomics of the synergy between waste and knowledge many business models could be developed as, for example, lines of services in the accounting and auditing businesses, because if statement (8) becomes a standard industrial observation, then knowledge and externalities should be accounted for in a completely different manner in financial terms. Maybe generally accepted accounting practices could also very well benefit from a retrofitting in the Blue Economy. From the high mesoscopic point of view of multinational firms, this would be a novel, readily available through untapped well of profitability, and boasting a genuinely operational "Blue" label (which could become the trade of ad hoc rating agencies) could end up a critical success factor for any publicly traded company. This is exactly the kind of mesoscopic effect that the Blue Economy should seek. From the lowest macroeconomic level then, that of national or regional administrations, one could observe that such a behaviour would encourage corporations to internalise their externalities as systematically as possible. Why externalise wastes indeed if these liabilities, once blended with the appropriate knowledge, end up being assets? These possible mesoscopic behaviours seem rather intuitive, but they are merely the outline of the more thorough economic model one should develop to systematise the synergy between nooconomics and industrial ecology.

5 Conclusion: In Terms of Stygmergies

Peer-pressure is a normative phenomenon. One of its emerging effects is not only social inertia, but also social stabilisation. Thus at the beginning of a revolution, peer-pressure is what encourages an idea to be considered ridiculous and dangerous, but then it is also what helps revolutions to move towards self-evidence. There is a tipping point between conservative peer-pressure and innovative peer-pressure.

A stigmergy is the self-organisation of a synergy, for example, an optimal collective pathway, usually at the mesoscopic scale (that of groups and groups of groups) through various forms of reinforcements. The self-organised stabilisation of the shortest pathway through pheromone signalling in anthills is the textbook example of a stigmergy. Economic systems typically display stygmergies as well, be they destructive (like a bank run), constructive, or neutral. The history of Silicon Valley was typically one of stygmergies, based on constructive peer-pressure, e.g. "if I succeeded, why would you not". The opposite, of course, destructive peer-pressure would be "if I failed, why would you succeed?".

The dynamic adoption of the Blue Economy poses the question of its entrepreneurial stygmergies, especially at the mesoscopic scale (that of small and medium enterprises). If the emergence of Silicon Valley was a typical "technopolis phenomenon" by which the enthusiasm of novel project holders self-organised along the fresh trails of earlier successful entrepreneurs, could there be an identical "Blue Technopolis Phenomenon"? How would the man-hill of blue entrepreneurs self-organise? Would it reach homeostasis, expand—by converting new entrepreneurs—or collapse?

Here I opinionated how the knowledge flow can alter the meso-economic dynamic of Man's interaction with wastes, along the general idea that, in the twenty-first century, Man will learn to bury not their wastes in either dirt or water but in knowledge. If waste + knowledge = asset (proposition 8), then the entire interest of the Blue Economy is to catalyse the conversion of wastes into "kwastes", that is, appropriately paired bits of wastes and bits of knowledge. Surely crowd-sourcing could benefit such a global catalysis, especially considering that human knowledge is intrinsically collegial. The crowdsourcing of the collectivisation of knowledge (Wikipedia being the clearest example of it) can encourage stygmergies and the emergence of a common sense. Should an equivalent "kwaste" platform be established, to maximise the interaction between wastes and knowledge (and appropriately called "Wikiwaste")? Nature's way to deal with wastes is to surround them with trial-and-error. Man's way of dealing with wastes should be to surround them with knowledge. As the No Free Lunch theorem establishes that there is no more efficient optimisation strategy than a random search over all types of optimisation problems, maybe trial-and-error is actually the very best way for the ecosphere to incorporate novel molecules into its many cycles. Should it also be that of Man?

References

1. Benton D, Hazell J, Hill J (2015) The guide to the circular economy: capturing value and managing material risk. DoSustainability, Bradford
2. Benyus JM (2009) Biomimicry. HarperCollins, New York

3. Benyus JM, Barberis JC (1992) The secret language and remarkable behavior of animals. Black Dog & Leventhal Publishers, New York
4. Benyus J, Pauli G (2009) Nature's 100 best: world-changing innovations inspired by nature. Chelsea Green Publishing Company, White River Junction
5. Capra F, Pauli G (1996) The challenge. F Capra and G Pauli, New Delhi, Response
6. Capra F, Pauli GA (1995) Steering business toward sustainability. United Nations University Press, Tokyo
7. Clark CW (2010) Mathematical bioeconomics: the mathematics of conservation. Wiley, Hoboken
8. Corning P (2010) Holistic Darwinism: synergy, cybernetics, and the bioeconomics of evolution. University of Chicago Press, Chicago
9. Ellen MacArthur Foundation (2012) Towards the circular economy. Ellen MacArthur Foundation, Cowes
10. Georgescu-Roegen N, Mayumi K, Gowdy JM (1999) Bioeconomics and sustainability: essays in honor of Nicholas Georgescu-Roegen. Edward Elgar, Cheltenham
11. Keller RP, Lodge DM, Lewis MA (2009) Bioeconomics of invasive species: integrating ecology, economics, policy, and management. Oxford University Press, Oxford
12. Kopnina H, Blewitt J (2014) Sustainable business: key issues. Taylor & Francis, Abingdon
13. Lovins A, Braungart M (2014) A new dynamic - effective business in a circular economy. Ellen MacArthur Foundation Publishing
14. Mathews JA, Tan H (2011) Progress toward a circular economy in China. J Ind Ecol 15:435–457
15. Moxnes E (1998) Not only the tragedy of the commons: misperceptions of bioeconomics. Manage Sci 44:1234–1248
16. Nabudere PDW (2013) From agriculture to agricology: towards a glocal circular economy. STE Publishers, Johannesburg
17. Ning D (2001) Cleaner production, eco-industry and circular economy [J]. Res Environ Sci 6:000
18. Pauli G (1995) Industrial clusters of the twenty-first century. In: Steering business toward sustainability. United Nations University Press, Tokyo
19. Pauli G (1996) Breakthroughs: what business can offer society. Epsilon Press Limited, London
20. Pauli G (1997) Zero emissions: the ultimate goal of cleaner production. J Cleaner Prod 5:109–113
21. Pauli G (1998) Technology forecasting and assessment: the case of zero emissions. Technol Forecast Soc Change 58:53–62
22. Pauli GA (1987) Services: the driving force of the economy. Pergamon, Oxford
23. Pauli GA (1998) Upsizing. Greenleaf, Saltaire
24. Pauli GA (2010) The blue economy: 10 years, 100 innovations, 100 million jobs. Paradigm Publications, Boulder
25. Rougerie J, Civard-Racinais A, Fuchs A (2010) De Vingt mille lieues sous les mers à SeaOrbiter. Democratic Books, Paris
26. Slavid R (2009) Extreme architecture. Laurence King, London
27. Yuan Z, Bi J, Moriguichi Y (2006) The circular economy: a new development strategy in China. J Ind Ecol 10:4–8
28. Zhou Q (2011) Applied economics, business and development: international symposium, ISAEBD 2011, Dalian, China, August 6–7, 2011, Proceedings. Springer, Berlin

Exploring the Synergy Between Motorists and Motorcyclists in Urban Mobilization

Juan C. Correa

1 Looking for the Synergy of Road-Users in Urban Traffic

Recent evidences show some reasons that explain why inhabitants of highly dense cities prefer the use of "risky" means of transportation (motorcycles) instead of "safe" vehicles [1]. Economic factors such as the low-prices for acquiring and maintaining a motorcycle are obvious advantages. Yet, other non-trivial factors are relevant for road-users who face urban traffic congestions on a daily basis. The practice of "motorcycle lane-sharing" occurs when motorcyclists ride in between the lanes of stopped or slow-moving vehicles, despite the fact that this practice might jeopardize traffic safety on the roads. This practice of mobilization, induced by density and economic factors, presents us a behavioral interaction between motorcyclists and motorists (i.e., drivers of all kind of four-wheeled vehicles) who act as elementary components of a phenomenon which has its own peculiarities at both the macroscopic and the microscopic scale.

The initial approaches of traffic provided us a set of tools to analyze the relationship between its macroscopic variables (i.e., speed, flux, and density) to deal with traffic management issues. However, other complementary approaches have been developed to offer us a viewpoint which enables us to understand why road capacity can be benefited from the synergistic effects of motorcycle lane-sharing. These approaches are intended to understand the dynamics of traffic from the behavioral interaction that takes place when motorists share their lane with motorcyclists [2].

It has been noticed, for instance, that the speed of motorcyclists is systematically higher than the speed of motorists and their maximum flow rates and critical speeds decline with an increase of maximum speed deviations [3]. It has also been observed

J.C. Correa (✉)
Fundación Universitaria Konrad Lorenz, Bogotá, Colombia
e-mail: juanc.correan@konradlorenz.edu.co

P. Bourgine et al. (eds.), *First Complex Systems Digital Campus World E-Conference 2015*, Springer Proceedings in Complexity, DOI 10.1007/978-3-319-45901-1_32

that the presence of motorcycles enlarged the capacity of the road, due to the fact that motorcyclists are able to progress by filtering when the movements of cars are constrained by vehicular density. This behavior, however, can be affected as the number of motorcycles increase; so the introduction of more motorcycles will inevitably impair the cars' movement, producing a lower traffic flow in certain conditions [4].

The complexity of traffic and the intrinsic relationships between its macroscopic and microscopic properties can be hard to tackle in natural settings due to the amount of instrumentation that is required for data gathering purposes. Fortunately, nowadays we have several tools that can be applied to overcome some of these difficulties. Computerized simulations in general and agent-based simulations in particular are just a couple of these tools especially suited to study urban traffic as a complex system [5]. But, how can we observe the synergy in urban traffic by using these tools? What are the variables that we need to collect and analyze in order to detect such synergy?

1.1 The Elements of Synergy in Urban Traffic

Synergy, roughly understood as the creation of a whole that is greater than the simple sum of its parts, is a ubiquitous phenomenon which has been observed in several fields like particle physics, chemistry, genetics, ethology, and psychology, just to mention but a few [6]. In urban traffic, however, it appears to be a subtle phenomenon closely related to the aforementioned practice of motorcycle lane-sharing. As part of my doctoral thesis [1], I designed an agent-based simulation to examine the quantitative effects of motorcycle lane-sharing on motorists' behavior in terms of their average speed and the frequency of acceleration/deceleration and lane-changing maneuvers. The simulation experiment consisted in comparing these metrics with and without the presence of motorcyclists on a virtual double-lane road created in a multi-agent simulation platform. In the simulation, motorcyclists choose a lane to share, they never attempt to change their chosen lane and drive in it at a speed which is higher than those of the motorists. The simulations showed that under these specific circumstances the motorists get almost confined to their lane, and were forced to increase up to four times their acceleration/deceleration maneuvers without significantly affecting their average speed. These results, although very specific for motorists, say little about the synergy in urban traffic. But the view of the whole system changes when the analysis focuses on examining what happens with the simultaneous mixed traffic of motorists and motorcyclists; also known as "heterogeneous traffic."

1.2 Heterogeneous Traffic Flow as an Indicator of Synergy

The heterogeneous traffic is characterized by the interaction of motorcycles and four-wheeled vehicles driving without lane discipline. The absence of lane discipline consists in vehicular movement that is influenced by the presence of vehicles in the front as well as on the sides. This led to a complex traffic behavior and it cannot be analyzed by using conventional traffic variables such as time headway and space headway (i.e., the "bumper-to-bumper" distance that separates two vehicles). Consequently, it is desirable to have a metric that can represent the heterogeneous nature of traffic. Since varying vehicle dimensions and road-users' behaviors are attributed to the lack of lane discipline, a modified measure of occupancy termed as "area occupancy" has been recently proposed [7].

Area occupancy is based on the concept of "occupancy" which is defined as the percentage time any arbitrary road section is occupied by a vehicle over a given period of time. The calculus of area occupancy is as follows:

$$\rho_A = \frac{\sum_{i=1}^{N} O_i \times w_i \times d}{T \times W \times d} \tag{1}$$

where ρ_A is area occupancy; O_i is occupancy time of the ith vehicle in seconds; w_i is the width of the ith vehicle; W is road width; d is the length of the road section under consideration; and T is observed time period in seconds. In Eq. (1), the numerator value takes care of the occupancy time (similar to occupancy); the amount of time a vehicle with a given area is spending on the road section under consideration. Its value will depend on the traffic composition and speeds of the vehicles. It must be noticed that the practice of motorcycle lane-sharing allows the motorcyclists to have a faster occupancy time due to the possibility of percolating through stopped or slow-moving vehicles. Thus, the purpose of measuring heterogeneous traffic flow presents us an interesting opportunity to examine the synergy that takes place in urban traffic when motorists and motorcyclists interact on the road.

Some years ago Indian researchers tried to estimate the heterogeneous traffic flow by using a sophisticated video analysis software to collect traffic volumes and composition by automatically identifying types of vehicles on the road [8]. The data was collected on the National Highway-1 (NH-1), connecting New Delhi and Amritsar. In this setting they employed two reference lines (similar to the speed trap) the time instances at which the vehicles of all types (motorcyclists, cars, trucks, busses, etc.) were entering and leaving the road section were noted down. By considering these time instances and length of the road section, the speeds and occupancy times were calculated for all the vehicles during a period of 30 min. Their results showed minimum fluctuations of speed. Yet, the interesting part of their results was associated with the observed differences between the occupancy and the area occupancy. It turns out that apart from speed, ρ depends on the vehicle length and ρ_A depends on the vehicle area. When more short vehicles with low vehicle area or vehicles with large areas were traveling on the roads, there shall be some difference between the values of ρ and ρ_A.

Since ρ_A is encompassing area of the vehicle and its speed under prevailing traffic conditions, its value is different from that of ρ. The study of area occupancy gives an insight into the behavior of heterogeneous traffic. The results obtained from this study have major ramifications on concepts of roadway capacity and level of service under real heterogeneous traffic conditions. The concept of area occupancy can also be used as a measure of performance in converting heterogeneous traffic stream into an equivalent homogeneous traffic stream, which is easily analyzed by observing the fundamental diagrams that depict the relationship among speed, flow, and density. However, this measure needs further validation under various traffic regimes (i.e., free-flow, congested, etc.). Since vehicle dimensions are very crucial in realizing the area occupancy, there is a lot of work to be done in this direction. Another remarkable effort was done by Lee [9] who compared the effects of traffic composition on the overall traffic. As part of his doctoral thesis he developed an agent-based model to simulate the heterogeneous traffic with different proportions of motorcycles on a virtual road of 300 m long and 10 m wide with three lanes in one single direction. In particular, his study compared the density-flow relationship in five different scenarios (i.e., 100 % passenger car flow; 25 % motorcycle flow; 50 % motorcycle flow; 75 % motorcycle flow, and 100 % motorcycle flow). Among the results it was reported that the worst traffic corresponded to a 100 % passenger car flow, while the best one corresponded to the scenario of 100 % motorcycles and the different compositions occupied an intermediate position between these two "homogeneous traffics." These results are consistent with those that indicate the benefits of splitting the heterogeneous traffic by creating a dedicated motorcycle lane [9]. Clearly then, the interaction between motorists and motorcyclists produces a traffic flow synergy that can be exploited in natural settings.

2 Final Comments

In this paper I summarized the experimental results of three different studies that show the possibilities to observe and model the heterogeneous traffic and the synergy that takes place on the road when motorcyclists and motorists interact. This synergy consists in the rise of the road capacity and the overall traffic flow, due to the fact that the motorcycle can exploit the road space by filtering through a slow-moving flow using the clearance between two parallel cars or the possibility to weave in and out of a stationary flow via the safety margins between vehicles. The aforementioned results required a considerable amount of work to model the behavior of the heterogeneous traffic and its corresponding validation with field observations. These incipient efforts were intended to observe, analyze, and simulate some complexities of the heterogeneous traffic that still require further studies to eventually improve the way we deal with real traffic management.

In computer simulations we have seen that the overall traffic flow improves when the traffic composition is not completely homogeneous. Thus, a fruitful exploration in the near future should take into account when is convenient (for the whole traffic system) to restrict the practice of motorcycle lane-sharing. Another fertile venue in studying the synergy of motorists and motorcyclists is to examine under what specific conditions works this synergy; namely to examine the effects of lane-width, number of lanes, etc.).

References

1. Correa JC (2015) Simulaciones y observaciones de la filtración de motos en el tránsito heterogéneo. PhD Thesis, Universidad Simón Bolívar
2. Mallikarjuna C, Ramachandra Rao K, Naga Venkata SKS (2010) Analysis of microscopic data under heterogeneous traffic conditions. Transport 25:262–268
3. Lan LW, Chang C-W (2005) Inhomogeneous cellular automata modeling for mixed traffic with cars and motorcycles. J Adv Transp 39:323–349
4. Lan LW, Chiou Y-C, Lin Z-S, Hsu C-C (2010) Cellular automaton simulations for mixed traffic with erratic motorcycles' behaviours. Phys A Stat Mech Appl 389:2077–2089
5. Bazzan ALC (2007) Traffic as a complex system: four challenges for computer science and engineering. In: Proceedings of the XXXIV SEMISH, Citeseer, pp 2128–2142
6. Corning PA (2003) Nature's magic: synergy in evolution and the fate of humankind. Cambridge University Press, New York
7. Mallikarjuna C, Rao KR (2011) Heterogeneous traffic flow modelling: a complete methodology. Transportmetrica 7:321–345
8. Mallikarjuna C, Rao KR (2006) Area occupancy characteristics of heterogeneous traffic. Transportmetrica 2:223–236
9. Lee T-C (2007) An agent-based model to simulate motorcycle behaviour in mixed traffic flow. PhD Thesis, Imperial College

Synergetic Economics. Scientifical Approaches of Synergistic Economic Networks and Systems

Laura Melinda Stan

1 Introduction

The present competition between the active entities of the socio-economic environment (in the context of shortages of resources, competencies, and required knowledge) is governed by the imperative of the deep collaborative efforts' increasing efficiency. For the past 10 years, the use of the micro- and macrosynergism's specific prefixes (*co-:* cooperation, coordination, coworking, co-opting and *con-:* conglomeration, conglobing, etc.) has been expanding in literature. On these lines, the intrinsic issue with literature became the understanding of the meaning of synergy [1] and the understanding of the way in which this phenomenon can amplify the competitive success of networks and systems represented at different levels.

Synergetics is still treated as a new science, unconventionally applied in Economics; furthermore, the statistics continue to be extremely vague in this phenomenon's highlighting. Synergetics (and implicitly Synergetics studies applied in Economics) is focused on the study of collaborative behavior of entities of different natures and on the study of the way in which these ones manage their complexity in order to increase their value. On these lines, *synergy* is assigned to proactive and super efficiency-oriented entities with strategic collaborative excellence; synergy is a property of all entities that turn to combined increased efforts, with multiplicative effects, through reinvented and repositioned[1] tools.

[1]În the maximum efficiency positions, with maximum utility for network or system.

L.M. Stan (✉)
Faculty of Economics and Business Administration, West University of Timişoara, Timişoara, Romania
e-mail: stan.melinda@gmail.com

P. Bourgine et al. (eds.), *First Complex Systems Digital Campus World E-Conference 2015*, Springer Proceedings in Complexity, DOI 10.1007/978-3-319-45901-1_33

Forasmuch as synergy emerges through extremely varied dynamic frames and contexts, Synergetic Economics accordingly is an interdisciplinary science, in flux, as the economic and social entities become more open, more dynamic, and interconnected with the elements of the environment in which they operate.[2] This branch of Synergetics has in the foreground the possibility of creating added value through comprehensive collaborative relationships—in the collaborative social systems' case as well as in the economic or other types of systems' case.

2 The Evolution of Synergetics Studies in the Economic Field

From the etymological point of view the "synergy" concept was used for the first time around the seventeenth century, originating from the Greek "synergos" term which defines the action of working together or of cooperate. "If during this period the synergy concept was mainly used in the theological field (describing 'the cooperation of human effort with divine will'), in the nineteenth and twentieth century, 'synergy' was promoted in physics and biochemistry" [2].

In economy, "synergy" was approached for the first time in the 1960s–1970s, within the studies of open economic systems, in the context in which the transition towards open-natural systems was noted. But the concept has been sluggishly integrated in management and economy; as a result, the verb "*to synergize*"[3] was introduced in English only in 2009–2010 (in the USA) [3]—concurrently with the amplification of the synergy phenomenon in the global economic crisis' context.[4] Researching the systematic "soundings" of the synergy concepts, I identified at least six evolutionary phases for the study of active economic entities' synergy (see Fig. 1):

I. *The precursory phase of Synergetics development as an interdisciplinary science applicable (also) in Economics (1963–1970).* R. Buckminster Fuller (futurist and systems theorist) approaches in an innovative way the open

[2]See the transition from *closed-rational systems,* with rigorous collaborative behaviors (1900–1930) to *closed-natural (social) systems*—theory Y of integration—(1930–1960), to *open-rational systems,* based on the interfering of internal systems' components/mechanisms with those of the environment—according to the contingency theory—(1960–1970), and to *open-natural (social) systems*—according to the organization theory, the institutional theory, the social ecological theory, and the resource dependence theory—the specific actions of these entities (mergers, joint ventures, alliances, co-options, diversifications, political involvements, etc.), being specific elements of synergistic structures (1970–present).

[3]Which means acting with/through synergy, manifesting synergy, cooperating with others to remedy something, and determining synergistic actions.

[4]When the added value's creation through comprehensive collaborative efforts has become an imperative for the active economic and social entities and the emergence of negative synergies required a better knowledge of the value creation chain's mechanisms.

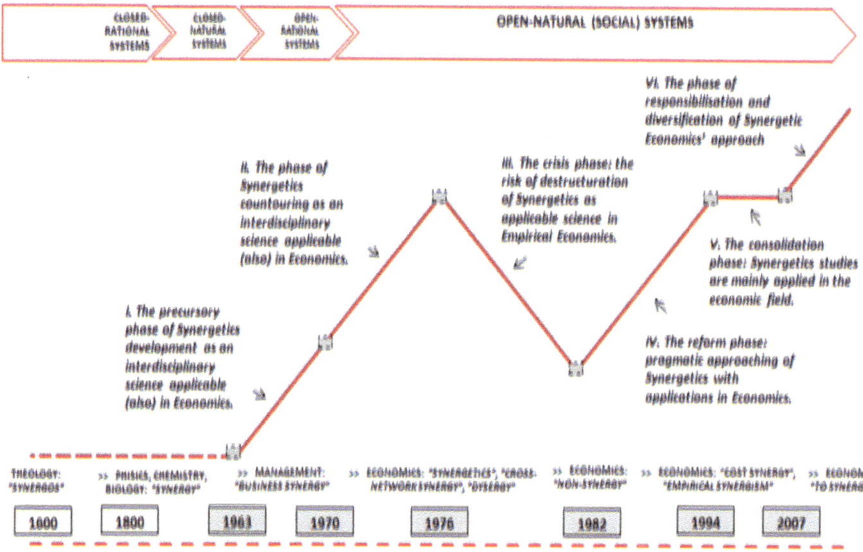

Fig. 1 Representation of the evolution of Synergetics studies in the economic field

systems, from the perspective of the fact that these entities—as a whole—constitute results of synergy effects insured by the constituent subsystems' deep pulling together [4]. Therefore, he will get to lay the foundations of a future new science, the Synergetics (unframed in the respective period), paying attention to its multiple implications in the field of exact and social sciences; as a result, Igor Ansoff, the father of strategic management, introduces synergy for the first time in the study of the open economic systems (1965), but approaching the concept (even in the next two decades) only from the management perspective—at business level and rarely at industries level [5–7].

II. *The phase of Synergetics contouring as an interdisciplinary science applicable (also) in Economics (1971–1976).* As a result of the introduction in the USA of the synergy concept in the study of Economics, Synergetics began to be treated as an interdisciplinary science, correlated with the economic life from the *ephemeralization* perspective, ≪doing more and more with less and less resources≫, especially through the technological advance's ability [8] (therefore creating exponentially abounding and diversified goods and services). Thus, new correlated concepts come to light, such as technological convergence or information flows between the component subsystems of an economic entity [9]. Due to the diversification of approaching, in Great Britain, A. Coulter comes to conclude that "synergy is *of universal applicability*. It can be applied not only to human mind, but also to human groups, to organizations of all kind, to industrial enterprises, *to entire economic systems*" [10].

III. *The crisis phase: the risk of destructuration of Synergetics as applicable science in Empirical Economics (1977–1982).* As the oil crisis' domino effect has affected the entire Western economic system at the end of the 1970s, more and more caterwauls appeared against Synergetics as applicable science in Economics. At micro- and macro-level repeated failures in shaping synergy into practice were noticed, seeming that sophistries took the place of synergism applicability. One of the reasons was the fact that synergy *became an abused concept* in the economic speech and, consequently, misinterpreted. It was believed that through the simple combination of capacities and efforts, synergy is obtained anytime/anyway, without taking into account the redundancies/false synergies/dysergies/non-existent collaborative strategies. "By the late 1970s, however, enthusiasm for synergy had waned. It seemed that synergy was a nice idea, but it rarely occurred in practice (. . .). Firms often used it only to justify the actions undertaken for other reasons" [11] (particularly with the aim of quickly attracting investments), which affected the adequate understanding and exploitation of synergistic potential.

IV. *The reform phase: pragmatic approaching of Synergetics with applications in Economics (1983–1994).* Despite the synergy's description in pretty vague terms in the economic and management literature, some entities steeply climbed in the global competitive hierarchy in the late 1970s and in the 1980s, benefiting from unexpected generous results, synergy producing exponential effects beyond the handy elements [12]; practically, these entities' (such as Microsoft Corporation, Apple Inc., the Japan's electronics industry, or the fax industry from Taiwan) activities were based on the principle of complementarity, innovation, and continuous transfer of know-how among units with different functions; furthermore, the receptivity for the external environmental opportunities and the propagation of cross-functional support have determined the detaching of these synergistic entities from the classic economic competitors. As a result, Synergetics "reinvented" itself, becoming in a renewable way a systematic knowledge assembly regarding *synergy based on empiricism* (see the works of R. Buckminster Fuller, W. Weidlich, G. Haag, P.A. Corning, M. Porter etc. from the 1980s–1990s).

V. *The consolidation phase: Synergetics studies are mainly applied in the economic field (1995–2006).* As a result of the review of Synergetics approach, generating added value through the profound combination of efforts, know-how or collaborative networks began to be studied mainly in economic and management fields; synergy's definitions were renewed in the Western dictionaries, the first explanation criterion of synergy phenomenon being the economic and management criterion [13, 14]. Between 1995 and 2006, the great part of the books published in the world on the Synergetics' theme, approached also its economic component, focusing on: (*a*) synergy resulted in the context of mergers, acquisitions, and alliances [15]; (*b*) organizational cultures explained through economic networks' synergizing [16] (that requires creative, strategic, and transformational thinking [17]); (*c*) cost reductions generated through the identification and optimal exploitation of the active economic system's interdependent resources [18]; (*d*) exploitation of collaborative networks' synergistic potential determining the creation of *a new level of economic performance* [19]; (*e*) spatial synergistic (urban) networks' development, which determined the regional/national clusters' consolidation [21], etc.

VI. *The phase of responsibilization and diversification of Synergetic Economics' approach (2007–present).* In the context in which a series of economic entities proved to be deeply affected by the global economic crisis, synergeticists began to be mainly concerned with (1) the diagnosis of economic systems which failed in the extended competitive environment, and (2) the way in which the synergistic networks are and should be generated and monitored in the crisis affected economic environment. These researches led to the rise of the era of synergistic potential's exploitation strategies [22], Synergetics being regarded as *one of the primary solutions for the exit from the <created chaos>*. The American researches drew attention to the fact that "synergy is not a unitary phenomenon" [23], there being *a variety of applications* of synergistic functions and possible synergy effects: positive and negative synergies, maximal and minimal synergies, etc. In Europe, the initiated studies *correlated* Synergetics *with cognitive economy* [24]; furthermore, they have laid the foundation for the horizontal and vertical synergistic networks theory, the vertical networks (based on the functions and capacities' complementarity) being the most complex ones, such as the type of conglomerates or polycentric urban regions, like the Ranstad region from the Netherlands [25]. The passage from an economic system's units' decentralization towards reconnecting dispersed units and capacities through consolidating geographic synergistic networks can be noticed, the tendency being to promote the most concentrated collaborative processes.

Related to the systems' failures in controlling the crisis' negative effects, J.A. Lybeck drew attention to the lack of the synergistic infrastructure's control. For example, the 50 State departments of the US financial institutions had separate or superposed jurisdictions during the outbreak of the economic crisis in 2007, this situation leading to the insurance companies' Federal non-surveillance [26] and to the transfer of the low yields from a unit to another interconnected entity. As a result, the Financial Stability Board identified "thirty financial groups worldwide that are considered to create systemic risk and should be supervised cross-border" [26].

The diversification of Synergetic Economics' research approach *also continues in the present*, concludent results of this stage being systematized as they represent the passage to another future phase. By that time, we can take into account the recent works of Q. Bai, S. Banerjee, E. Frehland, H. Haken, A. Liening, A.S. Puryaev, L.M. Stan, V.V. Yakimtsov, etc., focused on the Synergetic Economics' implication in the management of complex systems affected by the local, sectorial, and global crisis.

3 The Genesis and Evolution of Synergistic Network and System

The tangible and intangible support of the economic synergistic actions is constituted by *the synergistic network*, the assembly of capacities, and flows of different natures from which optimally exploited interdependence, the economic entity benefits from synergy effects. In contradiction to the economic system which is a complex construction of interdependent resources which constitute a whole, the synergistic network presents itself much more prominently with synergistic functions of links' facilitation. The literature emphasizes most of all the active economic networks' synergy, because these entities have as primary role the links' creation between two or more systems and subsystems; the scope is to proactively facilitate the optimum transfer of know-how, resources and other capacities needed by production or other type of systems. If the economic systems *also* depend on synergies, the economic networks depend *most of all* on synergies (having synergies' facilitation as primary role)!

The typical synergistic entities are the ones which reinvent the collaborative behavior when a disequilibrium is produced (in moment T): an apparent disorder/chaos in which the respective network/system is *induced,* resulting certain conflicts/opposing processes which require reorganization and profound collaboration between the present (new and preexistent) elements—see Fig. 2:

As a result of processes and collaborative flows' reorganization and repositioning of technological, informational, and financial resources, an increased <whole> as

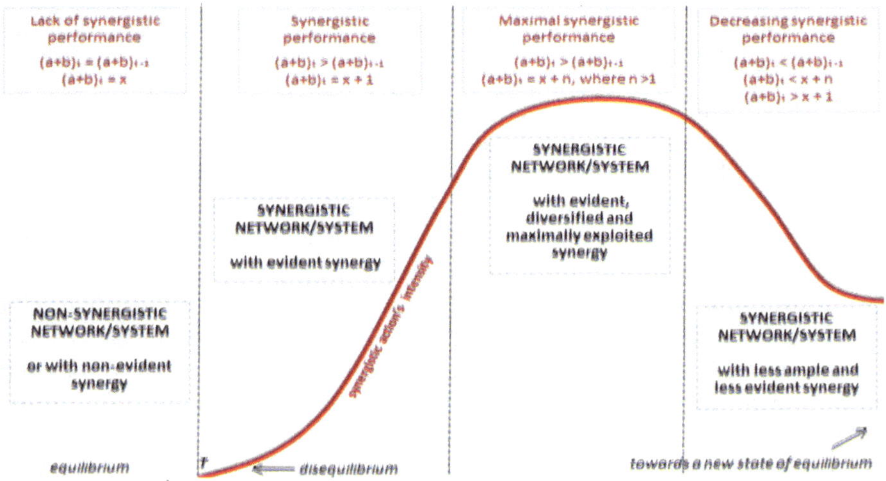

Fig. 2 The genesis and evolution of synergistic network or system and their undertaken synergistic actions' intensity. Examples of associated synergistic performances

value[5] is created (the collaboration between the a and b entities in the moment t having a greater value than the earlier collaborative results), tending to a diversification of included value chains. Subsequently, through the search and finding of a new state of equilibrium/symmetry (when the resources crisis is reduced), the synergistic actions' intensity decreases and the collaborative processes (by $a+b$ type) tend to deploy at a minimal, standardized level. From the moment of facing with a situation of <apparent chaos> (moment T), synergy may occur in a moment after ($T+1$ or $T+n$) because of the resistance to change/compromise/flexibilization of new required relationships or processes. If the produced disequilibrium is one scheduled (induced voluntarily) or prefigured in $T-1$ or $T-n$ moment, the resistance to change may take place "from time," the synergistic strategy itself being applied no later than the moment in which the disequilibrium, chaos, or *NEW* is confirmed in that entity's activities.[6]

3.1 The Synergistic Actions' Determinants in Economy

Due to the doctoral research of the premises and synergistic actions' determinants in economy, I got to outline the *synergistic gravity equation* (*SYNGEq*) that represents "a synthesis of the endogenous and exogenous factors which determine the private and non-private economic decision makers to call to actions of synergistic exploitation of the economic network in which they operate" [2]. This equation can also be applied to non-economic (but social) entities which are competing. *SYNGEq* consists of the following elements (in which the "imperatives" may mean "necessities") [2]:

$$\sum \mathrm{SYN.\ Act} = \sum \mathrm{R}^{-} \cdot \mathrm{I}\left(\mathrm{CRed} + \mathrm{Coop}^{+} + A^{\mathrm{Unimit.}}\right) \cdot V\left(\mathrm{Cust} + \mathrm{Info}\right) \cdot cc$$

where:

$\sum$SYN.Act = the sum of the synergistic actions adopted by the economic actor
$\sum \mathrm{R}^{-}$ = the amount of unpurchased but necessary resources of capacities
ICRed = the imperative for cost reductions (the global costs)
ICOOP^{+} = the imperative for deep cooperation (for the functional interdependence's increasing)
$\mathrm{IA}^{\mathrm{Unimit.}}$ = the imperative for purchasing unimitable competitive advantages
VCust = the necessity for customer value's increase
VInfo = the necessity for informational value's increase
cc = the specific environmental competitive conditions in which the economic actor operates.

[5]In terms of synergies with corresponding utility for the new needs of system/network.

[6]Examples of announced disequilibriums: scheduled mergers/alliances/relocations, the news of a direct competitor's appearance, the intention of connecting to the intermodal transport system, the change of legislation or of administrative regions, the liberalization of goods circulation, etc.

For example, SYNGEq helps to demonstrate the difference which exists between the prevailing synergistic actions' determinants in developed countries, with a national synergistic strategy of the cluster's system (the case of VInfo's prevelance in France) and the synergistic actions' determinants which prevail in less developed countries, without any national synergistic strategy (the case of $IA^{Unimit.}$'s prevelance to the right of less interconnected economic entities such as those from Romania). Concrete examples of SYNGEq's applicability will be presented in upcoming articles beside the post-crisis specific synergistic actions' determinants.

3.2 The Evolution of Active Economic System's Synergistic Potential and Its Degree of Exploitation

The undertaking of synergistic actions is generated, as we saw, by different endogenous and exogenous determinants, but the synergistic potential's degree of exploitation (SYNP) of an open economic system also changes depending on a series of social components. A very important factor to be taken into account is the type of system's organizational structure. SYNP is not exploited directly proportional to its evolution (in percentage) because the variables of synergistic exploitation's capacity[7] become the more numerous as the complexity of system/network increases, and this fact also requires specific synergistic strategies which limit to a certain extent the exploiting of the whole valid synergistic potential (see Fig. 3). The entities which have a higher capacity of SYNP exploitation are

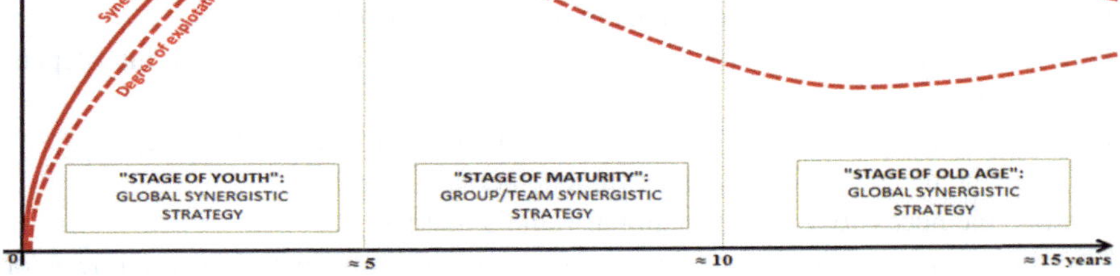

Fig. 3 The evolution of active economic system's synergistic potential and its degree of exploitation in terms of the evolution stage of the economic entity's general development. Specific synergistic strategies of these evolutions (Source: Stan L-M., 2011, p. 22, revised)

[7]The flow of resources, the internal bureaucracy, the number of interested groups/social partners involved, etc.

those which resort to *global synergistic strategies*, these ones being especially the young and less complex entities (small companies, newly established business incubators, clusters of zonal vocation, etc.); they have a higher synergizing ability of the internal and external capacities forasmuch "they gain perspectives across the entire business instead of a small functional area" [27], being able to prevent the possible synergistic counterperformances and to optimally place of know-how, human resources, technologies, etc., in pole position. For all that, the most prepared entities for complex synergistic strategies are the mature ones (for example, transnational corporations, clusters of national or international vocations), these structures benefiting from: a significant integration and use of ICT, a more advanced available know-how, and numerous and manifold human/material resources—all these generating mostly vertical synergies based on complementarity relations.

Nevertheless, as the system gets bigger and more complex the exploitation of the entire SYNP becomes more difficult, because of several factors: (*a*) the system's internal bureaucracy which is higher, (*b*) the reward systems which concern more the individual or team performances, generating unfair competition and poor communication between units, (*c*) the increasing of the risk of capacities/resources' wrong positioning among collaborative networks, and (*d*) the increasing numbers of interest groups with different expectations and requirements. Consequently, it will occur a more evident *emergence of the synergistic opportunities which are not (optimally) exploited*, of the efforts and capacities that are not (enough) combined and of increasing expenses for resources' requirements which could have been assured through synergy. Therefore, *SYNP is exploited even more as:* (1) its value is better identified/awarded within and outside the network; (2) the competitive conditions require the proliferation of added values generable through synergy; and (3) the undertaken synergistic strategies aim also at the combined increased effect at the entire level. When the entities start to be affected because of their age (because of the processes' rigidity, the excessive bureaucracy, and other elements related to an improper motivation of the collaborative intercessions), their SYNP size is reduced. But these disequilibriums often induce a comprehensive thinking of the value chains (when most of the mergers and alliances do take place), facilitating reorganizations, the creation of new collaborative relationships, and a wider exploitation of the available SYNP through a globally rethought synergistic infrastructure.

4 The Role of Information in Supporting the Synergistic Infrastructure and Adequate Exploitation of Synergistic Potential

The synergistic infrastructure of the open economic system (and not only economic) represents the assembly of tangible and intangible elements which correlated functions sustain the contouring of synergy effects necessary to the system; these elements belong also to other collaborative entities, the synergistic system/network

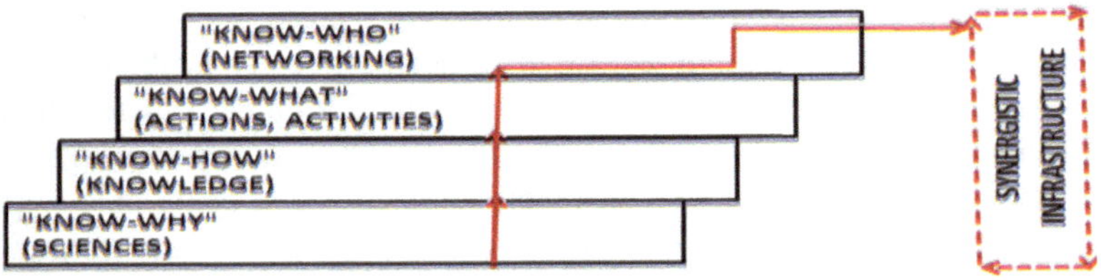

Fig. 4 The open economic system's synergistic infrastructure—product of an evolved economic situation. This infrastructure applicable or valid for other types of systems, too

continuously interacting with their environment. Hereby, true synapses (contact points) are built between strategies and internal and external capacities. The key elements which identify the synergistic infrastructure of any economic network or system—in the context of pursuing of a higher level of their competitiveness—are necessarily the following, too: *technology* (including innovation, information, and ICT) and *education* (especially the accessibility of knowledge and abilities, and their positioning in the maximum efficiency workstations)—elements that conditionate increases in the tangible and intangible values of a long series of interdependences. The synergism requires to directly *know-what* (the actions) towards *know-who* (networking with the most adequate partners) through *know-how* (practical knowledge to do these things), at the basis of all these standings being *know-why* (the sciences which sustain the collaborative activities of production entities) (Fig. 4).

The adequate identification of the synergistic infrastructure also helps us to better appreciate the real possibilities of SYNP exploitation and especially the role played by the collaborative flows in the actual exploitation of this potential. "In order to quantify the degree of the usage of the synergistic relational potential of the economic network or of an open economic system, the accounting of mutual information flows is essential, these ones being basic elements that facilitate the shaping of multiple and valuable synergies within the economic entity" [28]. Besides the number of activated network links (having/transporing mutual information flows), the formula of actually exploited SYNP also comprises *their intensity* (the number of links created in time) and *quality* (the quality of information and other qualitative aspects intended by the network coordinators) so as the shared information to correspond to the entity's needs and to guarantee the superiority in value of the network links. The satisfaction degree of information needs is expressed as a percentage because we deal with real (and not with ideal) economic networks and it may be known for a unique moment, t [28]:

$$\mathrm{SYNP}^t = \sum \frac{L_A^t}{L^t} \cdot qs^t.$$

If we take into account the current existent and not-yet-existent (but latent) links which are *necessary* and should be established (and also activated), it will result the total SYNP at moment t:

$$\mathrm{SYNP}_T{}^t = \Sigma \cdot qs^t$$

where:

$SYNP_A$ = the actual synergistic potential effectively exploited, used, which can be taken into account
$SYNP_T$ = the total, theoretical, nominal (of reference) synergistic potential, valid for the synergistic potential of the economic network in moment t
$L^t{}_A$ = the number of activated network links (that have/sustain mutual information flows) in moment t
L^t = the number of network links in moment t
q = quality parameter (parameter concerning the quality of information and other qualitative aspects intended by the network coordinators)
s^t = the satisfaction ratio of information needs (in the moment t) of the total information needs belonging to the network links.

Besides *the capacity of knowing and comparing the SYNP* exploited by different entities, this formula helps us understand that: (1) synergies suppose the procurement of some significant added values in contexts where information is complementary used, in an integrated, transparent, and coordinated manner, the dispersed information/knowledge's totality and utility being superior to the quantitative amount of the dispersals achievable through separate efforts [29]; (2) the information flows between the component subsystems of an economic entity represent integrated parts of the SYNP exploitation and in connection with which we can study the roles of network flows, information entropy, turning points in collaboration, etc.

5 Conclusions

Synergy represents a complex, *but real* experience. It means the condition and also the sign of living networks' vigor and dynamics, reflecting a series of undertakings and collaborative nurtured options in order to obtain a less conventional competitive advantage in the economic environment (and not only economic).

The current presentation has evidenced the necessity of an integrative manner of open economic system's synergy's approaching in the contemporary economy. As the economic entities pass towards the stage of natural (of more profound networking with the elements of the environment in which they operate), they implicitly become more interested in the synergy topic.

Synergetic Economics, framed in six distinctive phases of approaching, studies the specific phenomena of the economic systems in transformation, with a view to acquire dynamic equilibrium. It does not deal only with different activities that combine in order to achieve a common exponentially goal, but it also constitutes a science of self-(re)organization of open economic systems, on the basis of their constituent units' comprehensive collaborative relations, generating globally special behaviors.

The transitioning phase from the order to the disorder state (or from the state of equilibrium to the disequilibrium one) constitutes a critical point at which the systems or networks are becoming synergistic, *"the apparent chaos" having a profound role in enhancing the advantageous/superior reorganization* of the system's/network's elements through innovated or rethought collaborative processes.

Economic, informational, social, and cultural elements may be congregated in a set of determinants of the synergistic actions in economy: the case of the *synergistic gravity equation* (*SYNGEq*) which helps us compare the synergistic determinants of different systems/networks which are in competition.

The systems' synergistic potential (*SYNP*) amplifies as the complexity of these entities or of included networks grows; percentagly speaking, the SYNP exploitation's degree varies depending on the systems' organizational structure and their developing stage, being maximum in the first evolutionary period and minimum in the relative stability stage of these entities.

The economic system's *synergistic infrastructure* includes support elements for creating added value (sciences and knowledge) and elements that actively upkeep the synergistic intercessions in economy (actions and networking); such an infrastructure represents a part and parcel of the environment's competitive conditions in which that entity operates, directly influencing the intensity of synergistic efforts and the value of resulted synergistic effects.

The information flows (the quality and intensity of dispersed information, besides the number of activated network links and the satisfaction ratio of the network's or system's information needs) are essential to SYNP exploitation.

The synergistic network, as part of the open economic system, is an active network, oriented towards the internal possibilities and the requirements of the external environment in which it operates. Therefore, the *Synergetic Economics* is assisted by *varied interdisciplinary perspectives*, finding and emphasizing connections/links between seemingly disparate elements (technology, culture, ecology, social groups, etc.), which influence the intensification of this remarkable nonlinear phenomenon—synergy—with so many important effects in economy.

References

1. Sirower ML (2000) The synergy trap: how companies lose the acquisition game. The Free Press, Avenue of Americas, New York, p 13
2. Stan L-M (2011) The necessity to exploit the economic network's synergistic potential. LAP LAMBERT Academic Publishing, Saarbrücken 5. 33. pp 37–38
3. Merriam Webster Dictionary: http://www.merriam-webster.com/medical/synergize?show=0&t=1289571154
4. Buckminster Fuller R (1963) No more secondhand God: and other writings. Southern Illinois University Press, pp 123–161
5. Ansoff IH (1965) Corporate strategy: an analytic approach to business policy for growth and expansion. McGraw-Hill, p 167
6. Ansoff IH (1976) From strategic planning to strategic management. Wiley, pp 102–147
7. Ansoff IH, McDonnell EJ (1988) The new corporate strategy. Wiley, pp 62–97

8. Buckminster Fuller R (1971) Nine chains to the moon. Anchor Books, p 252
9. Bode HW (1971) Synergy: technical integration and technological innovation in the bell system. Bell Laboratories, pp 9–26
10. Coulter NA Jr (1976) Synergetics: an adventure in human development. Wichita Human Study Group, Prentice-Hall, p 15
11. Porter M (2001) Avantajul concurențial. Manual de supraviețuire și creștere a firmelor în condițiile economiei de piață. Editura Teora, Bucureşti, pp 300–301
12. Rappaport J, Swift C, Hess R (eds) (1984) Studies in empowerment: steps toward understanding and action. Routledge, pp 201–207
13. (2006) Le Robert MICRO. Dictionnaire de la langue française, Edition Poche, Pierre-de-Coubertin, Paris, p 1290
14. (2006) Macmillan English Dictionary for advanced learners. Bloomsbury Publishing Plc, p 1457
15. Faulkner D, Bowman C (2000) Elemente de strategie concurenţială. Editura Teora, Bucureşti, 7–8. 92
16. Harris PR (1998) The New Work culture: HRD transformational management strategies. HRD Press, Amherst, p 58
17. Dundon E (2002) The seeds of innovation. Cultivating the synergy that fosters new ideas. AMACOM, Broadway, New York, p 13
18. Dreikorn MJ (2004) The synergy of one. Creating high-performing sustainable organizations through integrated performance leadership. ASQ Quality Press, Milwaukee, Wisconsin, pp 15–21
19. Maxwell JC (2004) Winning with people. discover the people principles that work for you every time. Thomas Nelson, Inc., Nashville, Tennessee, pp 228–231
20. Buckingham M, Coffman C (2007) Manager contra curentului. Ce fac marii manageri altfel decât ceilalţi. Editura ALFA, Bucureşti, pp 32–35
21. Portugali J (2000) Self-organization and the city. Springer, Berlin, p 330
22. Aaker DA, McLoughil D (2010) Strategic market management. Global perspectives. Wiley, West Sussex, pp 140–141
23. Corning PA (2007) Synergy goes to war: a bioeconomic theory of collective violence. J Bioecon Springer 9:2
24. Ivancevic VG, Ivancevic TT (2008) Complex nonlinearity: chaos, phase transition, topology change, and path integrals. Springer, p 649
25. Meijers E (2007) Synergy in polycentric Urban regions. Complementarity, organising capacity an critical mass. Haveka, Alblasserdam, Netherlands, pp 14–25
26. Lybeck JA (2012) Istoria globală a crizei financiare (2007–2010). Editura Polirom, Bucureşti, 45. 266
27. Fung KV, Fung KW, Wond Y (2009) Concurenţa într-o lume plată. Cum să construim o companie într-o lume fără graniţe. Editura Publica, Bucureşti, p 121
28. Stan LM, Jivan A (2012) The exploitation of the open economic system's synergistic relational potential, JKMEIT Scientific Papers, II(2), France, 13.15–16
29. Jivan A (2000) Servicitate. Mai mult decât productivitate în economia de serviciu. Editura SEDONA, Timișoara, p 115

Synergy Drives the Evolutionary Dynamics in Biology and Economics

Klaus Jaffe

1 Introduction

The present paper does not pretend to present novel facts nor brand new theory. It aims at producing a synthesis of known facts that open novel windows that allow for fresh views on established knowledge, favoring the flux of ideas between areas of science that have developed quite independently from each other. New multidisciplinary ways to look at old facts broaden our understanding of nature by helping us rethink established dogma in search of Consilience [1]. Here I present a synthesis of a lifelong effort in building such an interdisciplinary window [2]. A version of this view, more focused on biology, was published in 2016

The theory of evolution, formulated by Darwin and Wallace some time ago, was built on the insight that heredity, natural selection, and variability interacted to produce biological evolution. The breakthrough in thinking was not the discovery of natural selection, or of heredity, or diversity. All these features were described in detail by Alexander von Humboldt, much cited by Darwin, and who lived a generation before Darwin and Wallace (Humboldt died the year Darwin published the *Origin of Species*). Humboldt had a working knowledge of selection and of the importance of the survival of the strongest, of heredity, and the logic of domestication of plants and animals by selective breeding, and was aware about diversity, describing detailed variations between species and among species [3]. The important contribution by Wallace and Darwin was the insight that evolution emerged from the synergistic interaction of these three features, and that this evolutionary dynamics could explain the emergence of species (top right cycle in Fig. 1). That is, the continuous interaction between heredity, variations produced by mutations and the environment, and natural selection, produce the evolutionary

K. Jaffe (✉)
Universidad Simón Bolívar, Caracas, Venezuela
e-mail: kjaffe@usb.ve

P. Bourgine et al. (eds.), *First Complex Systems Digital Campus World E-Conference 2015*, Springer Proceedings in Complexity, DOI 10.1007/978-3-319-45901-1_34

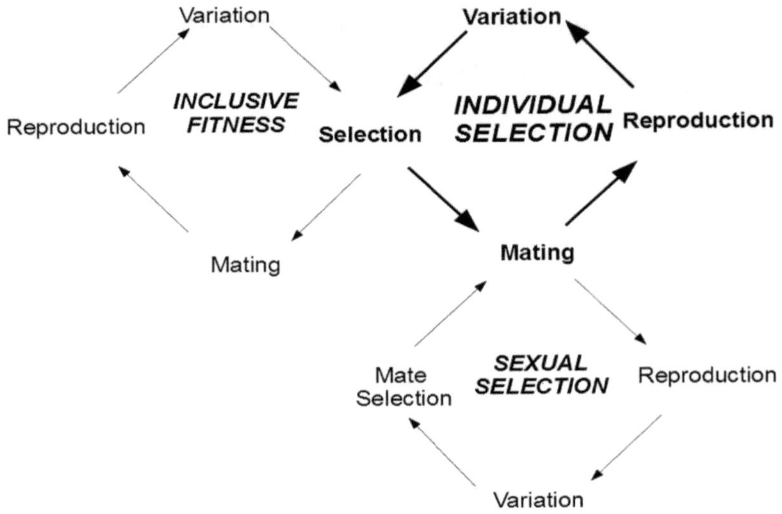

Fig. 1 Schematic representation of selected aspects or components of the network of relationships responsible for the dynamics of natural **Selection** driving biological evolution. **Individual Selection (i)** represents natural selection acting on the individual; **Sexual Selection (s)** that acts on mate selection strategies and intra-sex competition; and **Inclusive Fitness (o)** cycles represent the co-evolutionary effect on selection of the action of other organisms. **Variation** represents genetic mutations and phenotypic variations, **Reproduction** represents the reproductive and life-history strategies of individuals, **Mating** stands for sexual reproduction. Organisms suffer evolution through Individual Selection (*bold arrows*), which in turn is affected by at least two other cycles: Sexual Selection and Inclusive Fitness. Evolution among asexuals differs from this description [146], as no mating's occur

dynamics that allows species to adapt to their environments and eventually diverge in their evolutionary path producing new species. That is, natural selection operates through the differential reproduction of individuals, measured as fitness. Higher levels of fitness are achieved by higher rates of reproduction, which in turn may be enhanced by higher survival probabilities. This theory, however, had difficulties in explaining many mating displays using exceptionally bright colors, ornate plumage, and conspicuous forms among living creatures, which attract predators and thus decreased the odds of individual survival. To overcome this limitation, Darwin introduced the concept of sexual selection to complement that of individual selection to explain biological evolution (bottom right cycle in Fig. 1). Darwin described two conceptually different ways at looking at fitness: individual survival that favors the strongest and most able individual, and sexual selection that favors the most prolific in mates and descendants. Both selecting forces might work synchronously, or they might diverge. Although both, survival and reproduction, are parts of the individual's fitness, theory that looks separately at each of the two processes help us in gaining a deeper understanding of evolution.

Soon after Darwin, important advances in our understanding about how evolution proceeds emerged. Development of population genetics, mainly between 1918 and 1932, and the expansion of Mendelian genetics were incorporated together with a more detailed theory of natural selection and gradual evolution into a modern evolutionary synthesis. This modern synthesis, produced between 1936 and 1947, reflects the consensus that is still valid today [5–9], and others. The next important advance was a better understanding of cooperation and the emergence of societies that was not explained satisfactorily by Darwin [10] and Wallace [11] nor the just mentioned synthesis [12]. Cooperation is important in a number of settings, including behavioral interactions, biological evolution, sociobiology, cultural dynamics, and collective intelligence; yet, the features allowing it succeed are not well known and are still discussed today [13]. Inclusive Fitness Theory (IFT) as originally stated by Hamilton [14, 15] has been the most successful theory so far to provide explanations for the evolution of cooperation. Hamilton grasped that the effect of other individuals (con-specifics or not) affects the odds of survival of an individual. Specially among social species, the action of others could affect the fitness of an individual, so as to form a web of relations that affect the fitness of the participating organisms (the top left cycle in Fig. 1 represents just one cell of such a network).

IFT can be summarized by an expression quantifying the fitness costs of a cooperative interaction as $c<b\ .\ r$; where c is the fitness cost to the donor in a cooperative act, b the benefit to the receiver, and r as the probability that an allele in one individual will also be present in a second individual via common descent. This simple formula, centered on $c<b$, is a consequence of the law for the conservation of energy, or first law of thermodynamics, as applied to biology: in the long term, survival of organisms requires that its total energy expenditure must be equal or lower than its total income. In order for fitness to be positive, positive survival rates are required. Hamilton's proposition was to treat b as a quantity modulated by r. This IFT was initially misunderstood. Soon after rejecting Hamilton's original paper on IFT submitted to Nature [16], Maynard-Smith [17] introduced the Kin Selection Theory (KST) to explain the phenomena Hamilton described as IFT. This historical circumstance has obscured the relevance of IFT until today and favored that of KST which is more intuitive and easier to understand. The difference between KST and IFT is that the former only considers the genetic relationship between cooperating individuals as relevant for calculating fitness, whereas the later accepts that other factors are also relevant. KST simplifies IFT by assuming that r in the formula $c<b\ .\ r$ represents "only" the genetic relatedness between donor and receiver. This simplification, though, has become very popular. So much so that Google Scholar in April 2015 retrieved about 1.4 more papers using the term "kin selection" compared to "inclusive fitness." Many scholars today still do not distinguishable between both concepts (see [18–20], for example). This confusion between KST and IFT is augmented by histories of IFT that assign Haldane a primary role in it (see, for example, [21]). Yet, relating Haldane's [7] casual comments on how expanded parental care may be favored by selection, with a pioneering role in the development of IFT, is equivalent to calling Alexander von Humboldt the grandfather of Darwin's theory of evolution. Modern evolutionary theory assigns to the components affecting inclusive fitness that are nor related to kin a much more important role in explaining

evolutionary phenomena, including economics, than those considered by KST. The main shortcoming of KST is that it focuses on the actions between genetic relatives as drivers for the evolution of cooperation, whereas IFT also explains cooperative interactions between non-kin.

Substituting IFT with KST was never accepted by Hamilton ([16] and personal communication), and even Maynard Smith [22] eventually recognized its distinctiveness. IFT is a much more general theoretical framework than KST, which is a special aspect of the former. The focus on inclusive fitness rather than on kin selection allows for a finer understanding of population genetic dynamics. Inclusive fitness being >0 can be the right criterion for social behavior to be selected, also in models where kin selection is absent, and where assortment is brought about by something other than common descent. Inclusive fitness, in addition to the genetic relatedness between the actors in a cooperation, takes into account "the likelihood of sharing genes above random levels due to statistical effects in genetic population dynamics" [23]. This means that the effect of covariance on selection also determines the degree of assortation that may occurs between organisms [24]. IFT is an open theoretical framework, which might conceive as multipliers to b any means that increase the frequency of an allele in a population through social interactions, such as mutualism, synergistic cooperation, and others [25–27]. It is not necessary to refer any more to kin selection. Flecher and Doebeli [28] wrote: "The most fundamental explanation for how altruism (defined by local interactions) increases in a population requires that there be assortment in the population such that the benefit from others falls sufficiently often to carriers (and at the same time nonaltruists are stuck interacting more with each other). Nonadditivity if present can play a similar role: when collective cooperation yields synergistic benefits (positive nonadditivity) altruistic behavior can evolve even in the absence of positive assortment, and when there are diminishing returns for cooperation (negative nonadditivity) the evolution of altruism is hindered [25, 29]."

Independently to the theoretical development just described, robust tools for handling non-linear emergent phenomena in mathematical biology, such as numerical simulations, reached the same conclusions, confirming a central role for social synergy in the evolution of cooperation, specially among non-kin groups. That is, agent based computer simulations studying the evolutionary dynamics of inclusive fitness on haploids, diploids, haplo-diploids, and asexual and sexual organisms showed that social cooperation without social synergy is unable to emerge and sustain itself in scenarios for biological evolution [30, 31] and in scenarios of economic markets [32]. These simulations, validated by empirical experimentation [33–35], showed that both, biological evolution of social behavior and market dynamics, require social synergy for its working. Social synergy is defined here as the process by which emergent properties arise through social interactions. For example, cooperation in retrieving food by insects allows them to handle food that no single individual would be able to capture and retrieve alone, expanding opportunities to exploit novel niches to groups of cooperating foraging workers. Such type of cooperation seems to explain the evolution of social behavior among bees [36] and wasps [37]. Social synergy is not reduced to an abstract concept as it can be measured quantitatively and empirically in different settings

[38]. Cooperation were all interacting individuals benefit, are often refired to as Mutualism ([39–41], for example) and can be viewed as a special kind of social synergy.

EIFT is based on Queller's version of Hamilton's rule [25], as presented by Flecher and Doebeli [28], who formulated r, the modulator of b as a ratio of covariances (cov) so that:

$$r = \text{cov}\,(G_A, P_O) \,/\, \text{cov}\,(G_A, P_A)$$

where G_A measures the genotype or breeding value in each individual in the population (subscript A for actor), P_A the phenotypic value of each actor (e.g., 0 for defection and 1 for cooperation), and P_O is the average phenotype of others interacting with each individual actor (subscript O for others).

This formulation implies that the altruistic genotype represented by G_A increases in frequency if those with the genotype on average get more benefit from the behavior of others than they pay in cost for their own behavior. This relationship uses measures of assortment (covariance) between those with this focal genotype and the helping behaviors of others, scaled by the value of these behaviors. Taking the covariance over the whole population ensures that if this inequality holds for the helping genotype, it cannot simultaneously hold for the alternative nonaltruistic genotype. Therefore, when Hamilton's rule is satisfied, carriers on average have higher direct fitness than the population average (for details see [28]).

Here I propose a slightly different formulation that facilitates its application to human economic dynamics. This expanded theory allows to bridge conceptual divides between biological and economic sciences. Very recently, Corning [19] presented a similar bioeconomic approach to cooperation giving a preponderant role to synergy in evolution. His approach differs somewhat to the one developed here, as it focuses on multi-level and group selection [19] and to "synergistic selection" in the context of the emergence of complexity [42]. Corning defines synergetics as the study of synergy or the collaborative behavior of entities of different natures and on the study of the way in which these ones manage their complexity in order to increase their value. The present proposition differs from these conceptions but is complementary to Corning's approach, and is not a substitute for it.

2 Expanding Inclusive Fitness Theory

A dynamic narrative that includes both biological and cultural evolution requires three assumptions, definitions, or concepts in order to formulate an EIFT.

1. The first is to refer to agents as a concept that includes organisms and social structures. In biology, mating is described as a cooperation between two agents to produce offspring; whereas in economics, cooperating agents are productive units which can be individuals or aggregates such as companies. Cooperation is

at the heart of any business and thus the basis of economic dynamics. Using agents as the unit for dynamic studies is getting more common in biology, sociology, ecology, and economics as shown by the ever increasing literature (some examples are: [43–47], in addition to the literature cited so far).
2. Another definition that requires modification is reproduction. Reproduction should be viewed as reproduction of information, which includes diffusion and multiplication of information. This information can be of the genetic kind in biology or, in economy it might mean memes [48], information attached to productive systems [49], or scientific knowledge quantifiable with scientific papers [50], etc.
3. The third definition is a synthesis of the concept of evolutionary fitness function with economic utility function [51]. If we assume that natural selection molded behavior of extant organisms, including humans, so that they maximize their capacity for adaptation, then behaviors that aim to optimize the fitness of an individual will take into account the same natural selection forces working in evolution. As evolutionary fitness functions and economic utility functions are the product of the same natural forces, they will have equivalent structures dealing with the same constraints [52], including those formulated by IFT.

As represented in Fig. 1, the fitness of an individual has at least three aspects or components:

1. The selection acting on the survival capabilities of the individual that relate to its capacity to manage and respond to its environment (i),
2. The abilities to mate and reproduce that can be grouped under sexual selection (s), and
3. The inclusive fitness or fitness affected by the presence and actions of other individuals with which it interacts or which it bestows upon others (o).

The total fitness of an individual (f) is a composite function which includes the fitness conferred by the phenotype of the individual, which in turn depends on its individual survival capabilities (i) and its reproductive success (s). In addition, f depends on the consequences of interactions with others (o), so that:

$$f = f(i, s, o)$$

The component o has at least three parts to it. (1) The likelihood that a gene is present in another individual due to genetic relatedness or the kin selection component (k), (2) The probability that a gene is shared due to assortation (a), (3) The probability that a gene will favor the fitness of another due social synergies or economic considerations that emerge from the presence of specific alleles in each individual (e). Therefore:

$$o = f(k, a, e)$$

Assortation includes the concepts of kin selection, as preference for cooperating with kin is a specific kind of assortation. Thus $o = f(a, e)$. The fitness of the individual f can be summarized as the product of two related networks of relationship: f_i or factors affecting individual fitness directly; and f_o or factors affecting the individual fitness via the action of others:

$$f = f_i\left(i,\ s, f_o\left(a,\ e\right)\right)$$

This formulation converges with the formulation proposed by Queller [25, 27], Fletcher and Doebeli [28], and others, in that is treats assortation (a) and social synergy or non-additive benefits (e) as the most important features determining the evolutionary viability of cooperation. Of all these terms, a and e are the less well understood and will explained below.

2.1 Assortation

Assortation refers to the fact that similar organism attracts each other. This is described in phrases such as "birds of one feather flock together" and is also referred to as homophily: love for things similar to oneself; or narcissism: love of oneself. The term assortation was already used by Hamilton [53] and he helped to show its relevance to IFT motivating George R. Price to develop a mathematically useful formulation [24]. This paper showed that assortative mating can increase the frequency of an allele. This effect was shown to be so fundamental that it also works in mate choice in sexual reproduction [54]. Complementing these findings, computer simulations showed that without some kind of assortative mating, sexual reproduction is unlikely to emerge among complex diploid organisms [30].

The working of assortation in favoring the success of cooperative strategies seems to be associated with the possibility of forming globular clusters, as is the case of some network structures [55]. Assortation is favored by tags or a green beard effect [15], consisting of signals, behaviors, or other features that allow agents to discriminate among potential cooperators and regulate the type of agents that will interact cooperatively [56, 57]. A very basic form of cooperation often occurs among sexually reproducing mates. But assortation or homophily evolves in many other cooperative interactions [58]. Many behaviors of modern humans, such as the choice of mates and pets, can be explained as a result of assortation. For example, humans select mates based on visual perception of their faces [59], or of their pets [60], and friends [61] assortatively. In addition, homophily is very common in social settings [62, 63]. Assortation or homophily has particular interesting effects on the evolutionary dynamics of cooperation, even beyond what can be explained with IFT. They reduce error thresholds of mutations [64], and accelerate the speed of evolution [30], favoring the emergence and maintenance of cooperation [54].

Assortation has been studied extensively in assortative mating and assortative cooperation (see review in [54]). Empirical evidence for assortation has been mounting. Here, just few random examples: Evidence among vertebrates includes studies showing that chimpanzee friendships are based on homophily in personality [65]; the existence of assortative mating in lesser snow geese [66] and in blue tits [67]; assortation among humans in games of experimental economics [68, 69]; from anthropological and archeological studies [70, 71]; and of course, from ethology [59, 60].

2.2 Social Synergy

Much work on cooperation has centered on altruism. Indeed, the ultimate sacrifice of ants and bees for the good of their colonies is an impressive feat. But eventually, all sustainable social behaviors involve interactions that are beneficial to all intervening parts [28]. Interactions where all parties gain, the so-called win–win interactions (Fig. 2), are very much known among economist [73–75]. One important concept is social synergy, i.e., non-additive benefits and positive feedback of social behavior

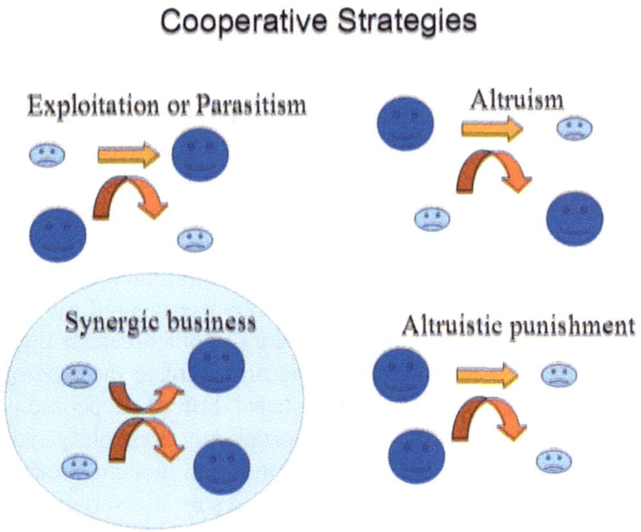

Fig. 2 Effect of different cooperative strategies on the fitness of the actors. In interactions involving exploitation or parasitism, one organisms benefit at the expense of the other, increasing its fitness (*bigger dark blue faces*) and reducing that of the other (*small light blue face*). In altruistic interaction the reverse holds. The altruist reduces its fitness and the other increases it. Interactions where altruists punish individuals not complying with social norms, both the altruist and the other reduce their fitness. In synergistic business both actors win, increasing their fitness, not necessarily by the same amount [72]

that affect individual fitness. Social synergy refers to synergies triggered by social cooperation that increase economic and other benefits to social individuals favoring its evolution [26, 27, 30, 32, 38, 76]. Synergies that emerge from social interactions can be quantified [38, 77], and are fundamental in explaining the maintenance of complex societies [32].

Although not ignored, social synergy has been little studied quantitatively among living creatures other than humans [38]. The father of IFT already recognized that several different mechanisms are needed to explain the prevalence of social cooperation among extant species [78]. Studying bees, Michener [36] demonstrated the existence of several different evolutionary routes leading to sophisticated societies that benefited all or most of its members. To understand these evolutionary dynamics, economic and ecological considerations are more important than genetic ones [37, 79].

The economic forces unleashed by human cooperation have been studied by political economist for many years ([80], for example). More recently, social synergy has been mentioned when studying pay off matrices, altruistic punishment, benefits of social life, and cooperation. The cost/benefit ratio of cooperation might reveal the existence of this synergy. Cost/benefit ratios have been shown to be important for the evolution of cooperation in different settings ([54, 76, 81–86].

From where does this synergy that produces win–win situation arises? Economics has an answer to this question. Long ago, among others, Aristotle recognized that division of labor enlarges and elicits innate human differences [87], which allow the existence of complex society. Adam Smith recognized the existence of a special synergy working in the markets as an "invisible hand," but neither he nor others focused on the specific mechanisms that allowed its working. Adam Smith [88] writes in The Wealth of Nations "The greatest improvement in the productive powers of labor, and the greater part of the skill, dexterity, and judgment with which it is anywhere directed or applied, seem to have been the effects of the division of labor ... It is the great multiplication of the productions of all the different arts, in consequence of the division of labor, which occasions, in a well governed society, that universal opulence which extends itself to the lowest ranks of the people". Friedrich Hayek [89], specially when tackling the "Economic Calculus," and many others ([90], for example) hint to the emergence of synergistic effect in social interactions due to the existence of differently specialized actors. Division of labor has also been associated with social life in insects and many other animals [12] and is correlated with the degree of order in ant societies [91, 92]. Thus, a very important source of social synergies in the economic dynamics of markets and in biological cooperation is the division of labor. The force behind Adam Smith's invisible hand of the market that triggers the complex market dynamics intuitively described by Hayek is social synergy as described here. This can be evidenced using computer simulations [93] and robot swarms [94]. It is the specialization of labor that allows complementary interactions to produce ever stronger synergies that confer non-linear economic advantages to societies that allow and foment individual liberty and division of labor [72].

Not all social synergies arise from division of labor. Many other mechanisms are possible. Economies of scale, for example, also allow the individual to achieve higher fitness or economic gains among humans and other animals [38, 77, 95]. Social intelligence can also be viewed as an emergent phenomena [96]. Again, economic science has explored these issues much more and/or differently than biology.

Among economists, the existence of non-linear dynamics in wealth accumulation has been recognized long ago. Karl Marx [97], for example, when he described the surplus value and attempted to balance wealth across a society, recognized that more than simple additive arithmetic's was required. More sophisticated thermodynamic approaches to study non-linear dynamics in economics were initiated by Georgescu-Roegen [98] and developed further by many others (see [99], for example) which eventually lead to a systematic use of the concept synergy in economics. Underlying the concept of social synergy in economics is the fact that some actions and the exploitation of some resources are only possible after a certain threshold size of social aggregates has been reached, producing a non-linear or emergent effect on wealth aggregation.

Examples of synergy used to simulate social evolution in biology can easily be applied to human economics. The example of two wasp mothers that attend their brood communally, each one investing 50 % of their time in brood care, achieving to protect their brood 100 % of the time, reducing the odds of losing their brood to zero with the same cost to parents [30], can be expanded to human societies and institutions in charge of communal security [85].

A simple example of a synergistic view of the relationship between increased utility and increased wealth is that if a wealthy donor gives a poor recipient a blanket, the recipient will get a much higher utility from the blanket than the donor, but there is no net increase in wealth. But if the object donated is a sewing machine, which is used in the rich donor's house as decoration, but the poor receiver uses it to produce blankets to sell, then there is a net increase in wealth. The first case illustrates a synergistic increase in utility, the second type one of wealth (Libb Thims personal communication).

The more we look at synergy in economics and business management, there more we find meaningful examples. Examples include: the impact of acquisitions on merging and rival firms [100, 101], economic development [102, 103], mergers and acquisitions [104–106], and evidence that certain type of competition over personal resources can favor contribution to shared resources in human groups [107].

2.3 From Biological Evolution to Economic Dynamics

How relevant is EIFT for our understanding of evolution? Cooperation among non-kin is as or more important than between kin. For example, symbioses are far more important in biological evolution than hitherto recognized [42, 108]. Theoretical evolutionary theory needs to digest this fact. In addition, recent reviews provide ample theoretical and empirical evidence justifying the extensions to IFT. For example, Van Cleve and Akcay [109] showed that the interaction between behavioral responses (reciprocity), genetic relatedness, and synergy interact are fundamental in understanding the richness of social behavior across taxa. The review by Bourke [110] on "comparative phylogenetic analyses show that cooperative breeding and eusociality are promoted by (i) high relatedness and monogamy and, potentially, by (ii) life-history factors facilitating family structure and high benefits of helping and (iii) ecological factors generating low costs of social behavior." The last factor is of course the mirror image of social synergy: Environments provide selection pressure to which organisms evolving cooperative strategies producing social synergy have to adapt. Many unequivocal examples of social synergy as a factor in determining the evolutionary success of social behavior have been reported. The best known example is probably the evolutionary history of social behavior among bees [36]. In the case in wasps, Hamilton's preferred species, social behavior generates indirect benefits by enhancing the productivity or survivorship of non-kin more often than that of kin ([111–114], for example).

Biologists are not the only ones interested in social evolution. The features that influence the dynamics of cooperation have been studied using different theoretical frameworks with different specific assumptions. The theoretical framework of studies of social dynamics by biologists, sociologists, economists, physicists, mathematicians, game theorists, computer scientists, and others differs in the concepts they use despite the fact that all are studying the same basic phenomena, making interdisciplinary communication of this issues difficult. However, all these disciplines have used applications of game theory, and specifically the Prisoner's Dilemma, to pursue their quest for answers in their fields. Thus, a common language bridging the concepts between these disciplines seems possible.

An important difference between biology and economy is that biology focuses on genetic evolution whereas economy studies cultural processes. This difference is much less important than the homologies in dynamic processes. For example, Manfred Eigen [115] insists that Darwinian evolution is not merely the organizing principle of biology but a law of physics that should be responsible for many phenomena in nature. Genetic evolution is based on vertical transmission of information, from parents to offspring, whereas cultural transmission includes in addition to the vertical kind a horizontal transmission of information. The overall evolutionary dynamics of both processes, however, is indistinguishable [116]. Both processes produce a continuous dynamics that may induce divergence or specialization [117].

The evolutionary dynamics in biology is centered on genes and organisms, whereas the economic dynamics is centered on business, enterprises, and companies. In biology, mating or cooperation between the sexes is fundamental for the survival of the population; whereas in economy it is cooperation among different type of laborers or companies that allows production of wealth. In both cases, the dynamics driving information, innovation, and social synergy is similar. EIFT formulas the equation: $\mathbf{c} < \mathbf{r} \,.\, \mathbf{b}$ using $r = f(a, e)$ or a function of the probability of the individual to possess a gene that confers it advantages in social interactions with others and the social synergy triggered by this interaction. In the case where the socially advantageous gene is shared between interacting organisms, we speak of assortation. If $f(a, e)$ has evolved by natural selection, economic utility functions are expected to include these same factors. Any comprehensive utility function in economics then has to include besides direct benefit/cost considerations, benefits to kin and group members, to individuals which might reciprocate positively in the future, and any entity that might synergize the individual actions. That is the utility function (u), analogously to the fitness function defined above, has to have at least three different components: $u = f\,(i,\ a,\ e)$. To express it in a way familiar to economists, physicists, and mathematician we can write the same formula as:

$$\upsilon = \phi\,(\iota,\ \alpha,\ \varepsilon)$$

implying that any general utility function in economics (υ) has to take into account the benefits accrued to the agents directly (ι), through interactions with others (α) and through synergies triggered by its behavior (ε).

The bioeconomic insight of EIFT is that the biological fitness function and the economic utility function have the same form, as both are the product of natural selection. The formulation proposed here allows to apply the analytical tools developed by Queller [25–27] for biological evolution to economics.

2.4 Benefits of an Extended Inclusive Fitness Theory?

The challenge of EIFT is to explain in more detail how biological and economic systems produce synergies by favoring specialization and division of labor, conferring the individuals in a cooperative society with fitness benefits that are much higher compared to a solitary life (see [93], for example). More experimental approaches in economics are required to address these issues (see [118], for example).

EIFT considers that factors other than genetic relatedness affect the cost benefit balance of cooperation and that fitness functions and utility function have to consider the direct effects on the individual as well as indirect benefits an individual achieves through assortation and synergistic interactions. These factors have been studied with different emphasis by biologist and economist. The most important factor often overlooked so far is probably the social synergy that emerges from cooperative interactions, such as synchronized division of labor. An important conclusion from empirical studies in economics, that try to assess the effect of social synergy or economic benefits that derive from social life, is that synergy is probably the most important driver in the evolution of cooperation, and that assortation or genetic relatedness are neither necessary nor sufficient for the emergence of cooperative phenomena (see also [19]). The same conclusion is reached when exploring the dynamics of cooperation in the repeated prisoners dilemma game [119]. Here social synergy is more important than assortation, which in turn is more important than kin selection, in fomenting cooperation. This suggests that an expanded version of IFT is required for a better understanding of the dynamics. Focusing only on kin selection is not enough. An insight into the economics of the cooperation is fundamental in understanding it. However, little quantitative empirical research on social synergy has been produced in biology (but see [38, 79, 120]), though it is recognized as of primary importance in the economics and business literature. The latest reviews of the empirical literature in biology confirm that a more economic view explains the descriptions of societies found in nature better [109, 110]. Even on co-evolution, the review by Ivens [121] shows a pattern among farming mutualism of ants and their domesticated species that seems to produce stability of these successful mutualisms: The component of inclusive fitness in the evolutionary dynamics (Fig. 1) dwarfs the sexual selection component. Most of these mutualisms are characterized by reduced symbiont dispersal and diversity (often in association with asexual reproduction and vertical transmission), promoted by specific ant behaviors of the ants, such as creation of protective environments. Co-evolution, viewed in this new light (see [12–124], for example), can easily explain many symbioses. Even extravagant proposals such as the one stating that host–microbe interactions influence brain evolution and development in mammals [125], can now be explained. An EIFT makes it unnecessary to treat symbioses and social cooperation as different phenomena as done by Corning [19], as both are considered in $o = f(a, e)$.

The central insight from recent empirical studies is that economic factors and assortation in its different forms determine social behavior. Social behavior cannot be understood without taking account of all of them. A synergistic interchange of theoretical knowledge between economics and biology looks promising for a novel attempted to deepen our understanding of social dynamics and should help to bridge the gaps in studies of evolution of social cooperation between economist, physicists, biologists, and others, providing for a common language in the quantitative assessment of the importance of specific features that aid social evolution.

A theory that helps us to look for the relevant features in the evolution of social behavior, dynamics of cooperation, and evolution of society might be useful. That is, more important than kin relationships are assortation and social synergy for understating social cooperation. Assortation is important in a number of fundamental instances of human cooperation [54, 126, 127] and may emerge in many other circumstances if we look for it. The most relevant potential contribution of this theory is that it might allow social science to profit from both economics and biology. It might help develop complexity sciences aiming to improve our understanding of social synergies unleashed by cooperation are of the fundamental forces driving the evolution of societies. These phenomena should be empirically observable. Three examples might help convince the reader about the empirical usefulness of this theory.

1. Many butterflies have associations with ants. They can either be mutualistic, exploitative, or parasitic. Quantitative phylogenetic analysis revealed a large prevalence of cooperation over competition in the symbiotic relationship [79].

As no possibility of genetic flow between ants and butterflies exist, there is no doubt here that social synergy is the driving force for cooperation. An impressive large number of symbioses are known to exist [19, 128]. This unified treatment of social synergy can be expanded to address the spontaneous commerce and cooperation networks that arise from the working of competitive advantages between nations [129] and firms [130] in economics, as originally conceived by David Ricardo [131].

2. Empirical evidence shows that different forms of division of productive activities in an economy account for differences in its capacity to produce and accumulate wealth [49]. This is linked to division of intellectual labor [132] in contemporary human society. For example, the division of labor in academic research accounts better for differences in relative economic success among nations than any other variable studied so far [50, 133]. These examples show that arrangements that affect social synergies, such as division of labor, are the key to understand contemporary economic development, including the working of finance [134]. An EIFT stimulates the exchange of analytical tools between economists and biologists for a novel view of the working of division of labor ([135], for example).
3. From a biological point of view, division of labor in ants is related to increased economic gains of social behavior [92] and at the same time, more sophisticated social behavior is related to a decreased individual complexity [136]. This is an example of social synergy driving social evolution at the expense of individual selection, easily explainable with the EIFT. In economics, we accept that societies confer energetic benefits to all individuals involved in both ants and humans [38, 77]. Thus cities allow synergies to emerge that provide non-linear benefits to society [137, 138]. These synergies are practically everywhere [128] and can even be detected in basic physical architectural arrangements [139] and therefore might be present in many as jet unsubscribed situations [140]. An EIFT might be better able to develop analytical tools to understand how and why synergies emerge from division of labor [72].
4. The study of the interactions between synergy and violence has been fruitful in bioeconomics [141], so might EIFT. Inclusive fitness theory shows that altruism is favored by natural selection if it directly or indirectly benefits the social group of the altruist, or if this social investment eventually benefits the altruist in the future. This mechanism is shown to work smoothly with simulations [32, 142]. Extant instincts and behavior in all plants and animals, including humans, are the product of natural selection. Extending these insights in building economic utility functions helps to understand behaviors that maximize individual benefits directly and indirectly through assortation and synergistic interactions. By doing this, EIFT explains features of modern life that has escaped explanations based on classical economic theory. Terrorism, for example, is a feature that is ever more important in contemporary society. Motivations to commit terrorist acts, however, are driven by both, biological and economic stimuli. The branch of biology studying animal and human behavior, ethology, tells us that aggression enhances group cohesion, that poor odds of survival or of alternative routes to

increase ones fitness (or utility) function increase the likelihood of aggressive interactions, and that large differences in individual strength, or low odds of retaliation, favor aggression [143, 144]. The consequences of a behavior determine its evolutionary success. The economic view tells us that the consequences of a behavioral action can be assessed by the ratio of benefits (b) to costs (c). EIFT shows that if b/c is high, even if b is dependent on a network of feedback mechanisms, biological and economic evolution will favor this behavior. Sometimes, b tends to infinity allowing for the existence of supernormal stimuli [145, 146]. Religion strongly favors such super-stimuli as shown by simulations [85, 147]. Relatively very large material benefits also induce supernormal motivations [148]. Thus, when pursuing heavens, avoiding hell, or aiming at paradise on earth, expected benefits tend to infinity. These irrational aims are a consequence of a simplified utility maximization function selected by evolution. If human dreams extend the benefits of its actions to life after death, self-immolation and extreme heroism may become an economically rational choice. These motivations, together with technological means that allow a single individual to inflict harm to many people, and to create diffusely defined but strongly inter-linked groups, allow terrorism to prosper. Modern communication helps creating fantasies that are disconnect with everyday reality, favoring these extreme b/c scenarios. Religion helps, but is not necessary in achieving these dreams. This bioeconomic view suggests that terrorism might be dealt with better by influencing the perceptions of the benefit/cost balance in the extended utility function. That is, perceived benefits of aggressive ants have to be reduced and cost increased, and benefits of alternative behaviors have to be increased and their costs decreased. By doing this, terrorism motivated by religion [147] can be shown to be very different from that motivated by politics, illegal business, or individual grievances. This insight improves our understanding of the emotions of potential terrorist and the role of inclusive groups, eventually reducing social aggressions by other means than by responding with more violence.

These examples show how a synthetic view of the dynamics governing cooperation might help achieving a better understanding between biology, sociology, economy, complex system sciences among others, eventually unleashing synergies that might advance our understanding of nature in important ways. Low hanging fruits might be found by economist exploring the working of assortation, which might improve our scant understanding of the interactions between family and business [149]. Homophily in human society achieves less diverse but more harmonious economies [150], suggesting a role for assortation hitherto overlooked in economics. Assortation viewed in the light of the present theory might, for example, explain the ubiquity of corruption among human societies, and help biologists to better understand economic synergies found in the social phenomena they study, opening our interdisciplinary world view in a consilient and quantitative way [151].

Acknowledgments Thanks are due to Michael Hrncir, Cristina Sainz, Emilio Herrera, Rodolfo Jaffe, and Peer 1038 of Peerage of Science. The paper was written as a response to questions and discussions triggered by my plenary talk honoring the 50 years of Bill Hamilton Inclusive Fitness Theory at the International Ethology meeting organized by the Sociedade Brasileira de Etologia in Mossoro in 2014. A non-quantifiable number of people helped crystallize the ideas presented during the last few decades. These include long evenings with Bill Hamilton during the eighties in the Americas and later in his house in Oxford; intensive conversations with John Maynard Smith in the University of Sussex a few months before his death; conversations and correspondence with David Queller; suggestions from Peter Corning; among many others.

References

1. Wilson EO (1999) Consilience. Vintage Books
2. Jaffe K (2016) Extended inclusive fitness theory: synergy and assortment drives the evolutionary dynamics in biology and economics. Springerplus 5:1092. doi:10.1186/s40064-016-2750-z
3. Humboldt A (1807) Le voyage aux régions equinoxiales du Nouveau Continent, fait en 1799–1804, par Alexandre de Humboldt et Aimé Bonpland. Paris
4. Jaffe K (1996) The dynamics of the evolution of sex: why the sexes are, in fact, always two? Interciencia 21:259–267, 22:48
5. Dobzhansky T (1970) Genetics of the evolutionary process, vol 139. Columbia University Press, New York
6. Fisher RA (1958) The genetical theory of natural selection
7. Haldane JBS (1932) The causes of evolution. Harper & Row, New York
8. Huxley J (1942) Evolution: the modern synthesis. Allen and Unwin, London
9. Mayr E (1963) Animal species and evolution. Animal species and their evolution. Google Books
10. Darwin C (1859) On the origin of species by means of natural selection. http://ns1.ias.ac.in/resonance/Volumes/14/02/0204-0208.pdf
11. Wallace AR (1870) Contributions to the theory of natural selection. Google Books
12. Wilson EO (2000) Sociobiology: the new synthesis. Harvard University Press
13. Skyrms B, Avise JC, Ayala FJ (2014) In the light of evolution VIII: Darwinian thinking in the social sciences. PNAS 111:10781–10784
14. Hamilton WD (1963) The evolution of altruistic behavior. Am Nat 97:354–356
15. Hamilton WD (1964) The genetical evolution of social behaviour I and II. J Theor Biol 7:1–52
16. Segerstrale U (2013) Nature's oracle: the life and work of W.D. Hamilton. Oxford University Press
17. Maynard-Smith J (1964) Group selection and kin selection. Nature 4924:1145–1147
18. Allen B, Nowak MA, Wilson EO (2013) Limitations of inclusive fitness. Proc Natl Acad Sci U S A 110:20135–20139
19. Corning PA (2013) Rotating the Necker cube: a bioconomic approach to cooperation and the causal role of synergy in evolution. J Bioecon 15:171–193
20. Gardner A, West AA, Wild G (2011) The genetical theory of kin selection. J Evol Biol 24:1020–1043
21. Hölldobler B, Wilson EO (2009) The super organism. Norton & Co, New York
22. Maynard Smith J (1983) Models of evolution. Proc R Soc B 219:315–325
23. Price GR (1970) Selection and covariance. Nature 227:520–521
24. Price GR (1971) Extension of the Hardy-Weinberg law to assortative mating. Ann Hum Genet 35(4):485–490

25. Queller DC (1985) Kinship, reciprocity, and synergism in the evolution of social behavior. Nature 318:366–367
26. Queller DC (1992) A general model for kin selection. Evolution 46:376–380
27. Queller DC (2011) Expanded social fitness and Hamilton's rule for kin, kith, and kind. Proc Natl Acad Sci U S A 108:10792–10799
28. Fletcher JA, Doebeli M (2006) How altruism evolves: assortment and synergy. J Evol Biol 19:1389–1393
29. Hauert C, Michor F, Nowak MA, Doebeli M (2006) Synergy and discounting of cooperation in social dilemmas. J Theor Biol 239:195–202
30. Jaffe K (2001) On the relative importance of Haplo-Diploidy, assortative mating, and social synergy in the evolution of social behaviour. Acta Biotheor 49:29–42
31. Jaffe K (2004) Sex promotes gamete selection: a quantitative comparative study of features favoring the evolution of sex. Complexity 9(6):43–51
32. Jaffe K (2002) An economic analysis of altruism. Who benefits from altruistic acts? J Artif Soc Soc Simulat 5:3
33. Chuang JS, Rivoire O, Leibler S (2010) Cooperation and Hamilton's rule in a simple synthetic microbial system. Mol Syst Biol 6(1):398
34. Crespi BJ (2001) The evolution of social behavior in microorganisms. Trends Ecol Evol 16:178–183
35. Harcombe W (2010) Novel cooperation experimentally evolved between species. Evolution 64(7):2166–2172
36. Michener CD (1969) Comparative social behavior of bees. Ann Rev Entomol 14:299–342
37. Silva ER, Jaffe K (2002) Expanded food choice as a possible factor in the evolution of eusociality in Vespidae. Sociobiology 39:25–36
38. Jaffe K (2010) Quantifying social synergy in insect and human societies. Behav Ecol Sociobiol 64:1721–1724
39. Axelrod R, Hamilton W (1981) The evolution of cooperation. Science 211:1390–1396
40. Bronstein JL (1994) Our current understanding of mutualism. Q Rev Biol 69(1):31–51
41. Hoeksema JD, Bruna EM (2000) Pursuing the big questions about interspecific mutualism: a review of theoretical approaches. Oecologia 125(3):321–330
42. Corning PA, Szathmáry E (2015) 'Synergistic selection': a Darwinian frame for the evolution of complexity. J Theor Biol 371:45–58
43. Axelrod RM (1997) The complexity of cooperation: agent-based models of competition and collaboration. Princeton University Press, Princeton
44. Bonabeau E (2002) Agent-based modeling: methods and techniques for simulating human systems. Proc Natl Acad Sci 99(Suppl 3):7280–7287
45. Epstein JM (2006) Generative social science: studies in agent-based computational modeling. Princeton University Press. Macromolecules. Naturwissenschaften 58(10):465–523
46. Pepper JW, Smuts BB (2000) The evolution of cooperation in an ecological context: an agent-based model. Oxford University Press, New York, pp 45–76
47. Tesfatsion L, Judd KL (eds) (2006) Handbook of computational economics: agent-based computational economics, vol 2. Elsevier
48. Dawkins R (1989) The selfish gene, 2nd edn. Oxford University Press, p 192, ISBN 0-19-286092-5
49. Hausmann R, Hidalgo CA (2014) The atlas of economic complexity: mapping paths to prosperity. MIT Press
50. Jaffe K, Caicedo M, Manzanares M, Rios A, Florez A, Montoreano C, Davila V (2013) Productivity in physical and chemical science predicts the future economic growth of developing countries better than other indices for knowledge. PLoS One 8(6), e66239. doi:10.1371/journal.pone.0066239
51. Kenrick DT, Griskevicius V, Sundie JM, Li NP, Li YJ, Neuberg SL (2009) Deep rationality: the evolutionary economics of decision making. Soc Cogn 27:764–785, Special Issue: Rationality Restored

52. Jaffe K (2009) What is science?: An interdisciplinary perspective. University Press of America
53. Hamilton WD (1975) Innate social aptitudes of man: an approach from evolutionary genetics. In: Fox R (ed) Biosocial anthropology. Malaby Press, London, pp 133–153
54. Jaffe K (2002) On sex, mate selection and evolution: an exploration. Comments Theor Biol 7(2):91–107
55. Kuperman MN, Risau-Gusman S (2012) Relationship between clustering coefficient and the success of cooperation in networks. Phys Rev E 86:016104
56. Riolo R, Cohen M, Axelrod R (2001) Evolution of cooperation without reciprocity. Nature 414:441–443
57. Kim J (2010) A tag-based evolutionary prisoner's dilemma game on networks with different topologies. J Artif Soc Soc Simul 13:3
58. Fu F, Nowak MA, Christakis NA, Fowler JH (2012) The evolution of homophily. Sci Rep 3:845
59. Alvarez L, Jaffe K (2004) Narcissism guides mate selection: humans mate assortatively, as revealed by facial resemblance, following an algorithm of "self seeking like". Evol Psychol 2:177–194
60. Payne C, Jaffe K (2005) Self seeks like: many humans choose their dog-pets following rules used for assortative mating. J Ethol 23(1):15–18
61. Christakis NA, Fowler JH (2014) Friendship and natural selection. Proc Natl Acad Sci U S A 111(Suppl 3):10796–10801. doi:10.1073/pnas.1400825111
62. Centola D, Gonzalez-Avella JC, Eguiluz VM, San Miguel M (2007) Homophily, cultural drift, and the co-evolution of cultural groups. J Conflict Resol 51(6):905–929
63. Kossinets G, Watts DJ (2009) Origins of homophily in an evolving social network. Am J Sociol 115(2):405–450
64. Ochoa G, Jaffe K (2006) Assortative mating drastically alters the magnitude of error thresholds. In: Runarsson TP et al (eds) Lecture notes in computer science LNCS 4193, Parallel Problem Solving from Nature - PPSN IX. Springer, Berlin, pp 890–899
65. Massen JJM, Koski SE (2014) Chimps of a feather sit together: chimpanzee friendships are based on homophily in personality. Evol Hum Behav 35(1):1–8
66. Cooke F, Finney GH, Rockwell RF (1976) Assortative mating in lesser snow geese (*Anser caerulescens*). Behav Genet 6(2):127–140
67. Andersson S, Andersson M (1998) Ultraviolet sexual dimorphism and assortative mating in blue tits. Proc R Soc Lond Ser B Biol Sci 265(1395):445–450
68. Bowles S, Gintis H (2005) Can self-interest explain cooperation? Evol Inst Econ Rev 2(1): 21–41
69. Hamilton IM, Taborsky M (2005) Contingent movement and cooperation evolve under generalized reciprocity. Proc R Soc B Biol Sci 272(1578):2259–2267
70. Apicella CL, Marlowe FW, Fowler JH, Christakis NA (2012) Social networks and cooperation in hunter-gatherers. Nature 481(7382):497–501
71. Durrett R, Levin SA (2005) Can stable social groups be maintained by homophilous imitation alone? J Econ Behav Organ 57(3):267–286
72. Jaffe K (2014) The Wealth of Nations: complexity science for an interdisciplinary approach in economics. Amazon Books, Paperback November 18, 2014
73. Dolfsma W, Soete L (eds) (2006) Understanding the dynamics of a knowledge economy. Edward Elgar
74. Lewicki RJ, Saunders DM, Barry B (1985) Negotiation. Irwin, Homewood, pp 166–169
75. Liu HB, Huang ZH (2007) An empirical research on the synergy effects in the integration of industry chains—the perspective of circular economy theory. J Technol Econ 9:007
76. Taylor PD (2013) Inclusive and personal fitness in synergistic evolutionary games on graphs. J Theor Biol 21:76–82
77. Bettencourt LMA (2013) The origins of scaling in cities. Science 340:1438–1441
78. Hamilton WD (1996) Narrow roads of gene land, vol 1: evolution of social behavior. Oxford University Press, Oxford

79. Osborn F, Jaffe K (1997) Cooperation vs. exploitation: interactions between Lycaenid (Lepidopera: Lycaenidae) larvae and ants. J Res Lepidoptera 34:69–82
80. Krotopkin P (1902) Mutual aid: a factor of evolution at Project Gutenberg
81. Baranski B, Bartz-Beielstein T, Ehlers R, Kajendran T, Kosslers B, Mehnen J, Polaszek T, Reimholz Schmidt J, Schmitt K, Seis D, Slodzinski R, Steeg S, Wiemann N, Zimmermann M (2006) High-order punishment and the evolution of cooperation. Proceedings of the Genetic and Evolutionary Computation Conference 2006: 379–380
82. Nowak MA (2006) Five rules for the evolution of cooperation. Science 314:1560–1563
83. Nowak MA, Sigmund K (1998) The dynamic of indirect reciprocity. J Theor Biol 194:561–574
84. Nowak MA, Sigmund K (2005) Evolution of indirect reciprocity. Nature 437:1291–1298
85. Jaffe K, Zaballa L (2010) Co-operative punishment cements social cohesion. J Artif Soc Soc Simul 13(3):4
86. Ohtsuki H, Hauert C, Lieberman E, Nowak M (2006) A simple rule for the evolution of cooperation on graphs and social networks. Nature 441:502–505
87. Aristotle (IV BC). Politics: a treatise on government. Project Gutenberg
88. Smith A (1776) The Wealth of Nations
89. Hayek F (1949) Individualism and economic order. University of Chicago Press
90. Becker GS, Murphy KM (1994) The division of labor, coordination costs, and knowledge. In: Becker GS (ed) Human capital: a theoretical and empirical analysis with special reference to education, 3rd edn. NBER Books, Cambridge, pp 299–322
91. Jaffe K, Fonck C (1994) Energetics of social phenomena: physics applied to evolutionary biology. Il Nuovo Cimento 16:543–553
92. Jaffe K, Hebling-Beraldo MJ (1993) Oxygen consumption and the evolution of order: negentropy criteria applied to the evolution of ants. Experientia 49:587–592
93. Jaffe K (2015) Visualizing the invisible hand of markets: simulating complex dynamic economic interactions. Intell Syst Account Fin Manage 22(2):115–132
94. Ferrante E, Turgut AE, Duéñez-Guzmán E, Dorigo M, Wenseleers T (2015) Evolution of self-organized task specialization in robot swarms. PLoS Comput Biol 11(8):e1004273
95. Hamilton WD (1971) Geometry for the selfish herd. J Theor Biol 31(2):295–311
96. Woolley AW, Chabris CF, Pentland A, Hashmi N, Malone TW (2010) Evidence for a collective intelligence factor in the performance of human groups. Science 330:686–688
97. Marx K (1867) Das Kapital. DC Books (2008)
98. Georgescu-Roegen N (1970) The economics of production. Am Econ Rev 60(2):1–9
99. Beinhocker ED (2006) The origin of wealth – evolution, complexity, and the radical remaking of economics. Harvard Business School Press, Boston
100. Chatterjee S (1986) Types of synergy and economic value: the impact of acquisitions on merging and rival firms. Strateg Manage J 7:119–139
101. Dopfer K (1991) Toward a theory of economic institutions: synergy and path dependency. J Econ Issues 25(2):535–550
102. Evans P (1996) Government action, social capital and development: reviewing the evidence on synergy. World Dev 24(6):1119–1132
103. Ostrom E (1996) Crossing the great divide: coproduction, synergy, and development. World Dev 24(6):1073–1087
104. Bradley M, Desai A, Kim EH (1983) The rationale behind interfirm tender offers: information or synergy? J Financ Econ 11(1):183–206
105. Eun CS, Kolodny R, Scheraga C (1996) Cross-border acquisitions and shareholder wealth: tests of the synergy and internalization hypotheses. J Bank Financ 20(9):1559–1582
106. Larsson R, Finkelstein S (1999) Integrating strategic, organizational, and human resource perspectives on mergers and acquisitions: a case survey of synergy realization. Organ Sci 10(1):1–26
107. Barker JL, Barclay P, Reeve HK (2013) Competition over personal resources favors contribution to shared resources in human groups. PLoS One 8(3):e58826. doi:10.1371/journal.pone.0058826

108. Kiers ET, West SA (2015) Evolving new organisms via symbiosis. Science 348:392–394
109. Van Cleve J, Akcay E (2014) Pathways to social evolution: reciprocity, relatedness, and synergy. Evolution 3(68):2245–2258
110. Bourke AF (2014) Hamilton's rule and the causes of social evolution. Phil Trans R Soc B 369:1471–2970
111. Gadagkar R, Gadagkar R (2009) The social biology of *Ropalidia marginata*: toward understanding the evolution of eusociality. Harvard University Press
112. Itô Y (1993) Behaviour and social evolution of wasps: the communal aggregation hypothesis. Oxford University Press (OUP)
113. Jaffe K, da Silva ER (2002) Expanded food choice as a possible factor in the evolution of sociality of Vespidae (Hymenoptera). Sociobiology 39:1–12
114. Strassmann JE, Queller DC, Solis CR, Hughes CR (1991) Relatedness and queen number in the Neotropical wasp, *Parachartergus colobopterus*. Anim Behav 42(3):461–470
115. Eigen M (1971) Selforganization of matter and the evolution of biological macromolecules. Naturwissenschaften 58(10):465–523
116. Jaffe K, Cipriani R (2007) Culture outsmarts nature in the evolution of cooperation. JASSS 10/1/7
117. Jaffe K, Flórez A, Manzanares M, Jaffe R, Gomes CM, Rodríguez D, Achury C (2014) On the bioeconomics of shame and guilt. J Bioecon 1–13
118. Tollefson J (2015) Can randomized trials eliminate global poverty? Nature 524:150–153
119. Montoreano C, Jaffe K (2013) Relative importance of social synergy, assortation and networks in the evolution of social cooperation. http://arxiv.org/abs/1311.4407
120. Smith J, Van Dyken JD, ZeeJune PC (2010) A generalization of Hamilton's rule for the evolution of microbial cooperation. Science 328:1700–1703
121. Ivens ABF (2014) Cooperation and conflict in ant (Hymenoptera: Formicidae) farming mutualisms – a review. Myrmecol News 21:19–36
122. Dawkins R, Krebs JR (1979) Arms races between and within species. Proc R Soc Lond Ser B Biol Sci 205(1161):489–511
123. Jaffe K, Osborn F (2004) Simulation of coevolution in Batesian Mimicry. Naturalia 28:9–18
124. Zaman L, Meyer JR, Devangam S, Bryson DM, Lenski RE et al (2014) Coevolution drives the emergence of complex traits and promotes evolvability. PLoS Biol 12(12):e1002023. doi:10.1371/journal.pbio.1002023
125. Stilling RM, Bordenstein SR, Dinan TG, Cryan JF (2014) Friends with social benefits: host-microbe interactions as a driver of brain evolution and development? Front Cell Infect Microbiol 4:147
126. Jaffe K (2008) On the adaptive value of sex. In: Unifying themes in complex systems IV. Springer, Berlin, pp 213–221
127. Weisel O, Shalvi S (2015) The collaborative roots of corruption. Proc Natl Acad Sci U S A 112:10651–10656
128. Corning PA (1983) The synergism hypothesis: a theory of progressive evolution. McGraw-Hill, New York
129. Porter ME (2011) Competitive advantage of nations: creating and sustaining superior performance. Simon and Schuster
130. Grant RM (1991) The resource-based theory of competitive advantage: implications for strategy formulation. Knowl Strat 33(3):114–135
131. Ricardo D (1891) Principles of political economy and taxation. G. Bell and Sons
132. Jaffe K, Florez A, Grigorieva V, Masciti M, Castro I (2010) Comparing skills and attitudes of scientists, musicians, politicians and students. Interciencia 35(7):545–552
133. Jaffe K, Rios A, Florez A (2013) Statistics shows that economic prosperity needs both high scientific productivity and complex technological knowledge, but in different ways. Interciencia 38:150–156
134. Jaffe K, Levy-Carciente S (2004) Simulating barter, money and financial economies. http://atta.labb.usb.ve/Klaus/art156b.pdf

135. Jaffe K (2014) Social and natural sciences differ in their research strategies, adapted to work for different knowledge landscapes. PLoS One 9(11):e113901
136. Jaffe K, Perez E (1989) Comparative study of brain morphology in ants. Brain Behav Evol 33(1):25–33
137. Florida R (2005) Cities and the creative class. Routledge
138. Haken H, Portugali J (2003) The face of the city is its information. J Environ Psychol 23(4):385–408
139. Fuller RB (1975) Synergetics. Pacific Tape Library, pp 372–434
140. Haken H (1973) Introduction to synergetics. In: Synergetics. Vieweg+Teubner Verlag, pp 9–19
141. Corning PA (2007) Synergy goes to war: a bioeconomic theory of collective violence. J Bioecon 9(2):109–144
142. Jaffe K (2004) Altruism, altruistic punishment and decentralized social investment. Acta Biotheor 52:155–172
143. Eibl-Eibesfeldt I (1979) Biology of peace and war. Thames and Hudson
144. Lorenz K (1963) Das Sogenannte Böse, Zur Naturgeschichte der Aggression first published 1963 by Verlag Dr Borotha-Schoeler, Vienna, Austria. English version: (2002). On aggression. Psychology Press
145. Mirás B, Cabrera A, Jaffe K (2007) Mate selection in the moth *Neoleucinodes elegantalis*: evidence for a supernormal chemical stimulus in sexual attraction. Anim Behav 73:727–734
146. Tinbergen N, Perdeck A (1950) On the stimulus situation releasing the begging response in the newly hatched herring gull chick (Larus a. argentatus Pont). Behaviour 3:1–38
147. Jaffe K, Zaballa L (2009) Cooperative punishment and religion's role in the evolution of prosocial altruism. Chapter 13. In: Feierman JR (ed) The biology of religious behavior: the evolutionary origins of faith and religion. Praeger, 301 pp
148. Levitt SD, Stephen J, Dubner DJ (2006) Freakonomics: a rogue economist explores the hidden side of everything. HarperCollins, New York
149. The Economist (2015) Family companies: to have and to hold. April 18. The Economist Newspaper Limited
150. Wang C, Steiner B (2015) Can ethno-linguistic diversity explain cross-country differences in social capital?: a global perspective. Econ Rec. doi:10.1111/1475-4932.12195
151. Jaffe K, Febres G (2016) Defining synergy thermodynamically using quantitative measurements of entropy and free energy. Complexity. doi:10.1002/cplx.21800

A Case Study in the Emergence of Recursive Phrase Structure

Emília Garcia-Casademont

1 Phrase Structure Grammar

Phrase structures combine words, phrases, and both, into phrases. Human languages exhibit phrase structure, though it is unclear why, and this issue can be approached from various directions. The aim of this study is to explore the origins of phrase structure from an evolutionary point of view [1–3, 6, 8, 11].

We build further on analytical arguments where we proposed concrete selectionist criteria and showed that phrase structure could be motivated by the need to avoid combinatorial search in parsing and semantic ambiguity in interpretation, while always keeping communicative accuracy [4, 7, 9]. We introduce an operational minimal model of communication (a specific language game [10]), together with a language acquisition device [5], to explain how phrase structure can be achieved and acquired at the individual and collective level.

Surprisingly, the grammars evolved by this operational model also exhibit recursion.

2 Recursion

Human languages exhibit recursion. Recursion typically refers to *syntactically structural recursion*, i.e., the presence of syntactic constituents containing elements of the same syntactic type as the whole constituent. See Fig. 1,

where the subindices just indicate different occurrences of the syntactic types.

E. Garcia-Casademont (✉)
Institut de Biologia Evolutiva (UPF-CSIC), DEXS, UPF, Barcelona, Spain
e-mail: emilia-maria.garcia@upf.edu

P. Bourgine et al. (eds.), *First Complex Systems Digital Campus World E-Conference 2015*, Springer Proceedings in Complexity, DOI 10.1007/978-3-319-45901-1_35

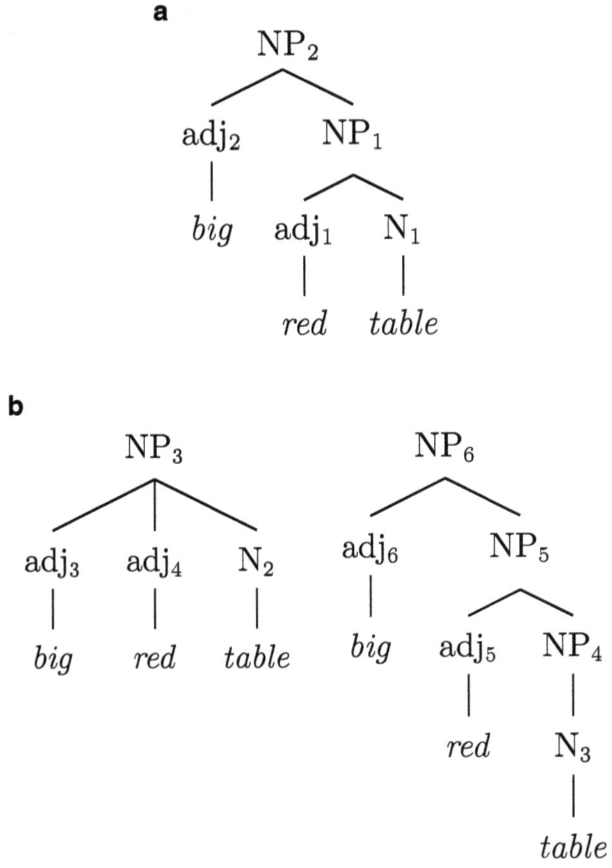

Fig. 1 Example of syntactic recursion: $NP \rightarrow adj + NP$

However, the linguistic analysis (the tree) depends obviously on the formalism and rules used to generate the linguistic tree. The same phrase could be analyzed otherwise in the following two ways, as well as in various other ways.

While the left tree in Fig. 1b doesn't exhibit recursion, the right tree exhibits it two times, both NP_4 and NP_5 are syntactically structural recursive.

Considering a sequence of rules that generates a linguistic tree, we define *processing recursion* as the fact of a rule applying to an input containing an element which is on its own the result of the application of this same rule. Among the previous examples (Fig. 1a, b), only the right tree in Fig. 1b exhibits processing recursion: the rule $NP \rightarrow adj + NP$ applies to NP_5, which is on its own the result of applying $NP \rightarrow adj + NP$ to adj_5 and NP_4.

Besides *syntactically structural recursion* for linguistic trees and *processing recursion* for sequences of rules generating linguistic trees, we define *semantically*

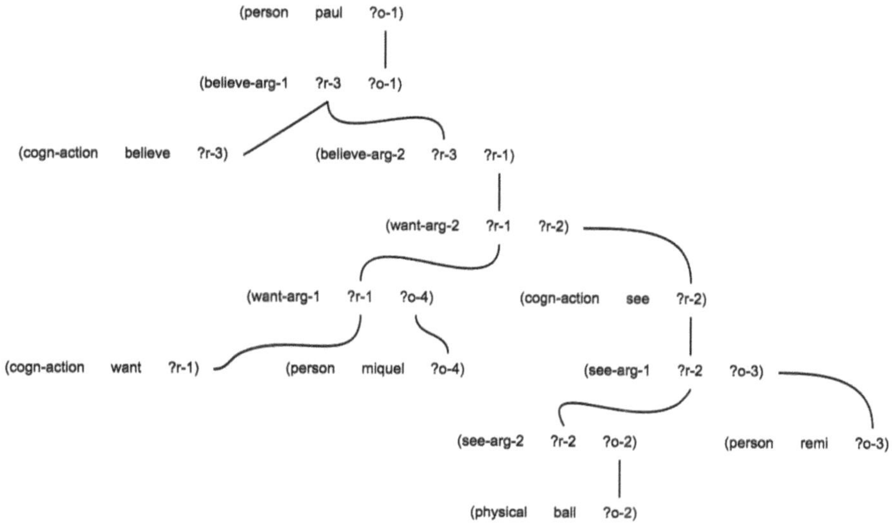

Fig. 2 Meaning representation of the meaning of the utterance: *Paul believes Miquel wants Remi sees ball*

structural recursion for meanings. Meanings are represented using predicate calculus. Each predicate has a variable introducing the entity to which the predicate applies, e.g., (cogn-action see ?r-2) represents that a predicate "see," which is of semantic type "cogn-action," introduces "?r-2" as the variable for the entity itself. Moreover, the arguments of a predicate have their own predications, each of which introduces another variable for the corresponding argument. For example, (see-arg-1 ?r-2 ?o-3) and (see-arg-2 ?r-2 ?o-2) are the predications for the first and second arguments of the entity represented by ?r-2, while ?o-3 and ?o-2 are the respective variables for the arguments. Finally, variable bindings are used to show various roles of the same entity, e.g., in Fig. 2, ?o-2 represents both the variable introduced by a predicate "ball" and the variable introduced by the second argument of a predicate "see," which means that the same entity is performing these two roles.

A meaning exhibits *semantically structural recursion* when the semantic type of one of the arguments of a predicate is the same as the semantic type of the predicate itself. In Fig. 2, "cogn-action" is the semantic type of "?r-3" and also the semantic type of "?r-1," which is the second argument of "?r-1". The same occurs between "?r-1" and "?r-2," therefore this meaning exhibits semantically structural recursion two times.

Summing up, we distinguish between three kinds of recursion: *syntactically structural recursion* (syn), *processing recursion* (pr), and *semantically structural recursion* (sem).

The table below represents how these recursions can be combined. $a_{ij} = +$ implies that Rec_i must occur together with Rec_j and $a_{ij} = +/-$ implies that Rec_i can occur both together with Rec_j and without Rec_j.

	syn	pr	sem
syn		$+/-$	$+/-$
pr	$+$		$+/-$
sem	$+$	$+(>1)/-$	

(3)

3 Presentation

We have run experiments where phrase structure grammars evolve and immediately exhibit the kinds of recursion defined previously. I will discuss in detail the three definitions of recursion and present a detailed example of the emergence of phrase structure grammar exhibiting at least one type of recursion. The presentation will be accompanied by a live web demo so that the potential of the system can be appreciated.

References

1. Beuls K, Steels L (2013) Agent-based models of strategies for the emergence and evolution of grammatical agreement. PLOS ONE 8(3):e58960
2. Bleys J (2010) Language strategies for the domain of colour. Ph.D. thesis, VUB
3. De Beule J (2007) Compositionality, hierarchy and recursion in language. A case study in fluid construction grammar. Ph.D. thesis, VUB
4. Garcia-Casademont E, Steels L (2014) Strategies for the emergence of first-order phrase structure. In: Evolang-X, Vienna, pp 50–57
5. Garcia-Casademont E, Steels L (2015) Usage-based grammar learning as insight problem solving. In: EAP CogSci conference
6. Maynard Smith J (1986) The problems of biology. Oxford University Press, Oxford
7. Steels L (1995) A self-organizing spatial vocabulary. Artif Life J 2(3):319–332
8. Steels L (ed) (2012) Experiments in cultural language evolution. John Benjamins, Amsterdam
9. Steels L, Garcia-Casademont E (2015) Ambiguity and the origins of syntax. Linguist Rev 32(1):37–60
10. Steels L, Garcia-Casademont E (2015) How to play the Syntax Game. In: ECAL conference
11. van Trijp R (2008) Analogy and multi-level selection in the formation of a case grammar. A case study in fluid construction grammar. Ph.D. thesis, UA

From "Traces" and "Human Trace" to "Human-Trace Paradigm"

Béatrice Galinon-Mélénec

1 Foreword

The question of trace as an object of research in the context of complex systems comes quite naturally in terms of multidisciplinarity. To articulate all these approaches, without merging them, we propose to gradually build an innovative form of research networks around the subject of "Trace". In the two previous conferences, I presented:

(a) the paper *The future of the "Homme-trace" A substantial societal challenge*, published in *Networks and Communication Studies,* vol. 28/201[1];
(b) at some events organized by the e.lab. Human-Trace complex System UNESCO (2013–2014).

Or the foundations of *a humanism of the trace*. Pr. Béatrice Galinon-Mélénec, full professor, UMR CNRS IDEES LH, (https://www.linkedin.com/pub/b%C3%A9atrice-galinon-m%C3%A9l%C3%A9nec/60/10a/931), Founder of the laboratory on The Human-trace DC Complex System UNESCO (https://en.wikiversity.org/wiki/Portal:Complex_Systems_Digital_Campus/E-Laboratory_on_human_trace). galinon@free.fr.

[1]Finally, we emphasised how the process of trace visibility worked: we concluded that, in our view,

B. Galinon-Mélénec (✉)
Normandie Université, UNIHAVRE, UNICAEN, UNIROUEN, CNRS, IDEES, 76600 Le Havre, France
e-mail: galinon-melenec@univ-lehavre.fr; galinon@free.fr
URL: http://www.umr-idees.fr/?l-homme-trace

P. Bourgine et al. (eds.), *First Complex Systems Digital Campus World E-Conference 2015*, Springer Proceedings in Complexity, DOI 10.1007/978-3-319-45901-1_36

This time, I'm putting forward an approach which aims to illustrate the Human-trace paradigm to a very diverse audience. In consequence, Scientific terms specific to the speciality[2] will be avoided.

2 Introduction

The explosion in multidisciplinary research works on trace has led to the precision here, in a language that is transversal to academic disciplines and players, of what we call the *ICHNOS-ANTHROPOS.* This new French anthropological paradigm is likely to found decisive anthropological semiotics when the veil must be lifted on the processes that produce what ordinary language calls "trace". This procedure would appear essential in a context where use of the term "trace" is spreading to all fields without a real meaning being fixed. Let us cite as an example of frequent usage: "digital traces"[3] or "cognitive traces" or "traces of pollution", etc. We have explored these different areas as well as the difference between *trace*, *imprint*[4] in different works and articles.

We will not take this up here. We invite interested parties to refer to our publications in CNRS editions[5] or *Intellectica.*[6]

it corresponds to processes of reciprocal excitation of traces, leading to what we call 'the echoing of signes-traces' which itself defines the relationship". See : https://en.wikipedia.org/wiki/Human_Traces#Other

In this chapter, we have seen how essential a role trace plays in this. As a consequence, we invited the reader to:

- question what is the relevance of a one-to-one correspondence between traces and signe,
- develop a wider understanding of interactions and relationships between trace and signe, the role of relationality and interdependence in any situation.
- establish how these interpretations of the trace correspond with complex interactions.

[2]More scientific terms? Please refer to the text of the e. session 2014 (open source); Galinon-Melenec [1].

[3]More scientific terms? Please refer to the text: Galinon-Melenec and Zlitni [2].

[4]We already said in the 4th International Conference on Complex Systems and Applications: "*A distinction must be drawn between the terms "trace" and "imprint". The origin of the term "imprint" (originally as "emprint" – late Middle English) is from Old French empreinter, based on Latin imprimere, from in- "into" + premere "to press", whose original meaning (1250) was "to stamp (a mark or outline) on a surface". The term "trace" covers a greater degree of general connotations and nuances. Our notion of trace includes the imprint that is a connoted trace of a more accentuated marking. On this basis, any imprint is a trace but any trace is not an imprint".*

[5]Galinon-Melenec [3].

[6]Galinon-Melenec Béatrice, "Des signes-traces à l'Homme-trace. La production et l'interprétation des traces placées dans une perspective anthropologique", dans Mille [4], pp. 89–113. (*From "Sign-Traces" to "Human-Trace"*, translated L. Brown, Hong Kong, 2013).

3 Trace

Human beings have sought, trace by trace, to go back in time to the origins of the universe and this search is evidence of the essential role held by the trace in the minds of all humans seeking to understand their place in Evolution.[7] Today, the quest culminates in the past, dating to between 10 and 15 billion years[8] when there was only unorganized matter, a mush burning at billions of billions of degrees. Current knowledge stops at this stage, known as the “Big Bang”. As there is no trace of what came before, the postulate tends to be the hypothesis that the Big Bang was the origin of the universe. The framework has been laid. The trace retraces time from trace to trace (Fig. 1).[9]

Derrida had already posited this idea in another way: “*Every trace is the trace of a trace. No element is never present anywhere (nor simply absent), there are only traces*” [5].[10,11]

---

[7]But, if this logic has enabled us to go back in time to the Big Bang, it does not prove that by using another form of logic, it would not be possible to gain access to what happened before it.

It is not because our logic has proved itself on earth that it is necessarily relevant for the whole of the universe.

[8]Estimations vary.

[9]We already said so in the 4th International Conference on Complex Systems and Applications: *“In March 2014, the astrophysicists who have been on a quest to find traces of a fundamental wave responsible for the origin of the universe think they have perceived a fossil vibration of the universe. It appears to be the oldest trace of the world’s early development to have been discovered. Il is important to take account of the fact that this was only made possible because scientists started looking for its existence, that is, from the moment it was assumed that the Universe began with the Big Bang, followed immediately by primordial waves. We will return later to the abduction process underlying this discovery. Thus, palaeontology and astrophysics are making considerable advances in what we know about humanity and the history of the universe. Their progress, in conjunction with observation of traces, shows how fundamental the question of trace is for the representation humans have of themselves and their place in the universe. This journey into history is equally relevant as it makes it possible to attempt to conceive human sustainable development. Furthermore, building on the ever increasing capacities of technology, research is making great strides in a number of major fields of knowledge including that of the human body (molecular genetics, genetic transmission, the brain, etc.)”.*

[10]Bennington and Derrida [6].

[11]We already said so in the 4th International Conference on Complex Systems and Applications: *“We had noted that the selection among traces of the same sign by two individuals might lead to a difference in signals according to how they are interpreted, i.e. to two distinct* signes-signaux. *In other words, the same* signe-trace *could be seen by two individuals as two completely different* signes-signaux.

The ex-post analysis of a communicative relationship can help to demonstrate that signes-traces *have echoed one another. This* echoing comes from interactions between signes-traces *that are entwined in their complexity. The nature of the relationship established cannot be reduced to the rational knowledge of the identity of individuals and communication frameworks. What underpins the relationship profoundly is taken to be a process whereby* signes-traces *are echoes of one another, the complexity of which is beyond explanation:* “If you press me to say why I loved him, I feel that I can say no more than it was because he was he, and I was I. There is, over and above all my discourse, and all that I can say in particular, some inexplicable and fatal force fostering this union.” (MONTAIGNE M., *“Of Friendship” in Essays, 1580).”*

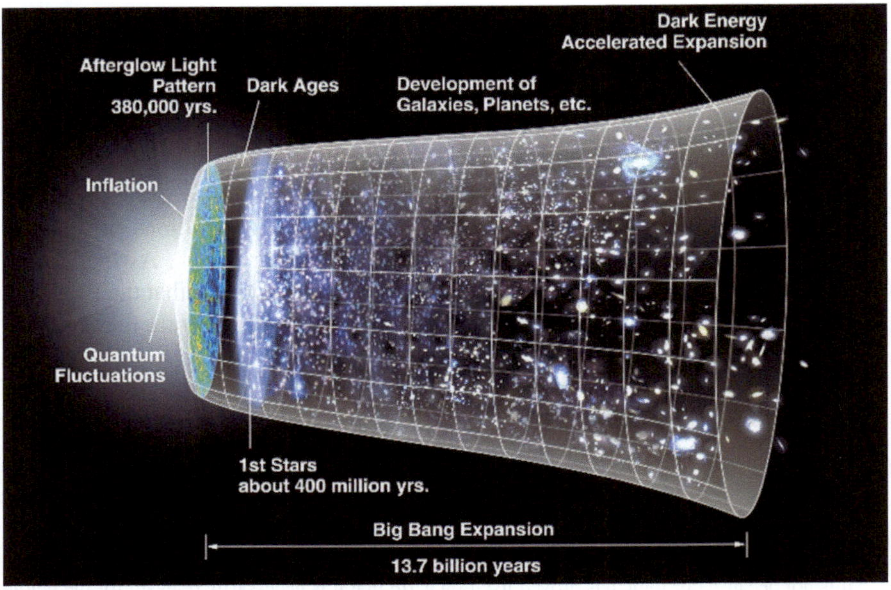

Fig. 1 Timeline of the universe, from the Big Bang to the present day.http://www.nasa.gov/

The complexity of the notion of trace comes from the fact that everything happens simultaneously in interactions, in multi-scale systems and the relationship. To obtain a clearer idea of the complexity, the processes that build the relationship between humans and reality and produce meaning must be made visible, must be taken apart, must be "*deconstructed*".[12]

4 From Trace to "Signe-Trace" and Processual Traces

We have used the term "trace" because it is a term in common usage[13] and considered as part of the collective perception which, although polysemous, can introduce the subject in question. But, in fact, we must distinguish the visible[14] trace which we call "*signe-trace*" (sign-trace) from the "*trace processuelle*" (processual trace). *By "sign-trace", we mean that what we can see at the present instance merits the term "sign" as much as the term "trace".*

---

[12]Cf. Derrida, op.cit.

[13]Cf. *Intellectica,* op.cit.

[14]In our paradigm, when the trace is identified as a trace, it becomes a *signe-trace* and, "If any sign is, in fact, a "signe-trace", a trace is not necessarily a sign".

We of course bear in mind Roland Barthes' warning that "*What happens in opposition to the signs and outside signs is quickly identified as a sign*" [7].[15] But we maintain that the sign[16] contains more than what it appears to contain, which is the process that constructed it and from which it issues: the "*processual trace*". The sign that we can see comprises an ambiguous "in-between" status[17] in which it is, in the present, the manifestation of a process of transformation. One might even consider that it is an implicit idea of association of the sign as a "processual trace" (and therefore of a "sign-trace") that the paleontologist or astrophysicist—to mention only these two scientific disciplines—has been able to reconstruct, "sign-trace" after "sign-trace", the history of humanity or the history of the cosmos.

The world in which we live is therefore constructed of "processual traces" and what we see are sign-traces. We posit that the principle of "processual traces" is apparent everywhere, in matter as well as in any life, at least since the Big Bang, given that we know of nothing before then. The conclusion is that if, since the Big Bang, the universe has become more and more complex, the present as a whole bears the "processual traces" of this complexity which continues its development. Hence the interest we have in locating the question of the trace within a complex context.

5 The *Human-Trace* (*Ichnos-Anthropos*)

Researchers observing contemporary society, its culture and techniques can also analyse it from the perspective of the trace.[18] We have opened up an avenue for research in positing that "*any human can be considered both as a construct of traces*

[15]Barthes [7].

[16]We already said in the 4th International Conference on Complex Systems and Applications: "*In our paradigm "L'Homme-Trace", we posit that the complexity of the real is inaccessible to the human mind. When a human distinguishes signs in a continuum of the real, it is the result of a sorting process; yet this process is linked to his/her history. An individual's history is an uninterrupted continuity of interactions with a human and non-human environment. These interactions produce a magma of traces inside the individual and they are themselves in the form of continuous and constantly renewed interactions. As a result, when humans perceive the world, they have the impression that it is a juxtaposition of emerging images of discontinuity whereas it is not the world itself (...).*

"*If reality (the entire universe) is defined as a complex referential inaccessible to humans as a whole, and if the universe is evolving (thus leaving in the matter and in the universe the* traces processuelles *of this evolution), it must follow that humans have access to only fragments of the real. The fragment selected by an individual represents what a human calls a "sign". However, the selection and interpretation of the sign is filtered and processed by the brain according to the "embodied traces" which are also* the traces processuelles *of its history*".

[17]The present, an indefinite time between the past and the future.

[18]Galinon-Melenec [3].

and as a producer of traces, the whole forming one system", which is what we have named *"the paradigm of the Human-Trace"*.[19] Here we can put forward a new illustration.

The twentieth century provided a great deal in the way of helping to understand our origins. The history of the world appeared more clearly to us and taught humans that they were the result of an evolution that began with the universe: *"The first particles, the atoms, molecules, stars, cells, organisms, living beings, right up to the curious animals that we are ... They all follow one another in the same chain, they are all carried along by the same movement*"[20]; with the precision: *"The elements that make up our bodies are those that formerly founded the universe"* (*Ibid.* p 9).

We base our argument on this breakthrough in science and on the inescapable corporeal dimension of human beings to illustrate our definition of the *Ichnos-Anthropos* (or *"Homme-Trace"*—Human-Trace).[21] This signifies that every human being's body includes traces, not only of the whole history of Evolution, but also those of all interactions with the environment since birth and even traces of those that precede it in his/her genealogy. In this sense, we can therefore say that humans are a "construct of traces". But they are not only that.

Everybody can observe that they produce, *a minima*, traces of respiration in space (and consequently modify it) and as soon as they move, they produce traces of their displacement. Obviously, the traces produced by human beings are more numerous than *"primary traces*". We can also distinguish *"secondary traces"*—among which there are the "technical ones", also called "artefact sign-traces"—and the *"tertiary traces"*—for example, the rules of law which are produced by humans with a view to regulating their relationships.

The definition of the Human-Trace (*Homme-Trace*)[22] functions perfectly well when one studies the history of humanity.[23] If one considers *Homo erectus*, he left traces in the way in which he occupied space (he adapted areas for sleep, others for meals, etc.) and mastered stone-splitting much better than his predecessors *homo habilis* and *homo ergaster.* When humans produced radically different "secondary traces", paleontologists qualified them as *homo sapiens*. One can distinguish two types: the "dextrous, creative" Neanderthal possessing an elaborate language[24] and *Cro-magnon*, who seemed "more culturally and biologically" equipped[25] than the Neanderthal and who survived him (Figs. 2 and 3).

[19]Open source: Galinon-Melenec [1].

[20]Reeves and Simmonet [8], pp. 9–10.

[21]This term has been translated into English by the *Human-Trace* and more generally by *Ichnos-Anthropos (ichnos = trace; anthropos = man).*

[22]*HUMAN-TRACE: "Man is both a producer of traces and a construct of traces, operating in a loop, a system in which each builds the other in a continuum"* [9].

[23]Cf. http://www.unesco.org/culture/humanity/.

[24]*La plus belle histoire du monde*, *op.cit*, p. 165.

[25]*Ibid*, p. 166.

Fig. 2 Ancient palaeolithic, Prehistory. From 2,500,000 to 300,000 year. http://www.histoire-en-maquette.com/

Fig. 3 Iron Age, ancient La Tène. From 400 BC to 200 BC. http://www.histoire-en-maquette.com/

The discovery of bodies makes it possible to observe that at each of these periods, the body has been modified.[26] The body has become taller and more upright (Fig. 4).

The brain has changed proportionally as tools have become more complex. This observation tends to confirm the definition of a systemic relationship existing

[26]In the 4th International Conference on Complex Systems and Applications (2014), We already posit that:

"- Humans and Milieu never cease to mutually influence one another through a system of interactions with traces;

- By observing a milieu, signes-traces *(signs of traces) of human existence can be identified.*

For us, these two dimensions are indissociable. Concerning technology, if it is recognised today that humans' use of it leaves traces in their bodily matter – in their brain – the notion of milieu *will be used when reference is made to its symmetrical and immediate influence on the technology itself, on its matter; both as continual and retroactive influences".*

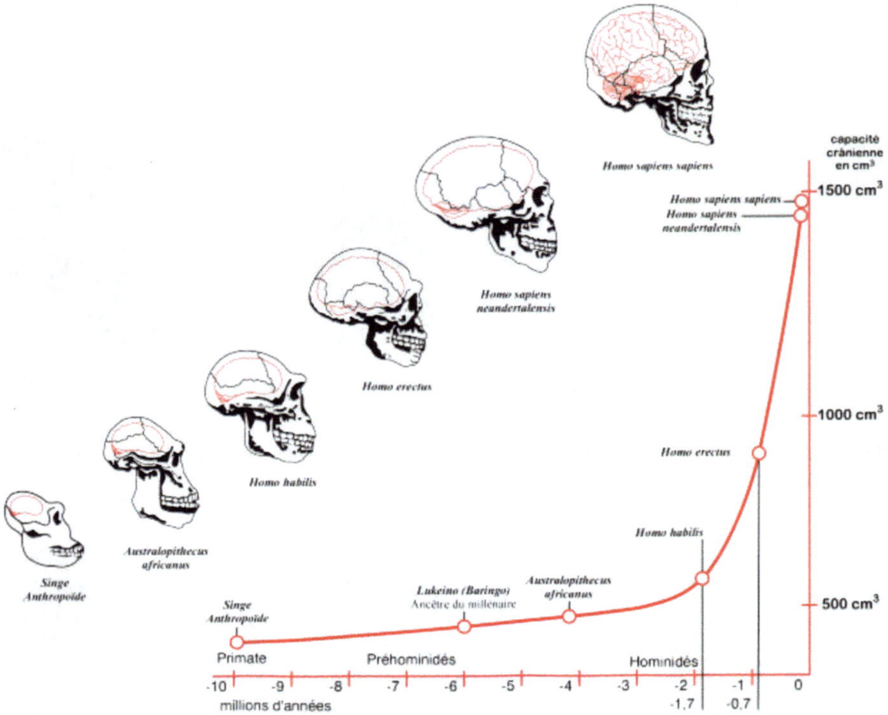

Fig. 4 Evolution of the brain. http://www.museummarseille.org/marseille_cerveau_evolution.htm

between the production of external traces (here the technique) and the production of internal traces (here linked to the ever greater activation of neuronal connections).[27]

Today, evolution continues. Technical, cultural and social mutations engender increasingly interdependent systems. A new world, an *artefact-world has appeared,* engendering *digital socialization.* The systemic dynamics of the Human-trace continues its work: with a more or less intensive use of digital technology, the brain continues its transformation (Fig. 5).

6 Traces of the Body

Aware now of the link existing between the evolution of the brain and the interactions of human beings with their environment, the twenty-first century researcher closely examines the plasticity of the brain.[28] Let us give the example of Dr Hugues

[27]Galinon-Melenec et al. [10].

[28]For more scientific terms please refer to:

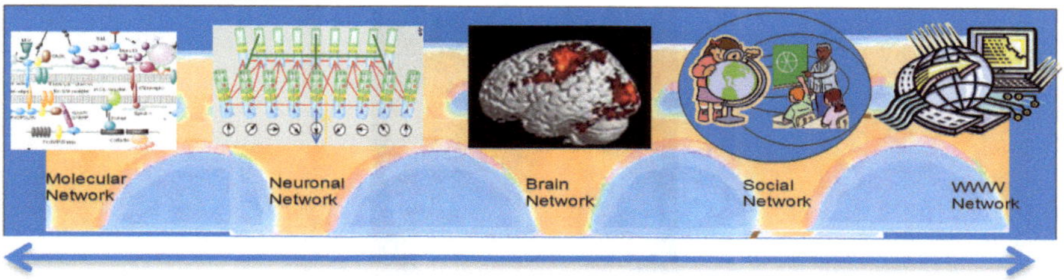

Fig. 5 Human traces/body. Example of interactions. Presented in E. laboratory human-trace UniTwin Complex System UNESCO, ICCS'14

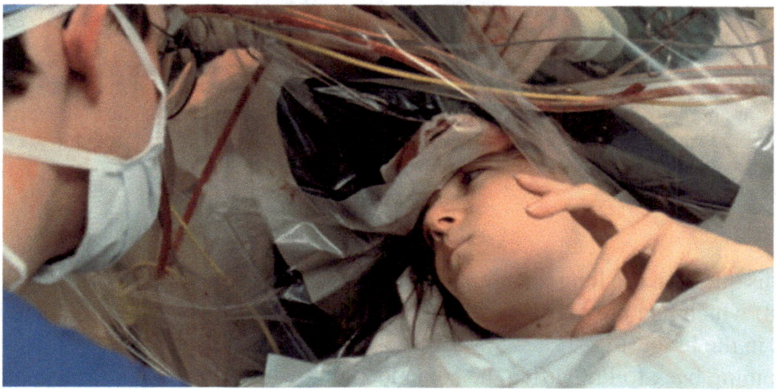

Fig. 6 © James King-Holmes/SPLY/COSMOS. Illustration: http://www.sciencesetavenir.fr/sante/20141009.OBS1652/video-operer-les-tumeurs-du-cerveau-sans-endormir.html

Duffau's study[29] which, based on *human brain mapping*[30] and what brain tumour patients feel, he operates on "the brain in total consciousness" (Fig. 6).

By making patients express what they feel when he stimulates different parts of the open brain, he was able to pinpoint that, while he was removing Broca's area (assumed to be that of speech) affected by the tumour, millimetre by millimetre, the patient continued to express himself in coherent sentences. He concluded from this that the relationship between "the cartography of the brain and tasks accomplished" should be called into question in the twenty-first century.

– Galinon-Melenec [11]

– Galinon-Melenec Béatrice, Martin-Juchat Fabienne, "Du genre social au genre incorporé: le 'corps genré' des SIC" in Bernard F. et Loneux Ch. (dir.), Recherches au féminin en sciences de l'information et de la communication, *Revue Française des Sciences de l'Information et de la Communication* n°4. En ligne sur http://rfsic.revues.org/857.

[29]Hugues Duffau, a University Professor at Montpellier, is a neurosurgeon and researcher in neurosciences ("a field which examines the neurobiological mechanisms which underlie cognition: motor function, perception, emotions, reasoning, language, memory, etc.". Cf. Duffau [12].

[30]Cartography of the human brain.

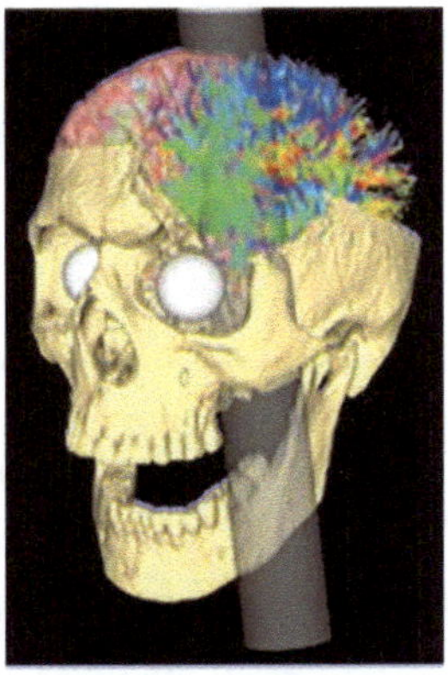

Fig. 7 Phineas Gage "On September 13, 1848, Phineas Gage, an American railroad worker, was injured in an explosion in which an iron rod passed through his brain. Against all expectations, he recovered from this accident, but his behaviour was radically altered. By studying his injuries, scientists gained a better understanding of the functions of the frontal lobe". Joan M.K. Tycko. Cf. http://thebrain.mcgill.ca/

This question is a long-standing one as demonstrated by the case of Phineas Gage (1848) which had previously led Dr Penfield to ask questions about the function of areas in the brain (Fig. 7).

Coming back to Pr. Duffau, he concluded that where the brain is concerned, one should not talk of "areas" but of "connections". These are capable of carrying out a reorganization [31] of what remains after the ablation. His conclusion was that there is a "*cerebral plasticity*" and that the distribution of the brain's functions is not fixed.

Post-operative imaging enabled him to see that the distribution of the brain's functions varied in the same patient, taken at two different moments in his life.[32] Imaging also enabled him to observe that it varied from one patient to another. This observation is very interesting in relation to our hypotheses. It confirms the idea that every individual's interactions with his/her environment are unique and leave specific traces within the person's body. The processes of interaction are performed constantly, the bodily traces of interactions evolving with them.

Certain traces of interactions are more striking than others—notably when the person is in a strong emotional state (Figs. 8 and 9).[33]

In this case, they can be transformed into imprints that are transmissible over several generations, as emerging research in epigenetics proves, which shows that

[31] Starting with *traces processuelles*? We suggest examining this hypothesis.

[32] DUFFAU H, *op cit,* p. 43.

[33] The role of psychic traces inscribed in the bodily matter should be stressed: embodied (in-body) interactions of humans with their milieu produce psychic traces. These are externalised in the milieu in the form of behaviours and practices, "*signes-traces*" of embodied psychic traces. The transformation into a *signe-trace* occurs as soon as another body perceives it.

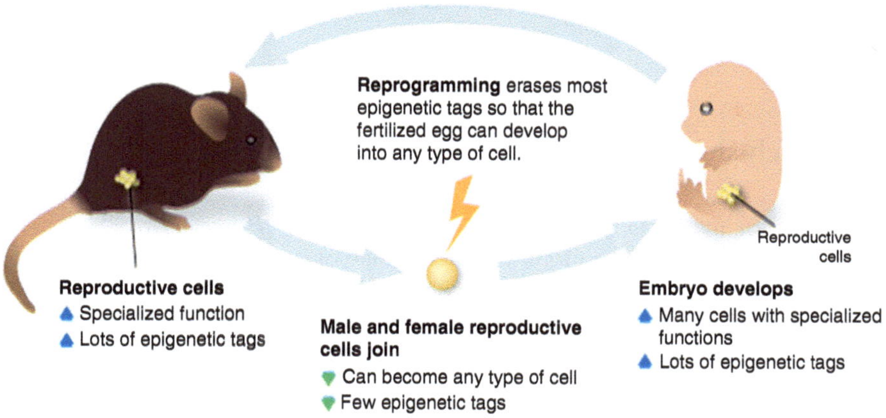

Fig. 8 Traces of stress. http://learn.genetics.utah.edu/

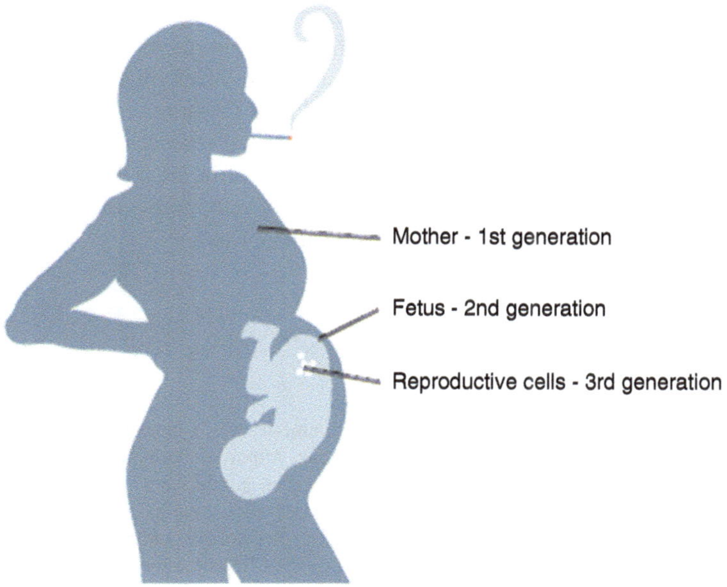

Fig. 9 Traces transmissible over several generations. http://learn.genetics.utah.edu/

the traces of traumatisms experienced three generations ago can be detected in the descendants' behaviour.[34,35,36]

7 Conclusion

In conclusion, the *Ichnos-Anthropos*—also known as the "*Homme-Trace*" (Human-Trace) to recall the French origin of the anthropological paradigm—encompasses the whole human species in which it is conceived as carrying within it the traces of history, not only the history of humankind but also that of the world. It therefore incorporates today's *sapiens sapiens* as well as those that preceded it: *homo habilis, homo ergaster*, *homo erectus* and *homo sapiens*. It also includes the humans of tomorrow who, travelling in space, will see their bodies being modified through interactions with a different environment than the one they are familiar with today.

The history of the Human-Trace is not over; each day that passes sees it evolving. It is impossible to predict its future but one can believe that by having become aware of the anthropological characteristics of *Ichnos-Anthropos,* it changes the perception of itself and its relationship with the environment: he becomes capable of understanding the role of the trace in sustainable development not only that of humanity but also that of the planet Earth within the Cosmos.

This is a considerable challenge. To take it up, all researchers working on the trace whatever their discipline are invited to take part in the e. laboratory Human-trace complex system UNESCO https://en.wikiversity.org/wiki/Portal:Complex_Systems_Digital_Campus/ELaboratory_on_human_trace.

References

1. Galinon-Melenec B (2014) The future of the "Homme-trace". A substantial societal challenge. Netw Commun Stud 28(1–2):107–130. En ligne sur : http://netcom.revues.org/1554
2. Galinon-Melenec B, Zlitni S (dir.) (2013) Traces numériques. De la production à l'interprétation, Paris, CNRS éditions, série L'Homme-trace tome 2, 274 pp
3. Galinon-Melenec B (dir.) (2011) L'Homme trace, Perspectives anthropologiques des traces humaines contemporaines, Paris, CNRS éditions, série L'Homme-trace tome 1
4. Mille A (dir.) (2013) De la trace à la connaissance à l'ère du Web. Intellectica n° 59
5. Bennington G, Derrida J (1991) Jacques Derrida. Paris, Seuil, Série les contemporains
6. Bennington G, Derrida J (1999) Jacques Derrida. University of Chicago Press, Chicago
7. Barthes R (1983) Empire of signs. Paperback
8. Reeves H, Simmonet D (1996) La plus belle histoire du monde. Points Seuil n° 897

[34] Francis [13].

[35] This example can also illustrate the fact that we apply the term "sign-trace" to behaviour.

[36] Fish et al. [14].

9. Galinon-Melenec B (2010) Traces internautiques et recrutement. Colloque L'Homme@ distance, Université du Havre
10. Galinon-Melenec B, Lienard F, Zlitni S (dir.) (2015) L'Homme-trace, Inscriptions corporelles et techniques, L'Homme-trace, tome 3, CNRS éditions
11. Galinon-Melenec B (2013) "Expérience incarnée, construction cognitive et jugement: Le rôle des "signes-traces" du corps dans la signification" dans. In: Boutaud JJ, Berthelot-Guiet K (dir.) "La vie des signes au sein de la communication : vers une sémiotique communicationnelle" Revue Française des Sciences de l'Information et de la Communication. En ligne sur http://rfsic.revues.org/487
12. Duffau H (2015) Le cerveau en toute conscience. In: Bourgine L et Bourgine J. (dir.) Ces Français qui révolutionnent la médecine. Editions de la Martinière
13. Francis RC (2012) Epigenetics - how environment shapes our genes. W. W. Norton & Company, Edition: Reprint (15 June 2012)
14. Fish EW, Shahrokh D, Bagot R, Caldji C, Bredy T, Szyf M, Meaney MJ (2004) Epigenetic programming of stress responses through variations in maternal care. Ann N Y Acad Sci 1036:167–180
15. Descola P (2013) Beyond nature and culture, Janet Lloyd (trans.). University of Chicago Press, Chicago
16. Eco U (1990) Introduction. In: Lotman YM (ed) Universe of the mind: a semiotic theory of culture. I. B. Tauris, London, pp vii–xiii
17. Hawking S (1998) A brief history of time, Paperback – Unabridged, September 1, 1998
18. Morin E (2008) La complexité humaine. Flammarion, Paris
19. Simondon G (1989) L'individuation psychique et collective. Aubier, Paris
20. Varela FJ, Thompson E, Rosch E (1991) The embodied mind: cognitive science and human experience. MIT Press, Cambridge

Foundation of CS-DC e-Laboratory: Open Systems Exploration for Ecosystems Leveraging

Masatoshi Funabashi, Peter Hanappe, Takashi Isozaki, AnneMarie Maes, Takahiro Sasaki, Luc Steels, and Kaoru Yoshida

1 Introduction

1.1 Environmental Problems of Agriculture

Humans have engaged in agriculture for over 10,000 years since the dawn of history, but that history has been inevitably based on the trade-off between agricultural production and environmental degradation [10, 62]. Modern agriculture is still situated along the same line, further loading the environment to realize physiological optimum in large-scale monoculture. The excess practice of conventional agriculture is now considered as a major factor of environmental degradation in both terrestrial and marine ecosystems that threatens the sustainability of our society by triggering irreversible global state shift of the environment [5]. Several International organizations provide scientific reports on the environmental degradation caused

All authors contributed equally to this article. The authors names are listed in an alphabetical order.

M. Funabashi (✉) • T. Isozaki • T. Sasaki • K. Yoshida
Sony Computer Science Laboratories, Inc., 3-14-13, Higahi-Gotanda, Shinagawa-Ku, Tokyo, Japan
e-mail: masa_funabashi@csl.sony.co.jp; masa314159265358979@gmail.com; isozaki@csl.sony.co.jp; sasaki@csl.sony.co.jp; kaoru@csl.sony.co.jp

P. Hanappe • L. Steels
Sony Computer Science Laboratory Paris, 6 rue Amyot, 75005 Paris, France
e-mail: hanappe@csl.sony.fr

A. Maes
OKNO, Koolmijnenkaai 30, 1080 Brussels, Belgium

So-oN, Vlaamse Steenweg 66, 1000 Brussels, Belgium
e-mail: annemie@okno.be

P. Bourgine et al. (eds.), *First Complex Systems Digital Campus World E-Conference 2015*, Springer Proceedings in Complexity, DOI 10.1007/978-3-319-45901-1_37

by intensive practice of conventional agriculture and associated market structure [34, 38, 79, 86, 87, 92, 93]. They commonly propose alternative models should take priority on small-scale local farming, environment-constructive method of cultivation, fair distribution, and equitable trade.

1.2 Sensing Technologies

Recent development of sensing technology has drastically changed the precision and scope of measurement from molecular to ecosystem levels. In the level of ecosystems, remote sensing technologies are extensively applied to the measurement in large-scale agriculture and assessment of ecological state. The application has a wide range such as international assessment of climate change [68], estimation of river inundation [72], weed control [80], prediction of drought [58], resource management [28], agroecosystem modeling [14], forestry planning [36, 55], estimation of crop biomass [85], and so on. The research trend is focused mainly on 2 areas, conventional agriculture systems and global observation of ecosystems: Sensing of agricultural ecosystems aims to gain specific information necessary for the prediction and control. While sensing of global ecosystems tries to capture global indicators associated with climate change in large scale.

There is, however, little application on the crossing area, especially on small-scale sustainable agriculture where biodiversity and cultural productivity are firmly related [8, 56]. Prediction model based on remote sensing data does not sufficiently incorporate biodiversity in its models. Small-scale farming between natural ecosystems and human disruption realizes rich ecosystems, which is similar to the biodiversity hotspot known as ecotone [90]. The importance of conservation effort of hotspots is due to its instability, which is considered to attribute to the chaotic behavior inherent in ecosystem dynamics [12, 65]. Such complexity requires methodologies of complex systems science besides the global measurement by remote sensing. We need to develop a framework to connect between diverse sensing methods and each ecosystem agent, which would provide rich indicators on the dynamics of biodiversity.

1.3 Biodiversity Database

Biodiversity is a key concept to assess the environment, evaluate the ecosystems services, and measure the effect of human disruption. Biodiversity comprises plural indicators describing the degree of variation, ranging over diversity within species, between species and of ecosystems from all sources [60]. Sensing of biodiversity in small scale still demands manual labor and classification of species by human. Important conservation issues based on the environmental assessment always require such precise information. Biodiversity and taxonomy database follows international initiative in view to describe the whole species existing on the earth

[17, 22, 23, 29, 44, 45, 47, 84]. These databases are complementary to remote sensing measurement, and both should be assembled to obtain a multi-scale interpretation of an ecosystem state. To establish a fine-grained measurement compatible to the concept of biodiversity, we have to proceed from several paths, such as refinement of sensing methods, models of biodiversity, and human observation, through an integrated development of technology, modeling, and practice.

Biodiversity is also finely related with local diet and health in traditional food, which is compatible to the growing needs in sustainable agriculture [50]. Foods are major factors of health amidst recent life style changes [31, 91]. It is important to explore the relation between environment, food, nutrition, and human health with the use of nutritional and biodiversity database.

1.4 Citizen Science and Virtual Platform

Development of information technologies and invention of mobile terminal such as smartphone has introduced a novel realm of interaction in citizen science. Diversity record related to geographical information has become possible to organize participation of individuals on site. The cloud sourcing of data collection is expected to be more accessible to local activities and essential to consider management on the diversity [26]. Integrated expression of environmental data on virtual globe software such as Google Earth is rapidly gaining its popularity. Participatory use of such platform can gather expert knowledge as well as lay-people observation, bridge the gap between professionals and amateurs, and is compelling to raise public awareness on the management including ethical discussion [71]. Virtual platform of environmental information with geographical expression can be a prominent platform to integrate sensing data, biodiversity database, and citizen observation.

1.5 Challenge of the e-Laboratory: Toward Citizen Prototype of Sustainable Agriculture—Exploring Ecosystem Agents as a Symbiotic Interface Between Human and Nature

Our e-laboratory gathers scientists, engineers, and artists to tackle the sustainability issues in food production and ecosystems management, ranging from urban gardening to agricultural production, from on-the-fly exploration to integrative simulation, from human to environmental health, and from data analytics to artistic expression [16]. Sensing technologies, databases, analytics, and citizen science will be combined in order to realize the leveraging of ecosystem services via human activity. The initial projects comprise six projects, namely Synecoculture (Sect. 2), P2P Food Lab (Sect. 4), Open Systems Data Analytics (Sect. 3), The Bee Laboratory (Sect. 5), Open Systems Simulation (Sect. 6), and One-Health Food Lab (Sect. 7).

Our challenge aims to yield the potentials of self-organizing nature of ecosystems with the bottom-up organization of each project. By exploring ecosystem agents as a symbiotic interface between human and nature, this challenge would consequently be led to develop diverse, parallelly distributed practices of sustainable food production and related scientific domains.

(Masatoshi Funabashi)

2 Synecoculture

2.1 An Alternative: Polyculture System with Ecological Optimum

Plants have been one of the main agents that transformed terrestrial environment through surprisingly intelligent adaptation and coevolution [27]. Ecosystems have been evolved by developing complex networks containing both competitions and symbiotic relations as a result of ecological optimum [66], which formed today's biosphere we live in. Indeed, natural ecosystems compatible to agricultural field such as grassland and forest is known to express symbiosis-dominant effect on biomass production with respect to the species diversity [2, 33]. Such productivity based on the ecological optimum is not yet sufficiently exploited in agricultural framework. The application of ecological optimum has been partially practiced in agriculture by perennial crops polyculture [24, 25, 43], no-till farming [37], natural farming [19], etc. The difficulty of cultivation method lies in the optimization of management cost and the productivity with a context of marketing. The challenge requires strong support of information technologies with open complex systems perspectives [21].

Considering the bottom-up emergence of biodiversity in ecological optimum, the measurement and control is difficult to achieve in a centralized top-down manner. We propose an alternative way, the organization of citizen science to collect and share the experience of diversified polyculture practice. By sharing the data and knowledge with the use of information technology, citizen science has a possibility to realize both diversity and flexibility of measurement and management, which may be more compatible to the dynamics of ecosystems [21].

2.2 Methodology of Synecoculture

As a polyculture system with ecological optimum, we introduced a novel system of agriculture based on the synthesis of ecosystems, namely "Synecoculture," mainly for the culture of vegetables and fruits [20]. Synecoculture is a high density mixed polyculture of edible species, that stems from the observation of natural ecotone and biodiversity hotspots. We associate plants according to their symbiotic interactions with soil, environment, and other vegetation, which augments the biodiversity of the culture beyond natural state. The fields overlap with a transition to mature secondary vegetation with human use [63]. The management of culture is based on the diversity and succession, similar to the forest renewal. We first plan the vegetation according to the environmental condition, and let the ecosystem self-organize the products with least human intervention: No tillage, no fertilizer, no chemicals, other than introduction of edible species. We harvest from the formation of ecological niche, in which thinning harvest from mixed and dense vegetation is effective for both year-round harvest and weed control. This system brings us quite diverse products all round year with extended culture seasons and thus suitable for local daily consumption.

2.3 *Experiments of Synecoculture*

Experiments of Synecoculture take place in several fields including professional farms and family gardens in Japan. Edible plant species numbers introduced in the proof of concept experiments are listed in Table 1. The edible species diversity introduced in Synecoculture fields, nothing but on monitored 0.3 ha, exceeds the item number of the traditional Satoyama agriculture in a regional scale, and is compatible in terms of cultivar variety. Although the growth of the plants are distributed in a long-tail and do not necessary survive after competition, human introduction of plant diversity interacts with both local flora and fauna, that creates a unique augmented ecosystem in ecological optimum [21]. This implies that

Table 1 Edible plant species diversity introduced during Synecoculture experiment in 2010/6–2014/5

Ise farms	Todoroki farm	Oiso farm	GIAHS Noto
134(133) species	173(157) species	170(158) species	173 items
–	467 varieties	426 varieties	705 varieties
–	215(196) species, 673 varieties		
Total 263(247) species			

Ise, Todoroki, and Oiso farms are Synecoculture farms in Mie-ken, Tokyo, and Kanagawa-ken in Japan, respectively. The total experimental surface of these farms are about 0.3 ha. Other citizen farms also exist with their own initiatives. Numbers in parentheses describe the number of species that survived more than 1 year after introduction or produced the seeds. For a comparison, GIAHS Noto is the UNESCO-certified globally important agricultural heritage system (GIAHS) in Noto peninsula, Japan, with 186,600 ha surface covering the northern half of Ishikawa-ken, renowned by its high biodiversity conservation value [32]

Synecoculture has a high potential to increase agro-biodiversity even in a small scale. Items produced were sold at a commercial farm (Ise) on site, at local restaurants, including on-line sales of 80 products all over Japan.

2.4 Exploration of Ecosystem Agents as an Interface

The productivity in ecological optimum is constrained by environmental conditions. Basic physical parameters such as temperature and precipitation are known to be the major determinants of vegetation [35, 48]. Sunlight also plays central role in photosynthesis. These parameters vary depending on local geographical condition and existing vegetation. At the same time, actual plant community varies depending on the seed bank, competition, vegetation succession, delayed action of past conditions [88], etc., within the same climate classification. The bidirectional dependencies between physical conditions and vegetation are an essential basis of ecosystems as complex systems. To treat the local diversity of plant community and make use of it in polyculture system, we need to know about the actual components and dynamics of ecosystems agents besides physical parameters of the environment.

Since ancient times, phenological events such as blossoming of flowers and certain animals' behavior were considered to be key indexes to detect the seasonal phase of ecosystem dynamics. With the development of ecological database, it is possible to link such phenological biodiversity observation with practical prediction and control of ecosystems management, including farming system with ecological optimum. With the aid of database, sensing, and communication technology, we can attribute each ecosystem agent to the property of interface for the management of ecosystems. Generally an interface is an element between two systems, which intermediates between inputs and outputs for certain objective. For example, harvest of edible species is the output from ecosystems, while planting, releasing, and introducing them correspond to input. One can consider each ecosystem agent as an interface to sense and make decision of input and output on ecosystem. The functional roles of plants and animals can be used as relational indexes for ecosystem management. With the development of interactive database between biodiversity observation and ecological knowledge, we try to explore and share practical framework to make use of ecosystems agents as an interface of sustainable agriculture.

As an example, Fig. 1 shows the comparison between different ecosystems with common species based on the database of Synecoculture [21]. Geographically distant ecosystems can be compared in terms of biodiversity with a time line, from which we can estimate what kind of species can be commonly introduced, and what is the condition that makes the difference. In this example, the earlier appearance of the species in Tokyo than in Paris may indicate a correlation with meteorological data such as temperature since spring comes earlier in Tokyo, which provides a candidate of effective physical parameters. In application, the species appearance can be interpreted as the biological reaction to environmental

Fig. 1 Comparison of species diversity between different ecosystems on Synecoculture contents management system. *Bottom*: Observed species in Paris, France on 2013/6/15. *Top*: Corresponding species observation in Tokyo Japan, whose appearance seasons range over several months

condition in total. When the culture condition is not equivalent and impossible to control each factor in an open environment in operation, the similarity of ecosystem reactions may consequently be an effective index for the management, as is the case in farming folk wisdom. In any region of the world, we can find diverse relations between climate, animals and plants that infer ecologically consistent structure. These include actual causal structure of ecosystems, and also the so-called pseudo-correlation in a massive data analysis, the correlation without direct causality but still applicable to the actual management of complex systems. These symbols and relations form certain universal structure in human cognition in various culture, which is pointed out to be essentially common to scientific thinking [52]. To determine the factors that support a species niche is necessary to diversify the polyculture in ecological optimum, which is a complex entanglement that depends on environmental condition, associated biodiversity, farming option, etc. The analysis is further developed in collaboration with Open Systems Data Analytics, which is described in Sect. 3.

(Masatoshi Funabashi)

3 Open Systems Data Analytics

There is a growing necessity for dealing with open systems such as computer networks, healthcare, and global climatic changes. The data related to ecology and agriculture are also exponentially increasing, such as biodiversity database, monitoring data of agricultural fields, buying history of products, nutrition profile, and geoenvironmental assessments. Such "big data" is gaining greater importance on the future of human society. In Open Systems Data Analytics, we are trying to develop novel analytical tools of massive data that have many variables dependent on each other, using probabilistic graphical models and methodologies of physics [40, 42].

In open systems, it is difficult to understand and control the whole systems due to their complex interactions and dependencies. Data oriented approaches are effective for understanding open systems when the subsystems are not explicitly defined. From the viewpoint of data analysis, we consider that causal relation inference from observational data is useful for understanding and managing open systems when possible practices of controlled experiments are limited during operation. Although the true cause-effect relationships are difficult to prove completely due to many unobserved variables, it is possible to make effective interventions as we augment the accuracy of the causal relationship inference in the systems [64]. A causal analysis may thus remain at the level of providing some suggestions for managing systems and repeatedly improve it with the introduction of possibly latent variables and new data.

We therefore apply Tokoro's open systems science [81, 83] with adaptation to a causal data analysis, and propose a methodology, which we call (a version of) Open Systems Data Analytics, as follows:

1. Define the problem and its domain.
2. Set variables and get their data.
3. Construct a causal model by causal inference algorithm from data.
4. If it is difficult to interpret causal relations in the model, important latent variables probably exist. Infer them and add data of the new variables to the dataset as far as possible.
5. Repeat the procedure (3)–(4) until a satisfactory result (or consensus) is obtained.
6. Do predictions, interventions, and/or construct theories.

More precisely, the definition of causal relationship is adopted from the Reichenbach's principle of causality on the sets of three (or more) variables [67]. It is based on the finding of directed Markov patterns in correlation networks for random variables by removing pseudo-correlations with statistical tests.

3.1 Analysis of Synecoculture Database

In Synecoculture project, polyculture with ecological optimum requires a huge amount of information on biodiversity, interactions, and vegetation succession to optimize as a productive system. Such information contains huge number of parameters, generally sparse, possibly biased, open-ended, etc., because it relies on human observation. Still, it can bring useful information and intriguing insight on the management if powerful algorithmic analysis is combined with appropriate human evaluation. Open Systems Data Analytics can combine massive data analysis tools with diverse feedbacks by humans working on site, so that to maximize the synergy between background knowledge, computational power, and human intuition. We apply the methodology to maximally augment educational effect of participants. The aim is different from conventional science that seeks for the strong reproducibility and predictability of phenomena with external observation, but rather to explore multiple choices for a better management from inside of open complex systems in operation, where internal observation is structurally inevitable.

We here demonstrate an example of the causal data analysis in Open Systems Data Analytics, the first step of the procedures (1)–(4). We analyzed biodiversity data obtained from Synecoculture farms and surrounding environment in Japan between 2011/4–2013/3. The data are the binary occurrence records of observations about plants, insects, date, and places, which defines the process (1) and (2). The analysis of the process (3) was performed using an algorithm called Combining Stage (CS) [41]. From 11,911 observations that comprise 1232 variables (1150 plants and animals, 3 years, 12 months, 67 places), 1131 sets of Markov dependency were detected on 611 variables.

Figure 2 Left shows the inferred causal model with a directed graph. We evaluated the suggestion with respect to the observers' experience and classified into three categories:

1. Trivial such as the causality between observation places and discarded from the analysis.
2. Unknown to the experience or other literatures, therefore candidates for further observation.
3. Known to the experience or other literatures, therefore validation instance of the model.

We then evaluated the information of 2, the suggestions unknown to experience by the following information quantity $I(c)$ for each unknown suggestion with a strength of inferred causality c:

$$I(c) = -\log\left\{\frac{\#(\text{known suggestions with inferred causality} \geq c) + 1}{\#(\text{known suggestions}) + 1}\right\},$$

where #() represents the number of suggestions, and the additions of 1 correspond to count a chosen unknown suggestion itself. The information $I(c)$ for each unknown

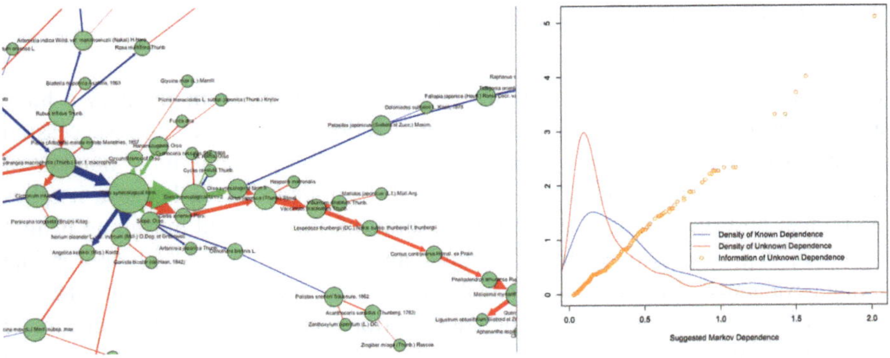

Fig. 2 *Left:* Graph visualization of the causality suggestion. The *circles* and the *edges* denote variables and Markov dependencies between two variables, and the *line thickness* denotes strength of dependencies quantified by mutual information. For variables A and B in the figure, if $A \rightarrow B$, A denotes a cause of B in the sense of Markov dependence. Colors of edges indicate that the suggestion was 1. trivial (*green*), 2. unknown (*red*), and 3. known (*blue*) to the observer's experience. An undirected edge means that there is a Markov dependence but without significant inference of causal direction. *Right:* Example density distributions of unknown (*red line*), known (*blue line*) causality suggestions and information (*orange circles*) distribution of unknown suggestions

suggestion is defined as the information of significance (p-value in statistical tests) with respect to the known suggestion distribution as a model. It represents the amount of information that the model would gain as a reward if the unknown suggestion was revealed to represent plausible dependence by further observation. The $I(c)$ is 0 when there is no known suggestion. It expands the value range as validated known suggestions increase, which reflects the plausibility of the model with respect to the accumulation of human observation and experience.

Figure 2 Right shows the example distributions of unknown, known causality suggestions, and distribution of $I(c)$ for each unknown suggestion with respect to the inferred causality. Unknown causality suggestions with high $I(c)$ exist, which are the preferential candidates for future observation. By incorporating these suggestions and by iterating the step 5 of Open Systems Data Analytics, it is expected to modify the observation framework by means of internal observation so that to achieve a better fit between causality analysis and observers' experience.

For further analysis, we plan to add environmental parameters such as temperature, humidity, precipitation, and hours of sunlight, in order to incorporate objective measures that do not depend on human subjectivity. Furthermore, since ecosystems inherently contain transient dynamics and farming options could change according to the objective of production, we expect to repeat and integrate the analysis year-by-year to investigate the degree of achievement in the steps 5 and 6.

(Takashi Isozaki and Masatoshi Funabashi)

4 P2P Food Lab

The goal of the P2P Food Lab project is to develop innovative and sustainable solutions for the growing, harvesting, transporting, consumption, and disposal of food. We evaluate new technologies that support new modes of crop production, in particular small-scale and micro-agriculture. As we are fully aware that technology alone cannot solve the current crises that our society is facing, including the challenges facing agriculture today [13, 15, 18, 73], P2P Food Lab includes a strong social aspect and aims to build alternative solutions using a bottom-up approach that involves all stakeholders, including citizens. The first phase of the P2P Food Lab project is to design an on-line/offline platform that groups the four main pillars—agriculture, communities, technology, and science—and to leverage the network effect to engage many people and increase the social impact of the platform.

4.1 P2P Food Lab Starter Kit

In our first experiment, during Summer 2014, we developed a "Starter Kit" for micro-agriculture that consisted of a small, Internet-connected greenhouse (Fig. 3). A sensor box was placed inside the greenhouse that took daily images of the crops and measured the air temperature, air humidity, and sunlight. The sensor box was built using standard off-the-shelf components such as Arduino, Raspberry Pi, and standard webcams. The goal of the connected greenhouse was to create an on-line

Fig. 3 *From left to right*: (1) The first P2P Food Lab greenhouse near Paris, (2) Subsequent version of the greenhouse in Brussels, (3) Sensor box with camera, (4) Screenshot from the web site, (5) Time-lapse of the radishes and weather data, and (6) Children from participating school near Paris

Fig. 4 *From left to right*: (1) The seeding calendar, (2) A 1.2 × 1.2 m plot with seedlings, (3) Matrix of photos uploaded by participants, and (4) Visualization of recorded environmental data

social network of participants and gardeners, and evaluate the use of sensors for the study of crop growing.

4.2 *The CitizenSeeds Experiment*

The Starter Kit still represented an entry point that was too steep to engage many people. We also found that we needed to give participants clearer guidelines in order to obtain reusable data. Consequently, we simplified the requirements for participation in Summer 2015. In the new experiment, called CitizenSeeds, participants only needed a camera-equipped mobile phone and a 1 m^2 plot of land (raised bed or plain soil) (Fig. 4). We also defined a fixed collection of seeds to plant and a fixed, shared planting schedule. These two elements greatly help in aligning the participants, comparing the data, and stimulating social interaction. To measure the environmental data (sunlight, air temperature, and soil humidity) the participants had the option to use the Flower Power device produced by the Parrot company. They also had the possibility to buy soil for their plot to normalize the soil used in the experiment, however, few participants chose this option.

Participants are asked to upload photos of their plot and the vegetables once a week. About 80 people registered of which about 30 contribute to the experiment. A single web page displays the states of the plots and the environmental data of the community.[1]

4.3 *Future Developments*

The Citizen Seeds project seems to have the right level of ingredients to evolve into an on-line platform for micro-agriculture. The main long-term developments that we envision are

- Distributed organization: Scaling up from tens of participants to thousands of participants will require the introduction of new organizational *scaffolding* structures [51].
- Citizen Science: Introduce well-defined protocols for planting and measuring in order to perform scientific experiments with the involvement of many amateur gardeners. Topics of interest to be studied include analysing the effects of crop intermixing [6, 11, 54, 57], measuring the evolution of soil biology [1, 4], tracing the phenotypical adaptation of varieties to local conditions.
- Increase plot size: Develop similar experiments and data sharing for larger plots and with (semi-)professional participants. The biggest interest is to study alternative farming techniques [49] that have been studied very little by agronomists until now, including Synecoculture, permaculture [7, 61], and bio-intensive

[1] See https://p2pfoodlab.net/CitizenSeeds/experiments/4.html.

micro-farming [46]. The platform also allows the community explore new technological tools and sensors to monitor and optimize these cropping techniques, using a collaborative approach [9, 39].

(Peter Hanappe)

5 The Sound Beehive Experiment

5.1 Overview

The Sound Beehive Experiment monitors the development of a bee colony on the basis of the sounds it generates, and creates artistic expression mediated by artificial intelligence. For this purpose, we developed a beehive that is equipped with sensors, microphones, and cameras. The Sound Beehive is installed in the Urban Bee Laboratory on a rooftop in the Brussels city centre. Data is streamed to central repositories and analyzed using statistical techniques and graphic visualizations.

5.2 Introduction

5.2.1 An Ethological Approach

Honeybees are bio-indicators. They provide a constant stream of information on the environment in which they forage, via their daily activity, and via the pollen and nectar they harvest. Environmental problems such as the use of pesticides can be detected by monitoring the colonies with audio and video tools and by scanning their daily activity over several years [59]. In nearly all industrialized nations, bee colonies are now threatened. The compromised state of the foraging areas for bees is worrisome. By using bees as bio-indicators and by translating the information into artworks, this project aims to make citizens aware of the increasingly negative effects of our lifestyle and methods of industrial production. AnneMarie Maes is a media artist collaborating with computer scientists and engineers to develop art-science projects. Interested in showing the hidden structures in nature, we try to use innovative technological methods to probe the living world.

5.3 Methods

5.3.1 Basic Elements

To study the bees in their natural environment, following the footsteps of von Frisch and other ethologists [89], we have built a customized "sound device." Microphones

Fig. 5 *Left*: Diagram of the instrumented Sound Beehive, *Right*: the Scaffolded Sound Beehive Sculpture

inside the beehive enable us to continuously monitor the colony's buzz. Together with outside and inside video monitoring it forms a non-intrusive scanning device for controlling the colony's health and development. We also installed a network of temperature and humidity sensors spread throughout the beehive. The annotated video and audio data are uploaded to our open source video database [74]. All time-stamped sensor data from the lab's weather station, as well as the temperature and humidity data measured inside the beehive, are made public on [75].

5.3.2 Instrumentation

Our custom-built audio, video, and sensor device are integrated in a Warré beehive (Fig. 5). It is a sustainable beehive in which the colony develops at its own pace. We started to customize our Warré beehive by putting electret microphones in the top cover and by attaching contact microphones in the frames of the brood box. All microphones are connected to pre-amps stored in the rooftop. They are powered by a battery that is located a few meters away from the hive to avoid the creation of electro-magnetic fields.

For recording the video images, we use Raspberry Pi computers. The Raspberry can be easily integrated in complex installations and is equipped with a series of USB and Ethernet connections to function in a network of devices. We integrated two small high-resolution cameras in our setup; one camera to record the activity on the landing platform and a second infrared camera to registrate the activity inside the brooding box.

The analysis of the images gives us information on the relation of the bees to the environment. A beecounter is integrated in order to determine the in/out flux and detect homing problems related to pesticide contamination. The images also give us information about the pollen supply and the development of the colony related to the activity level of forager bees, fanning bees, dead bees, and lazy bees on the landing platform.

5.3.3 Sensing of Bee Activity Related to the Environment

A bee colony is very responsive to the biotopes of which it is a part. The production of honey is dependent on the flowers we grow, the plants we like, and the garbage or pollution we produce. The colony is also very sensitive to environmental variables such as outside temperature, rainfall and humidity, the wind and hours of sunshine. We therefore compare the behavior of the bees and the development of the colony with the data from the weather station. In our rooftop field lab, we have installed a Libellium agriculture kit [53], including several environmental sensing devices. For example, the hours of solar activity, as well as the soil composition, determine the nectar flow of the flowers and their visits by the bees. Nectar secretion increases as pollinators visit the flower.

We set up a database of the pollen contained in the honey of our urban bee colonies and we started to determine the pollen source. By studying the pollen in a sample of honey, it is possible to collect evidence of the geographical location and genus of the plants that the honeybees visited [76]. As such, we start to trace green corridors through the city.

5.3.4 Data Processing

In January 2015 we started analyzing the recorded files. We scanned the sound files in terms of their brightness, loudness, and noise level. For the analysis of the video files, we made use of motion detection via the frame difference method. The analysis of the sound files is a complex matter. We therefore use techniques of Artificial Intelligence in collaboration with the Brussels Free University. We have recorded large amounts of data in order to investigate whether we can detect patterns. All together these data give us plenty of parameters to combine and to play with, to create models, and to compare different moments in time and thus to study the behavior of the colony relative to timeline/season and environmental parameters.

5.4 *Visual Expression of the Sound Beehive*

A video shows a graphical rendering of AI analysis of colony behavior combining real audio data with measurements of the microclimate inside the hive: temperature, CO2, and humidity. Another video shows 365 days of activity inside a real observation beehive, played back at high speed. The images were recorded with an infrared camera inside the hive and processed using pattern recognition, AI, and computer graphics algorithms. These images offer the stunning visual experience of a bee colony in action (Fig. 6). To create an immersive sound installation we analyzed the sound files recorded in the hive. The Scaffolded Sound Beehive (Fig. 7) is a wooden sculpture, constructed using open source digital fabrication and mounted on scaffolds of 2.5 m high. Visitors can enter this upscaled model of the Warré beehive and experience an auditory artistic interpretation of hive activity. We processed the recordings (made in the real beehive) using sophisticated pattern recognition algorithms and artificial intelligence analysis software, and edited the sound files by adding swirling electronic sound clusters to sonify the ebb and flow of swarm activity in the hive.

5.5 *Art Exhibition*

The Sound Beehive immersive installation has been shown at the Institute of Evolutionary Biology (IBE) in Barcelona (May–June 2015) [77], and at the AI and the Arts exhibition for the international conference of Artificial Intelligence in Buenos Aires, Argentina (July 2015) [78]. The enormously positive response of viewers shows clearly that the presentation of scientifically inspired art can have a strong impact and raises awareness of important societal issues, and also that art-inspired science can have a fruitful positive effect to push science in new directions.

(AnneMarie Maes)

6 Open Systems Simulation

6.1 *Motivation*

We as a modern society are facing many urgent yet unsolved problems, including the possible global shortage of natural resources and food, environmental pollution, economic instability, poverty, crisis of medical and health care systems, social insecurity, computer and network fragility, and so on. All of these are problems of open systems, which is literally a system that is open to the outer world and shows temporal development as interactions with the outer world progress [82]. As a consequence, we can neither have full control over the systems nor restart from

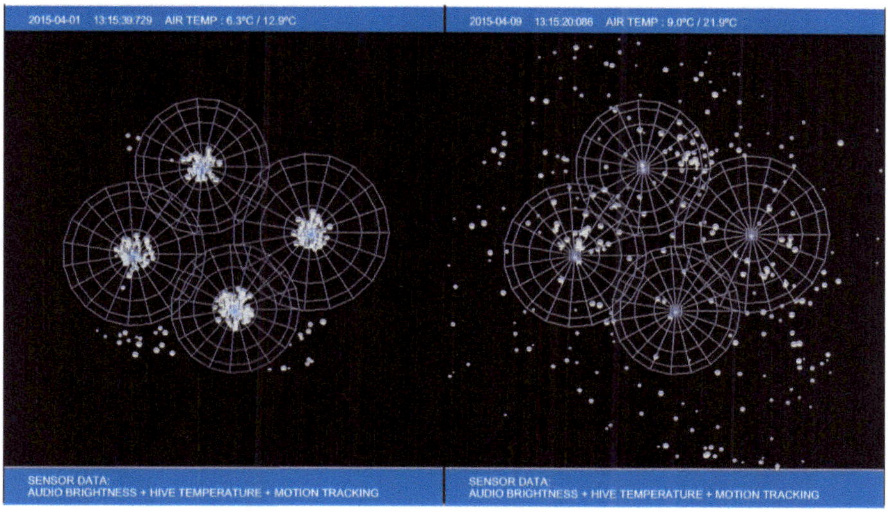

Fig. 6 Analysis of the data collected in the beehive, comparison between 2 days. The data used for visualization are: audio brightness (four sources), the temperature inside the beehive, and the motion tracking on the landing platform

the beginning to let them reproduce identical behaviors. Science has had difficulties in handling these kinds of problems because its methods rely on validation or falsification of hypotheses by observing reproducible instances. However, if simulation virtually reproduces an one-time-only problem, the boundary of applicable range of science can be pushed far beyond. Based on the above insight, an ultimate goal of the open systems simulation is to develop a framework and methods of simulation that can handle open systems problems in a constructive and an operational way.

6.2 Practice

Epidemics is an important factor of social-ecological systems that threatens productivity and sustainability. As a specific instance, we have been developing an integrated simulation of infectious disease, especially targeted for influenza [69, 70]. In order to understand the essential aspects of influenza epidemiology, it is not enough to simply look into a particular part such as the micromechanisms and processes of infection separately in a closed manner. We need to simultaneously take into account surrounding yet further-reaching factors, such as the structure and dynamics of human society, the population dynamics of humans, ecosystems (including those of other species), seasonal conditions, geographical constraints, and so on, in order to take an integrated and holistic view. As a first step of our endeavor, we have built a bi-layered model which interconnects macrosimulation of epidemic circulation among hosts and micro simulation of viral evolution driven

Fig. 7 *Top*: streaming activity on the landing platform. *Bottom*: video-analysis of the activity inside the beehive

by immunological interaction among hosts and viruses. In the model, we showed possible mechanisms that generate the limited diversity of viruses, which is one of the fundamental yet unexplained temporal behaviors observed in the evolution of real influenza.

(Takahiro Sasaki)

7 One-Health Food Lab

Under the slogan of "One World, One Health," the Wildlife Conservation Society convened public health experts from around the world to adopt the Manhattan Principles in 2004. The first principle is: "Recognize the essential link between human, domestic animal and wildlife health and the threat disease poses to people, their food supplies and economies, and the biodiversity essential to maintaining the healthy environments and functioning ecosystems we all require." It exemplifies the interdisciplinary movement toward solving the world's complex, interlocking problems, rather than simply taking reactive countermeasures against zoonotic diseases [30]. In One-Health Food Lab, we scientifically examine foods and medicinal herbs, which were produced in different growth conditions and placed in the market, to evaluate their quality and safety to the health of human and other creatures sharing the earth.

7.1 Activity in 2014

In 2014, we studied functions, compositions, and tastes of common vegetables in relation with their farming methods: Conventional (C), Organic (O), and Natural[2] (N). In one study on cabbage using a smart taste-sensing system for five different tastes (sweetness, saltiness, sourness, umami, and bitterness), bitterness was found to be the one that significantly differentiated N-cabbage from the others and also the most lasting one among the tastes [95]. In another study through absorption spectroscopy and metabolome analysis, the UV-absorption level and the phytochemical content were both detected to hold the order N>O>C in cabbage or C>O>N in carrots [94]. In addition, antibiotics were found in O-products; pesticides, carcinogens, and synthetic drugs were detected independently of the farming method. Through these studies, physiological differences of plants related with their growth environment and their effects on human health as foods and drugs have been partly elucidated.

(Kaoru Yoshida)

8 Conclusion

With the perspective of open complex systems, six projects have begun to collaborate in e-laboratory that independently interact and tackle the urging problems of food production in various aspects of social-ecological systems. The founding philosophy and initial steps of exploration are demonstrated as a progress report. Without any top-down restriction of methodology and expression other than open complex systems, our projects are in open collaboration inside and outside of the e-laboratory, expecting bottom-up convergence of new ideas that could be supported by plural independent scientific paths.

[2]Synecoculture products in Sect. 2.

We also expect that with this form of collaboration, new effective methodologies to treat open complex systems might be elaborated, not necessary constrained in food production but applicable to a wide range of open systems problems. When combined with interactive technologies and cloud computation, data analytics and simulation are traditional yet prominent domains to trigger such change in scientific methodologies.

The design of agriculture varies according to the local climate condition, selection of culture species, available resources, preference of consumers, health effect, disease risks, environmental preservation effort, economical state, community dynamics, means of distribution, etc., which are assigned parallelly in each project. We require an integrated approach to widen the choice of possible strategies in order to develop sustainable food production in an ever-changing environment. At the same time, fundamental question such as ecosystems function and health effect of food should maintain scientific objectivity without catering to a specific social activity. From each part to the whole, subsystems related to food production should be redefined while operating as open complex systems, intriguing effective social-ecological change with a scientific support.

This e-laboratory is a challenge that brings us back to the origin of agriculture and question de novo the design of food production ranging from urban gardening to natural state. The methodology and expression vary among projects, though we commonly share the mutual principle of challenge: Human activities from industrial production to citizen initiative should bring positive impacts on nature by augmenting the ecological state, and human society in return benefits from its ecosystem services. Human augmentation of ecosystems, or ecosystems leveraging, that exceeds the conventional reach of agriculture.

A possible outcome from this challenge is the derivation of a protocol that can infer and continuously ameliorate a suitable and sustainable food production system in a wide range of social-ecological condition. This protocol is not a mere ensemble of past data in various environment, but a trial-and-error tutorial that describes how to yield a concrete design of agricultural system in a given transient condition. This corresponds to a meta-algorithm that constantly integrates new data and provides wider choice of exploration in the practice of agriculture. The protocol should be extensively applicable in any arable and social condition, including future climate change. The tailoring of site-specific management system is expected to bring self-sufficient practice for the natural source management in highly variable and diverse farm conditions typical of resource-poor farmers in developing world, which is estimated to be about 1.4 billion people [3].

The databases, tools, and tutorials developed in each project will gradually be available in public including the CS-DC interfaces.

Acknowledgements We acknowledge Hidemori Yazaki, Kousaku Ohta, Tatsuya Kawaoka, Kazuhiro Takimoto, and Shuntaro Aotake who worked as research assistant in the Sects. 2 and 3, Kana Maruyama and Tatsuya Hiroishi in Sect. 3.

References

1. Abbott LK, Murphy DV (2003) Soil biological fertility. A key to sustainable land use in agriculture. Springer, Dordrecht
2. Aiba S et al (2005) Dynamics, productivity and species richness of tropical rainforests along elevational and edaphic gradients on Mount Kinabalu, Borneo. Ecol Res 20:279–286
3. Altieri MA (2002) Agroecology: the science of natural resource management for poor farmers in marginal environments. Agric Ecosyst Environ 1971:1–24
4. Bärlocher F (2005) Leaf mass loss estimated by litter bag technique. In: Methods to study litter decomposition. Springer, Houten, pp 37–42
5. Barnosky AD et al (2012) Approaching a state shift in Earth's biosphere. Nature 486:52–58
6. Bedoussac L, Justes E, Journet E-P, Hauggaard-Nielsen H, Naudin C, Corre-Hellou G, Prieur L, Jensen ES (2014) Intercropping 2014. An application of ecological principles to improve nitrogen use efficiency in organic farming systems. In: Bellon S, Penvern S (eds) Organic farming, prototype for sustainable agricultures. Springer, Dordrecht
7. Bell G (2005) The permaculture way. Chelsea Green Publishing, White River Junction
8. Burlingame B (ed) (2010) Proceedings of the international scientific symposium biodiversity and sustainable diets united against hunger, 3–5 November 2010. FAO Headquarters, Rome. http://www.fao.org/docrep/016/i3004e/i3004e.pdf
9. Campbell DC, Carlisle-Cummins I, Feenstra G (2013) Community food systems: strengthening the research-to-practice continuum. J Agric Food Syst Community Dev 3(3):121–138. http://dx.doi.org/10.5304/jafscd.2013.033.008
10. Carter VG, Dale T (1975) Topsoil and civilization. University of Oklahoma Press, Norman
11. Corre-Hellou G, Dibet A, Hauggaard-Nielsen H, Crozat Y, Gooding M, Ambus P, Dahlmann C, von Fragstein P, Pristeri A, Monti M, Jensen ES (2011) The competitive ability of pea–barley intercrops against weeds and the interactions with crop productivity and soil N availability. Field Crop Res 122:264–272
12. Cushing JM et al (2002) Chaos in ecology. Academic, London
13. De Schutter O (2014) Report of the Special Rapporteur on the right to food, United Nations General Assembly
14. Dorigo WA et al (2007) A review on reflective remote sensing and data assimilation techniques for enhanced agroecosystem modeling. Int J Appl Earth Obs Geoinf 9(2):165–193
15. EEA (2015) The European environment – state and outlook 2015: synthesis report, European Environment Agency (EEA), Copenhagen
16. e-laboratory website. http://www.elab-ose4el.net/ (2015)
17. Encyclopedia of Life. http://eol.org/ (2015)
18. FAO (2014). The state of food and agriculture, food and agriculture organization of the United Nations, Rome. http://www.fao.org/3/a-i4040e.pdf
19. Fukuoka M (1978) The one-straw revolution: an introduction to natural farming. Rodale Press, Emmaus
20. Funabashi M (2011) Fondation de la Synécoculture: Vers une agriculture de synthèse écologique et rentable, Actes du colloque Transversalités de l'Agriculture Biologique. English translation available at: https://www.sonycsl.co.jp/publications/result/?langmode=jp&bykeyword=funabashi
21. Funabashi M (2013) IT-mediated development of sustainable agriculture systems–toward a data-driven citizen science. J Inf Technol Appl Educ 2(4):179–182
22. Global Biodiversity model for policy support. http://www.globio.info/home (2015)
23. Global Biodiversity information facility. http://data.gbif.org/welcome.htm;jsessionid=F6C5F068ACF892631397FBB91CBE91D8 (2015)
24. Glover D et al (2010) Increased food and ecosystem security via perennial grains. Science 328(5986):1638–1639
25. Glover D et al (2010) Plant perennials to save Africa's soils. Nature 489:359–361

26. Goodchild MF (2007) Citizens as sensors: the world of volunteered geography. GeoJoumal 69:211–221
27. Graham LE (1993) Origin of land plants. Wiley, New York
28. Green EP et al (1996) A review of remote sensing for the assessment and management of tropical coastal resources. Coast Manag 24:1–40
29. Groombridge B, Jenkins MD (2002) World atlas of biodiversity: earth's living resources in the 21st century. University of California Press, Berkeley
30. GRF one health summit. http://onehealth.grforum.org/about/about-one-health/ (2015)
31. Harvard School of Public Health (2011) The global burden of non-communicable diseases. A report by the World Economic Forum and the Harvard School of Public Health
32. Hashimoto S, Nakamura S, Saito O, Kohsaka R, Kohriki C, Tanaka R (2014) Measuring ecosystem services of social-ecological production landscapes of Noto: toward the economics of Satoyama. In: Proceedings of international conference, sustainable management including the use of traditional knowledge in Satoyama and Other SELPS, pp 17–26
33. Hector A et al (1999) Plant diversity and productivity experiments in European grasslands. Science 286:1123–1127
34. HLPE (2013) Investing in smallholder agriculture for food security. A report by the high level panel of experts on food security and nutrition of the Committee on World Food Security, Rome (2013). http://www.fao.org/fileadmin/user_upload/hlpe/hlpe_documents/HLPE_Reports/HLPE-Report-6_Investing_in_smallholder_agriculture.pdf
35. Holdridge LR (1947) Determination of world plant formations from simple climatic data. Science 105:367–368
36. Homgren P, Thuresson T (1998) Satellite remote sensing for forestry planning. Scand J For Res 13:90–110
37. Huggins DR et al (2008) No-till: the quiet revolution. Sci Am 299(1):70–77
38. International Assessment of Agricultural Knowledge, Science and Technology for Development (IAASTD) (2009) Synthesis report with executive summary: a synthesis of the global and sub-global IAASTD reports. http://www.unep.org/dewa/agassessment/reports/IAASTD/EN/AgricultureataCrossroads_SynthesisReport(English).pdf
39. Introne J, Laubacher R, Olson G, Thomas M (2014) Solving wicked social problems with socio-computational systems, Kunstliche Intelligenz, February 2013.
40. Isozaki T (2012) Combining stage algorithm for discovering causal models. In: Proceedings of UAI workshop on causal structure learning
41. Isozaki T (2014) A robust causal discovery algorithm against faithfulness violation. Trans Jpn Soc Artif Intell 29(1):137–147
42. Isozaki T, Ueno M (2009) Minimum free energy principle for constraint-based learning Bayesian networks. In: Proceedings of ECMLPKDD 2009, Part1. LNAI, vol 5781. Springer, Berlin, pp 612–627
43. Jackson W (2002) Natural systems agriculture: a truly radical alternative. Agric Ecosyst Environ 88:111–117
44. Japan Integrated Biodiversity Information System. http://www.biodic.go.jp/english/J-IBIS.html (2015)
45. Japan Node of Global Biodiversity Information Facility. http://www.gbif.jp/v2/ (2015)
46. Jeavons J (2012) How to grow more vegetables (and fruits, nuts, berries, grains, and other crops) than you ever thought possible on less land than you can imagine, 8th edn. Ten Speed Press, Berkeley
47. KNEU-project (2014) KNEU-developing a knowledge network for European expertise on biodiversity and ecosystem services to inform policy making economic sectors. Biodiveristy-Knowledge. http://www.biodiversityknowledge.eu/
48. Kottek M, Grieser J, Beck C, Rudolf B, Rubel F (2006) World map of the Köppen-Geiger climate classification updated. Meteorol Z 15:259–263
49. Kremen C, Iles A, Bacon C (2012) Diversified farming systems: an agroecological, systems-based alternative to modern industrial agriculture. Ecol Soc 17(4):44
50. Kuhnlein HV et al (ed) Indigenous peoples' food systems & well-being. Food and Agriculture Organization of the United Nations. Center for Indigenous Peoples' Nutrition and Environment, Rome (2013). http://www.fao.org/docrep/018/i3144e/i3144e00.htm

51. Lane D, Maxfield R, Read D, van der Leeuw S (2009) From population to organised thinking. In: Lane D, van der Leeuw S, Pumain D, West G (eds) Complexity perspectives in innovation and social change. Springer, Berlin
52. Lévi-Strauss C (1990) La Pensée Sauvage. Pocket, Paris
53. Libellium agriculture kit. http://www.libelium.com/products/waspmote/ (2015)
54. Loreau M, Hector A (2001) Partitioning selection and complementarity in biodiversity experiments. Nature 412:72–76
55. Lynch J (2013) Choose satellites to monitor deforestation. Nature 496:293–294
56. Mader P et al (2002) Soil fertility and biodiversity in organic farming. Science 296:1694–1697
57. Malézieux E, Crozat Y, Dupraz C, Laurans M, Makowski D, Ozier-Lafontaine H, Rapidel B, de Tourdonnet S, Valantin-Morison M (2008) Mixing plant species in cropping systems: concepts, tools and models. A review. Agron Sustain Dev 29:43–62
58. McVigar TR, Jupp DLB (1998) The current and potential operational uses of remote sensing to aid decisions on drought exceptional circumstances in Australia a review. Agric Syst 57(3):399–468
59. Michels M (2011) A beehive monitoring system incorporating optical flow as a source of information. Masters thesis, Freie Universität Berlin
60. Millennium Ecosystem Assessment (2005) Ecosystems and human well-being: biodiversity synthesis. World Resources Institute, Washington, DC. http://www.unep.org/maweb/documents/document.354.aspx.pdf
61. Mollison B, Holmgren D (1978) Permaculture one: a perennial agriculture for human settlements. Trasworld Publishers, Melbourne. ISBN: 978-0-938240-00-6
62. Montgomery DR (2012) Dirt: the erosion of civilizations. University of California Press, Berkeley
63. Newbold T et al (2015) Global effects of land use on local terrestrial biodiversity. Nature 520:45–50
64. Pearl J (2000) Causality, models, reasoning, and inference. Cambridge University Press, Cambridge
65. Pool R (1989) Ecologists flirt with chaos. Science 243(4889):310–313; New Series
66. Putman RJ, Wratten SD (1984) Principles of ecology. University of California Press, Berkeley
67. Reichenbach H (1956) The direction of time. Dover Publications, New York
68. Rosenqvist A (2003) A review of remote sensing technology in support of the Kyoto protocol. Environ Sci Pol 6:441–455
69. Sasaki T (2013) Hierarchical multi-agent-based model for simulating the prevalence and evolution of influenza virus. In: Spring simulation multi-conference, pp 62–69
70. Sasaki T (2013) Simulating limited diversity in evolution of influenza. In: ECAL 2013: 12th European conference on artificial life, pp 470–471
71. Sheppard SRJ, Cizek P (2009) The ethics of Google Earth crossing thresholds from spatial data to landscape visualization. J Environ Manag 90:2102–2117
72. Smith LC (1997) Satellite remote sensing of river inundation area, stage, and discharge: a review. Hydrol Process 11:1427–1439
73. SOER (2010) The state of soil in Europe. A contribution of the JRC to the European Environment Agency's Environment State and Outlook Report, European Environment Agency
74. The Bee Laboratory open source video database. http://pandora.okno.be (2015)
75. The Bee Laboratory open sensor data. http://opensensordata.net (2015)
76. The Bee Laboratory pollen database. http://urbanbeelab.okno.be/doku.php?id=pollen_database (2015)
77. The Sound Beehive installation at the Institute of Evolutionary Biology (IBE) in Barcelona (May–June 2015). http://annemariemaes.net/presentations/bee-laboratory-presentations-2/bee-monitoring-devices-and-curious-observations/ (2015)
78. The Sound Beehive installation at the AI and the Arts exhibition for the international conference of artificial intelligence in Buenos Aires, Argentina (July 2015). http://annemariemaes.net/presentations/bee-laboratory-presentations-2/artes-ijcai/

79. The World Bank (2008) World development report, agriculture for development (2008). http://siteresources.worldbank.org/INTWDR2008/Resources/WDR_00_book.pdf
80. Thop KR, Tian LF (2004) A review on remote sensing of weeds in agriculture. Precis Agric 5:477–508
81. Tokoro M (2010) What is open systems science? In: Tokoro M (ed) Open systems science: from understanding principles to solving problems (the future of learning). IOS Press, Fairfax
82. Tokoro M (ed) (2010) Open systems science from understanding principles to solving problems. IOS Press, Fairfax
83. Tokoro M (2015) A new method for new challenges. Sci Am 499:64–66; Innovation Special Issue
84. Tree of Life web project. http://www.tolweb.org/tree/ (2015)
85. Tucker CJ (1980) A critical review of remote sensing and other methods for non-destructive estimation of standing crop biomass. Grass Forage Sci 35:177–182
86. United Nations Environment Programme (UNEP) (2009) The environmental food crisis - the environment's role in averting future food crises. http://www.unep.org/pdf/FoodCrisis_lores.pdf
87. United Nations Environment Programme (UNEP) (2012) Avoiding future famines: strengthening the ecological foundation of food security through sustainable food systems, United Nations Environment Programme (UNEP), Nairobi. http://www.unep.org/publications/ebooks/avoidingfamines/portals/19/UNEP_Food_Security_Report.pdf
88. Verbitskii VB, Verbitskaya TI (2007) Ecological optimum and effect of delayed action of a factor. Dokl Biol Sci 416:386–388
89. Von Frisch K (1953) The dancing bees: an account of the life and senses of the honey bee. Harvest Books, New York
90. Walker S, Wilson JB, Steel JB, Rapson GL, Smith B, King WMG, Cottam YH (2003) Properties of ecotones: evidence from five ecotones objectively determined from a coastal vegetation gradient. J Veg Sci 14(4):579–590
91. WHO (2003) Diet, nutrition and the prevention of chronic disease. Technical Report Series 916. WHO, Geneva
92. World Economic Forum (2010), Realizing a New Vision for Agriculture: a roadmap for stakeholders. http://www3.weforum.org/docs/WEF_IP_NVA_Roadmap_Report.pdf
93. World Economic Forum (2013) Achieving the new vision for agriculture: new models for action. http://www3.weforum.org/docs/IP/2013/NVA/WEF_IP_NVA_New_Models_for_Action_report.pdf
94. Yoshida K, Funabashi M (2014) Taste analysis on conventionally, organically and naturally grown cabbage. In: 1st IMEKOFOODS, October 12–15, Rome
95. Yoshida K, Shimizu-Yoshida Y, Funabashi M (2014) Functional and compositional characteristics of conventionally, organically and naturally grown cabbage and carrots. In: 4th International conference on food studies, October 20–21, Prato

Product Lifecycle Management (PLM): A Key to Manage Supply Chain Complexity

Imane Bouhaddou and Abdellatif Benabdelhafid

1 Introduction

Doing business globally involves very complex and increasingly dynamic processes and demands a high level of flexibility and adaptability from the companies involved [1].

Therefore, the complex management of an extended enterprise has increased needs for information exchange, sharing, and archiving.

A supply chain encompasses such entities as suppliers, manufacturers, assemblers, distributors, logistics centers, customers, and in some cases even competitors, which are interconnected among by flows of different resources, including goods, information, or finance, through physical and virtual network of their performance.

Managing increasing complexity in supply chains is absolutely necessary to companies to compete better in global market. Complexity in supply chains is associated with material and information flows between supply chain partners.

Product lifecycle management (PLM) enables a supply chain to become much more competitive by an effective management of the supply chain flows at various lifecycle stages of a product.

I. Bouhaddou (✉)
Ecole Nationale Supérieure d'Arts et Métiers, Meknes, Morocco
e-mail: b_imane@yahoo.fr

A. Benabdelhafid
Le Havre University, Le Havre, France
e-mail: Abdellatif.benabdelhafid@uni-lehavre.fr

P. Bourgine et al. (eds.), *First Complex Systems Digital Campus World E-Conference 2015*, Springer Proceedings in Complexity, DOI 10.1007/978-3-319-45901-1_38

2 Supply Chain Complexity

2.1 Characteristics of Supply Chain Complexity

There are some key characteristics of complexity occurring in a supply chain system which need to be discussed to understand the impact of these characteristics on the occurrence of complexity. These characteristics were detailed in [2]. We summarize them in the following chart (Table 1).

2.2 Supply Chain Flows

Complexity in supply chains is associated with material and information flows between supply chain partners. Traditionally, these flows are organized sequentially from supplier to customer. Today, information flows do not follow this linear form. Information flows rather now look like a simultaneous exchange, especially through electronic exchanges between all supply chain partners.

Table 1 Characteristics of supply chain complexity

Characteristic	Description
Number of components	Number of items (raw, manufactured, or end), products, processes, supply chain partners, relationships, interactions, goals, locations, etc. *Example*: A high number level of any components contributes increasingly in a complexity in a supply chain system
Diversity	Related with the homogeneity or heterogeneity of a system *Example*: A high level of diversity of any components (supplier, product, mean of transport) along the supply chain leads to system's heterogeneity and results a high level of complexity
Interdependency	Interdependence between items, products, and supply chain partners. Complexity increases in direct proportion to the increase of interdependence
Variety	Variety represents dynamical behavior of a system *Example*: A product or a process variety in supply chains leads to increase in complexity level over time
Uncertainty	Uncertainty represents all difficulties to be able to make a clear picture of a system due to the lack of information or knowledge The more uncertainty in a supply chain system is, the more complexity occurs in this system Proportion to the increase of interdependence

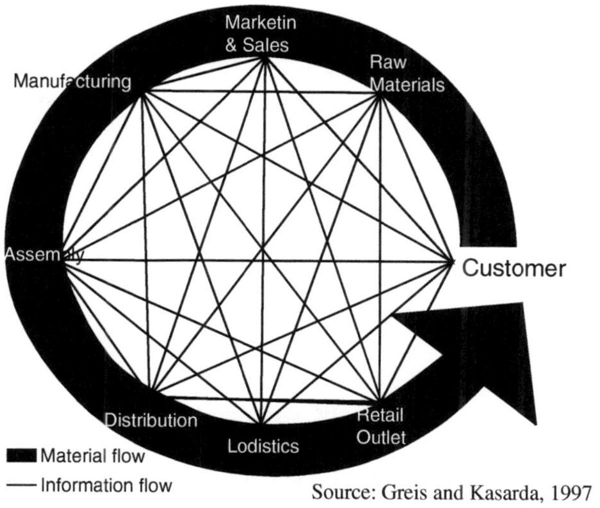

Fig. 1 Material and information flows in supply chain: new pattern [3]

Figure 1 is a schematic diagram of this new pattern. There are internal networks in the firm with simultaneous exchange between all units. Indeed, the emergence of information and communications technology facilitating exchange between all partners produced a change in the organization of the supply chain flows [4, 5].

In addition, different industrial activities with strong technological character generate and manipulate a lot of technical data that need to be exchanged, managed, and stored in a consistent and standardized manner. They have led to the developing of methods and systems to manage the product information throughout its lifecycle. It is in this context that the paradigm of PLM was born [6].

3 Product Lifecycle Management

3.1 Definition of PLM

To meet the challenges of today's global business environment, organizations with complex products and processes need to manage the entire product lifecycle from concept through end of life, working across functions, and supply chain partners. Additionally they must ensure that they are able to respond quickly to market opportunities as well as changes in technology and customer requirements [7].

PLM is an integrated approach that, with information technology aid, realizes an integrated cooperative and collaborative information product management during all the lifecycle [8]. PLM systems allow simultaneous and collaborative product development by taking into account all stages of the lifecycle [9].

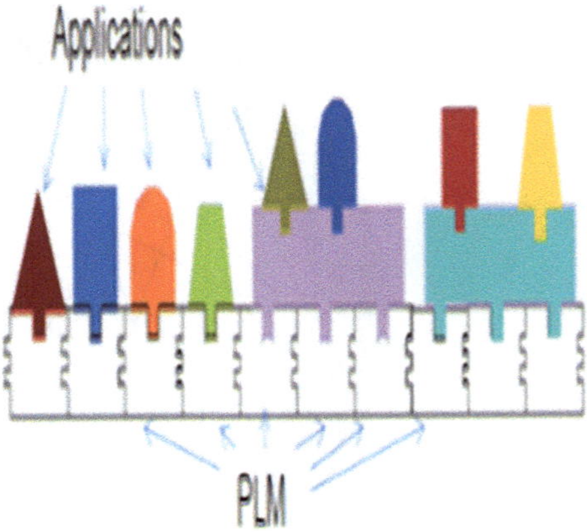

Fig. 2 PLM and other applications [11]

PLM is an application of manufacturing industry that allows to virtually simulate the complete product life [10]. PLM system should enable the flow of information between different business applications (Fig. 2). In fact, PLM enables organizations to develop consistent, repeatable processes that extend through the product lifecycle and enable collaboration across system disciplines and the supply chain. Figure 3 represents an analogy with human body. Indeed, informational connections in the human body are connected by the backbone. We consider that PLM plays the role of the backbone.

3.2 *PLM for Supply Chain Optimization*

Optimizing the supply chain and especially the simultaneous optimization of product and supply chain design is a difficult problem. Due to the complexity of the induced mathematical models, very few models address the integrated problems. They use a very large number of variables to model the problem in a comprehensive manner and are therefore difficult to solve [13].

Centralized approaches used to treat the problem of integrated design of the product and its supply chain generate complex mathematical models, we adopt an approach combining centralized decisions while integrating the constraints of the different supply chain partners during the product design and decentralized decisions when it comes to locally optimize each supply chain partner.

The decentralized approach reduces the complexity of solving mathematical models and allows the supply chain to respond quickly to the evolution of local conditions of each partner.

PLM will assure the integration of the different supply chain partners. Indeed, the information centralization by the PLM enables to take into consideration the dependence between these partners, improving therefore local optimization results.

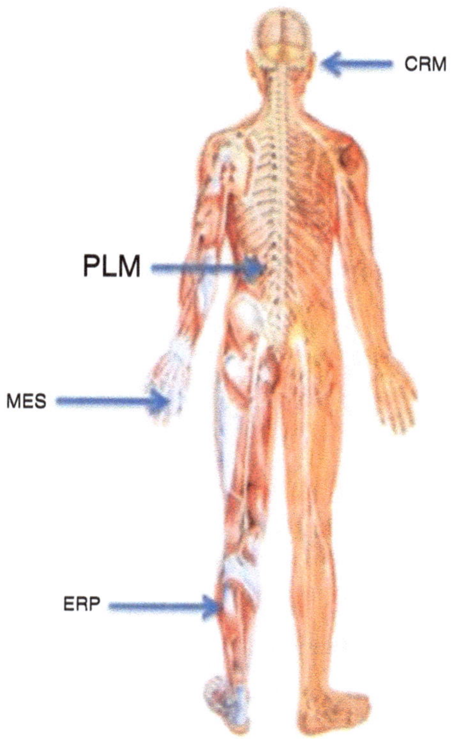

Fig. 3 Analogy with human body [12]

The local optimization allows to:

- Decrease the amount of data to be processed in each resolution process.
- Preserve the autonomy of each supply chain partners.
- Respond quickly to the evolution of local conditions of each partner.

With PLM, everyone is contributing to the product design.

The partners of a supply chain are linked by PLM. The PLM structures the design of the supply chain (Fig. 4).

The mathematical models to optimize the supply chain cost were detailed in [14]. They concerned a problem integrating supplying-production-storage-distribution which aims to minimize the supplying costs, production costs, storage costs, and distribution costs.

The models developed are related to strategic and tactical decisions for supply chain design.

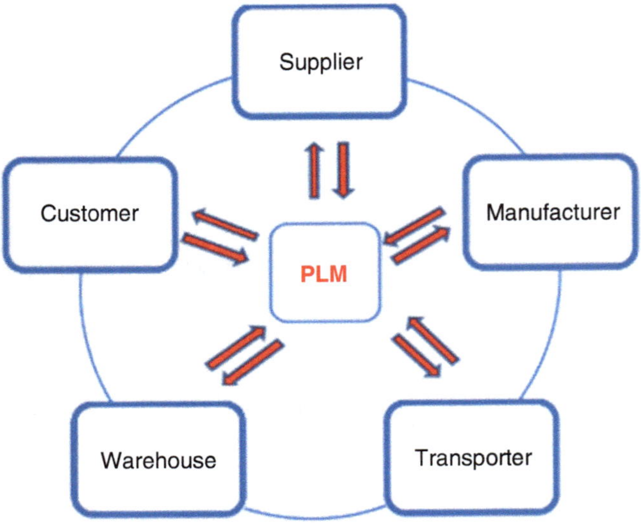

Fig. 4 PLM as an integrator of supply chain partners

4 Framework PLM/Supply Chain Optimization

The modelled framework using UML class diagram (Fig. 5) is based on PLM model [15] and the methodology for simultaneous design of a new product and its optimized supply chain proposed in our previous paper [16].

It includes the digital mock-up parameters (product in design) and the different variables needed for supply chain optimization.

This framework highlights the interactions between several factors heavily dependent around the product and consequently the many parameters to manage which illustrates the system complexity.

5 Conclusion

The work presented in this paper is related to the field of integrated engineering, specifically the integrated logistics in the early phases of the product lifecycle using PLM. The supply chain is a complex system and like any complex system, solutions are found by compromise.

The objective of the proposed approach is to solve formally the compromise between PLM and mathematical models to optimize the supply chain.

The approach is, on the one hand, centralized while integrating supply chain constraints and decentralized with a series of local optimizations which avoid the global supply chain optimization.

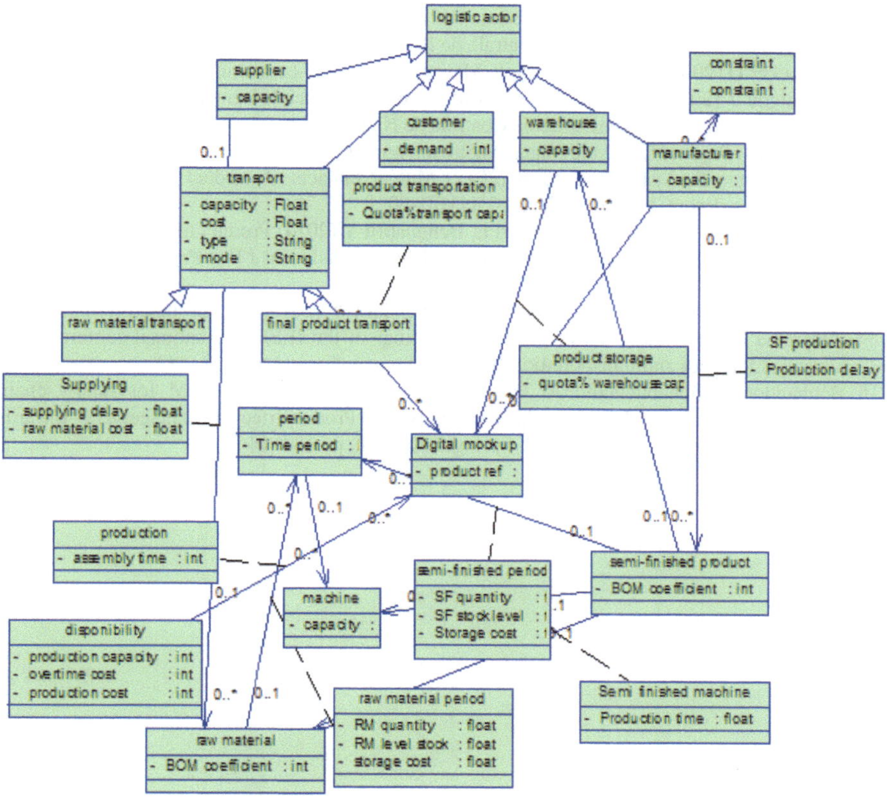

Fig. 5 Framework PLM/supply chain optimization

References

1. Denger A, Maletz M, Helic D (2010) Social computing: a future approach of Product Lifecycle Management. In: Proceedings of I-KNOW2010: 10th international conference on knowledge management and knowledge technologies, pp 451–457
2. Rimiene K, Bernatonyte D (2013) Supply chain management trends in the context of change. Econ Manage 18(3):596–606
3. Greis NP, Kasarda JD (1997) Enterprise logistics in the information era. Calif Manage Rev 39(3):55–78
4. Monteiro T (2001) Distributed management of cooperation between enterprises. PhD Thesis, Institut National Polytechnique de Grenoble
5. Handfield RB, Nichols EL (1999) Introduction to supply chain management. Prentice-Hall, New Jersey
6. Grieves M (2006) Product Lifecycle Management: driving the next generation of lean thinking. McGraw-Hill, New York
7. CIMdata (2003) Product Lifecycle Management, "Empowering the future of business"
8. Terzi S (2005) Product Lifecycle Management: definitions, open problems and referenced models. PhD Thesis, Nancy University I

9. Paviot T (2010) Methodology of resolution of interoperability in the field of Product Lifecycle Management. PhD Thesis, Ecole Centrale de Paris
10. Debaecker D (2008) Product Life-cycle Management. Vinci-consulting
11. Le Duigou J (2010) Modelisation of PLM Systems in extended enterprise. Application for mechanical SME. PhD Thesis, Ecole Centrale de Nantes
12. Aras (2013) Product Lifecycle Management and human body. Aras
13. Baud-Lavigne B, Agard B, Penz B (2012) Mutual impacts of product standardization and supply chain design. Int J Prod Econ 135(1):50–60
14. Bouhaddou I, Benabdelhafid A, Ouzizi L, Benghabrit Y (2014) Product lifecycle management approach to design a product and its optimised supply chain. Int J Bus Perform Supply Chain Model 6(3/4):255–275
15. Bouhaddou I, Benabdelhafid A, Ouzizi L, Benghabrit Y (2013) An optimized framework of integrated logistics based on PLM. In: Proceedings of the IEEE international conference on advanced logistics and transportation (ICALT), Sousse, Tunisia, pp 438–443
16. Bouhaddou I, Benabdelhafid A, Ouzizi L, Benghabrit Y (2012) PLM for supply chain optimization. In: Rivest L, Bouras A, Louhichi B (eds) Product lifecycle management - towards knowledge rich enterprises, vol 338. Springer, London, pp 134–146

Performance Improvement of Baggage Handling System in the Soekarno-Hatta International Airport

Dimas Novrisal, Nadia Hamani, A. Elmhamedi, and Tresna Soemardi

Globalization has big impact in the airports all over the world. Many international airports increased in large growth of passengers in recent years. There are 2.835 billion passengers of the airplane in the world in 2011 and International Air Transport Association (IATA) standard predicts that it will increased at 4.4 % in 2012 and 4.5 % in 2013 [1]. Worldwide international and domestic air transport passenger traffic grew 5.9 % to a new high of 5.2 trillion kilometers in the year of 2011 [2]. Nowadays, the globalization increases the importance of airports to worldwide connectivity. Passengers growth in the airport is still strong despite the economic slow in many regions in the world because global business and tourism depend on the air transport industry [2].

Air transport is the fastest, safest, and comfortable public transportation. Low cost airlines become popular in recent years which make prices affordable and inexpensive. Passengers are growing more and more effecting the airport with hectic rush and long queue in the peak hours. Passengers satisfaction depends on processing time in the system because the less time they spend in the system of the airport, the higher they are satisfied; however, airport must adhere to the standard of the security and safety regulation included proper identification, limited luggage weight, and safety procedures at the security checkpoint [3].

The airport management is a complex system related to the passengers, airplanes, and baggage. The passengers are the most important entities in the airport. The sustainability of airport depends on passengers because passengers are the users of

D. Novrisal • T. Soemardi
Faculty of Engineering, University of Indonesia, Depok, Indonesia

N. Hamani (✉)
University of Picardie Jules Verne, Amiens, France
e-mail: nadia.hamani@u-picardie.fr

A. Elmhamedi
University of Paris 8, Paris, France

P. Bourgine et al. (eds.), *First Complex Systems Digital Campus World E-Conference 2015*, Springer Proceedings in Complexity, DOI 10.1007/978-3-319-45901-1_39

airport service. Passengers' satisfaction has to be the main goal of the airport service. The airport must have adequate facilities and public area to serve the passengers properly. Passengers' satisfaction depends on the performance of airport in serving the passengers.

Performance of facility in the airport can be measured by simulation technique. Many researchers develop simulation technique to measure the capacity of the airport, utilization of each area, passengers waiting time, and passengers processing time in the airport system. Discrete-event simulation has been used to measure performance of the system in the departure terminal of SHIA. ProModel simulator version 7.5 of ProModel Corporation is used in order to simulate the system. The study consists of four steps: measurement of the performance of existing system, measurement of the performance of new system, and compare the performance of two systems. Several scenarios develop to see the reliability of new system with the growth of passengers until 2030 to the capacity of airport.

The results obtained through ProModel Simulator were summarized in two categories; Entity Activity and Utilization of Location. From the results, we conclude that the new system has a better performance than the existing system. The new system can reduce processing time and waiting time.

The scenarios result of new system shows that the number of passenger of SHIA in the year 2015–2020 has a little effect to processing time and waiting time. In the year 2025–2030 there is effect of number of passenger to waiting time in the airport. This effect will be related to processing time in the system.

After measurement of the performance of the existing system, further work will focus on measure qualitative indicator of baggage handling system at Soekarno-Hatta International Airport. Qualitative indicator related to perception of passenger to the service of the airport. We have to study the qualitative indicators to make a better analysis and understanding of passenger satisfaction.

References

1. PT Angkasa Pura II [Electronic Resources]: www.angkasapura2.co.id
2. International Air Transport Association Annual report 2012, 68th General Meeting Beijing, June 2012
3. Manataki IO, Zografos K (2009) A generic system dynamics based tool for airport terminal performance analysis. Transp Res C 17:428–433

Territory as a Narrative

Eddie Soulier

We will develop two ideas in this paper: a territorial assemblage is a narrative, it connects very heterogeneous ingredients. But as such, it is possible to live in it only if we are able to imagine a political mode of "being together" in a world-like-assemblages.

First we try to develop our first claim: a territorial assemblage is a narrative which organizes heterogeneous entities (characters, context, plot, tools, discourses, etc.) in a kind of complex hypernetwork.

1 In Which Sense a Territory Could Be a Narrative?

It is a pedagogical manner to clarify what is a territorial assemblage, because arranging is the action to arrange or the result of this action. But it is also an arrangement, or layout, an organization or way in which the things are laid out, arranged—here a territory—and the sequence of ongoing events which proceed there.

To demonstrate this point, we need to make three moves:

A first move considers territory as a kind of action. We oppose then territory as an object, like in positivist science, to territory as a process, like in pragmatics view.

In a second move we claim that action is a kind of narrative, in accordance with Ricoeur who considers "Text as action, action as text" in its article Meaningful Action Considered as a Text [1] or in accordance with Alisdair MacIntyre, who in his book *After Virtue* in 1981 speaks about "enacted narrative" as a basic form of social life.

E. Soulier (✉)
UMR CNRS 6281 "Institut Charles Delaunay", Université de Technologie de Troyes (UTT),
12 Rue Marie Curie - CS 42060, 10004 Troyes Cedex, France
e-mail: eddie.soulier@utt.fr

P. Bourgine et al. (eds.), *First Complex Systems Digital Campus World E-Conference 2015*, Springer Proceedings in Complexity, DOI 10.1007/978-3-319-45901-1_40

And the last move considers in short territory as a kind of narrative.

We use Assemblage's theory from Whitehead, Deleuze, Michel Callon, or Manuel de Landa for the conceptual modelling step, and the notion of hypernetworks, provided by Jeffrey Johnson [2], as a computational tool for territorial data analysis, in the implementation step. But we can't develop this part of our work in this introductive presentation [3–5].

So *the first move considers territory as an action.* This claim is grounded in a classic opposition between a territory or space as a portion of geographic space that is claimed or occupied by a group (here it is spatial dimension or administrative organization which are important) and concepts like territoriality or spatiality, as an aspect of social life; the "production" of territories; but also, on a subjective side, a feeling of belonging or to be excluded. Here territory includes all relationships of human activity with space.

Our position is closed to the idea of "The Social Construction of Human Space" promoted by Michel Lussault, instead of space as an objective and external reality, where "practicing places" are just here and there, "in a space." For example, Lussault and Stock [6] develop the idea of the spatiality of practice as "faire avec l'espace" (in French), doing with space (in English), where different spatial competences in the process of coping with space are fleshed out. In short, a territory gains to be studied from the lenses of action, practice, activity system, or, better, network of actions or actants, instead of thinking it as a substance.

In sociology, the semiotic term "actant" was incorporated into the actor-network theory by Bruno Latour and Michel Callon, the activity of which is described as "mediation" or "translation" (Wikipedia). In a more comprehensive way, an actant is something that acts or to which activity is granted by others. It implies no special motivation of human individual actors, nor of humans in general. An actant can literally be anything provided it is granted to be the source of an action. Examples of actants are a document, an utterance, a law, a budget, a bacterium, an artifact, an organization, or, of course, a human actor.

So, *the second move considers an action as a king of narrative.*

Let us avoid a first error of category or misunderstanding concerning narrative: it will not be said here that, on one side, there are reality or the action and, on the other side, what it is told about them or some representation of the reality they refer to. We will reject the theories known as dualistic theory of narration. We will adopt on the contrary a realistic theory, a kind of theory that considers "action like text, and the text like action," paraphrasing Paul Ricoeur. A territory, a city, and a landscape are, literally, a "story being done."

As Alasdair MacIntyre [7] affirms it also, the stories must be lived before being told, contrary to Louis O. Mink who considers that "Stories are not lived but told." Dualistic thinkers like Mink admit a divorce between the action (action strictly speaking) and the mental state of the subject (intention). A famous thinker like John Searle commits the same mistake. On one side, there will be the reality (the realm of causality, where stay the referent) and, on the other side, only the representation of reality (the domain of reasons to act, the fiction realm, where stays the signified).

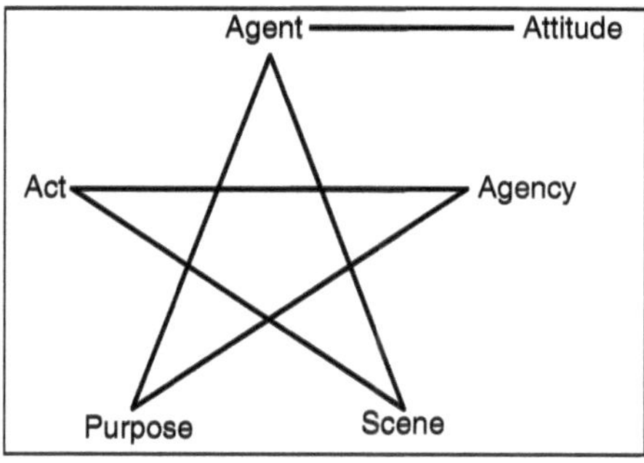

Act	What is being done
Agency	Mean by which an action occurs
Agent	Person doing the act
Scene	Setting/background for action
Purpose	Reason or rational behind action

Fig. 1 Burke's dramatistic pentad

But Monist thinkers say that there is only one plan of reality, not two (reality/fiction or action/speech).[1] For them, narrative form is neither disguise nor decoration or just a setting. "Stories are lived before they are told" (said MacIntyre). Enacted narrative is the most typical form of social life for him. Action itself has a basically historical character. We are "entangled in stories" as coined Whilhelm Schapp [8], a student of Edmund Husserl. Entanglement becomes an important philosophical category in assemblage philosophies and in some actual ontological debates.

The point is that analyses which unilaterally focus on "the actor," the anthropological subject, lose the fact that even a simple action only gains its meaning from the context (the setting, the social practice) in which it is embedded.

A well-known example of story as an assemblage of heterogeneous entities has been already given by Burke's Dramatistic pentad, in his 1945 book "A Grammar of Motives" (Fig. 1).

It is then necessary to return to a social vision of man, a neo-Aristotelian one, in opposition to a liberal point of view. As Michael J. Sandel from Harvard University claims it in his [9] paper, the image of what he called "the Unencumbered self," promoted by libertarian is a myth. The fact of becoming an individual is inseparable from our continuous participation in socially instituted practices. The subject is the effect of a process, not its end or its cause. It is necessary to consider agent behavior in the context of its previous activities and of its contexts. It is within the framework of practices that we are able to analyze the reasons to act (externalist thesis), and not in the spirit of the men (as claims mentalism).

So an analysis like Vladimir Propp or Algirdas Julien Greimas' ones makes the previous point clear, and makes easier to feel a territory as a kind of collective ongoing action, action which has a story-like existence and a story-like dynamics.

[1]Monist point of view advocates unity of origin of cause and reason, intention and action, action and narration (G.E.M. Anscombe in her 1957 Intention book said "intention belongs, together with action itself, in the category of process. To intend to act is already to be in the process of doing the action"). In sum, all is it about action, like in pragmatism.

For Greimas a narrative is composed of actants like CHARACTERS, who are constrained to adopt roles or figures according to the circumstances. They form alliances which are reinforced or not in the course of time. Let us not forget either the myriad of objects essential to very good narrative: what would become Cinderella without pumpkin or her court shoes? More seriously, and the realistic historians know it well, historical descriptions need to speak about objects or mediators. Without objects, no action is possible! The second category of entities is the SETTING. It is the backdrop of the characters and the plot, the context. If one thinks "territory," it is necessary to think of its environment, with factors likely to influence it (which reveals a SWOT analysis of the Strengths, Weaknesses, Opportunities, and Threats). The environment can be not only an opportunity but also a threat. We also need to think the objects or things of which any territory is made up. Finally, a narrative is a PLOT. Any story is made of a sequence of stages, a weaving punctuated by tests which structure alliances between characters. Bruno Latour called them "controversies." What is a project, an innovation, or a strategy if it is not a plot? To survive to these tests and controversies (and to hope to continue), characters must change their nature (as of the initiatory entry in the project), and are then to metamorphose (at the exit of each test). Reality is thus not something fixed and immutable. There are necessary TRANSFORMATION, CHANGE, and LEARNING. The metamorphosis is our condition. And a plot is more or less easy to follow, like our own life.

The difference that we try to promote with the concept of assemblage applied to territory is to give an absolutely realistic version of this story being done. The metaphor of the territory as a narrative is not one: a territory is really a story, and not a discourse. A territory is LITERALLY a story being done. This story is about a chain of experiments achieving itself, it is not a representation (presentation again) or a talk. One finds not only characters there, certainly, but also collections of objects, landscapes, tests, and complex plots. And especially, it is necessary to think this story like something absolutely material, very distributed, and at the same time very situated: an element of the territory is a very real and material thing, it is located here and now, but it is at the same time in relation to myriads of other elements. Elements can be people, equipment, documents, organizations, places, discourses, tools, or procedures. Let us admit moreover than these objects can enter in relation ones with the others according to multiple manners and dimensions, and not only according to the representation that we, human, can have. So in such circumstances this stone is an object in a museum, or a blunt object being able to be used for a crime, or then a reference in a note. But this stone is also in relation to wood, which draws reciprocal mechanical constraints, or with chemical agents, which opens then in a whole world of reactions. In short, far from limiting itself to the only relation of the man in the world, which traps us in the restricted field of representation, the relationality is on the contrary a property shared by all the beings.

So *the third move considers territory as a kind of assemblage.* Instead of thinking in terms of entities with only some dimension, limited to human realm, one is asked to think in terms of nodes that have as many dimensions as they have connections. And as entities are deeply and strongly heterogeneous, distributed, and weakly bound (see "the weak ties" from Granovetter, 1973), we have no simple unification principle to connect entities each others (cognitive, social or political, material, artifactual, or ecological relationship). Instead, "we have fibrous, thread-like, wiry, stringy, ropy, capillary character that is never captured by the notions of levels, layers, territories, spheres, categories, structures, systems" [10]. Assemblage thinking go further than actual actor-network theory (ANT): assemblage is a mode of ordering heterogeneous entities so that they work together for a certain time [11].

By introducing heterogeneity we lose any principle of unification of relations between entities, any center of crystallization of a unified meaning. The only thing we can do is to calculate the detail, the singularity, and the heterogeneous one, while following the metamorphoses of the entities, and the rich relational world in which they "are préhendent" mutually, to take the concept of Whitehead who saw in the relation of gripping between objects or events the basis of any reality and the basis of any coherent cosmology. Prehension means the action of grasping. The life (or the existence) is mainly a "logistic problem" between parts, but without Great Whole, God, Reason, or Nature to explain that.

So for us, assemblage is that: not only the narrative "making of" of the world or reality, but also a need for having an adequate tool to follow these networked ongoing actions. Because in this new ontology the world is not any more an object, consequently the territory either. The intelligence of the territory should be less dependent on a concept of representation than it should be an investigation allowing the follow-up of this making of this production, which is at the same time a performance (in a sense of performativity). A territory is a performative! As John Austin affirmed it, an act of language is performative if the statement carries out an action by the fact of its enunciation (like in the sentence like: "I declare you are husband and woman," "the priest declare the wedding". (When to say is to make or How to do Things with Words, [12]). In short, here, to name is to do things.

If one accepts the image which we give of the narrative as a "network of actions" or "action nest" [13], and not (only) a discourse about an action, then we rather need a more pragmatic method than semantic, pragmatic analytical to follow the continuous process of manufacturing the territory.

So it is the time now to explore the second part of our hypothesis, around a complementary issue.

2 Does the Territorial Assemblage Make a Viable World?

What was sometimes called "transpersonal multiplicity" in the vocabularies of Foucault, Deleuze, and others, or the "multitudes" (Hardt and Negri), "assemblages" (DeLanda or Callon[2]), and "pluralist world" (James) according to other traditions disconcert obviously people who conceive the territory like an object, a destiny

[2] See also Müller [14].

or an end, a set of actors well identified, and a certain form of social relations organization. However, the transpersonal multiplicities, which tend to escape from the stratifications and the well-established frameworks, are decisively becoming our today reality, at the time of social networks, of the digital society and the lateral power, as Jeremy Rifkin called it.

Territory as assemblages must have some regulatory rules. But how to speak about power, rules, decision, substances, or representations without adopting notions like levels, layers, territories, categories, structures, and systems? For people, territory is always a Euclidian space, composed with objects (like monuments, roads, etc.), with more or less well-identified people and groups, a history and destiny, and a perspective or a project, not a Rhizomatic ongoing process.

This brings us directly to the second point: those who adopt this idea of collective assemblage, which political design do they proposed finally? Is an assemblage a king of reality order which people can lived in? Which is the place for decision maker in these networks?

It is already possible to identify three logics at work inherent to the assemblage thinking. I call them:

- Apolitical or innovative perspective
- Autonomy and Self-governance
- And the third is diplomacy perspective.

Frequently the image of assemblage is carried by apolitical personalities. It is the first perspective. Let us think of Deleuze. Deleuze is interested in what is new, innovative, the creative movement of the spiritual and the material, the desiring strengths. Michel Serres too. Henri Bergson thinks in Becoming terms (not in the Being or Substantial terms). Whitehead thinks in Process terms. They are not interested by the basements of our social life. For them, Genesis (Process) is more important than Substance (Object).

Their formula could be: "*What is common is what is new*"!

The second attitude could be that of Felix Guattari, co-thinker of the concept of "rhizome" with Gilles Deleuze. Slavoj Zizek in its recent book on Deleuze said that: "The ontology of productive Becoming clearly leads to the Leftist[3] topic of the self-organization of the multitude of molecular groups which resist and undermine the molar, totalizing systems of power—the old notion of the spontaneous, non-hierarchical, living multitude opposing the oppressive, reified System." Example are "zadistes": A member of the "zone à défendre" (in French); and anti-austerity movement, also referred to the Indignants Movement; or what Pierre Rosanvallon called Counter-Democracy, in an Age of Distrust.

Their formula should be: "*What is common is what for which we fight*"! The fighting is the common ground of progressist assemblage, composed by those who make or support progress.

[3]Radical left.

A third attitude emerges, carried by Bruno Latour. Recently, Bruno Latour has been interpreted to be a radical democratic theorist, opening up and expanding the demos of democracy to include nonhumans (it is his idea of "A Democracy of Things"). But he also contests the Western modernity in its worldwide dominance, notably based on its mastery of science and hence of nature. He sees that the Western project can no longer proceed unchallenged because other points of view can no longer be ignored. Time is coming to consider that there are other points of view. You can only have diplomacy where you have an enemy (when other people or entities can't be simply obliterated.). You need to negotiate what you are in common with others, without any help from any transcendental points of view. Nature, science, reason, and so on are potential transcendental point of view, but they don't work anymore. So the unique alternative for Latour is the long-term reason of the diplomat.

In this case, the formula will be: "*What is common is a common mediator*" (Fig. 2).

There exists also a cybernetic form of political model in the networks, very emergent, around the idea that judgment results from an accumulation of myriad of small and scattered experiences, not by induction from a controlled or homogeneous position (or central situation). A kind of new casuistry, someway. And so come the big data! The massive or big data paradigm pushes this idea that every problem could be resolved from data. It is believed that one can control the activity from data and digitalization of the world. What is given (the data) is no longer built, but immediately available, in the digital reality. Big data encloses (digitalized) reality on itself. (New) Cybernetics abandon all forms of scale, benchmark, or hierarchy,

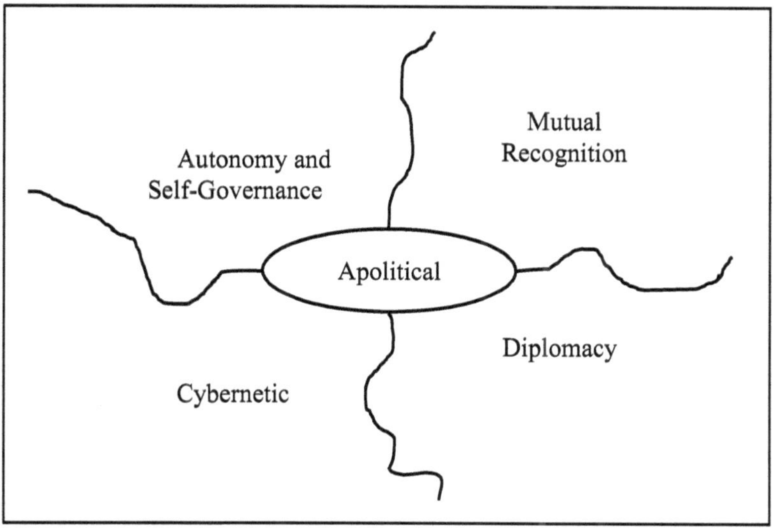

Fig. 2 Assemblage political models framework

in favor of an immanent normativity evolving in real time, minimizing all (real) confrontation with individuals. Rouvroy and Berns [15] in a recent paper called this movement, according to Foucault's terms, "Algorithmic governmentality."

The formula is: "*What is common emerge from statistics, focus on relations*". Quantified Self or Smart City concepts are typically based on this cybernetic view.

Finally we look at a fifth political model as essential, it is that of mutual recognition proposed by Honneth [16]. The recognition concept is derived mainly from G.W.F. Hegel's early social philosophical works. Axel Honneth argues that "the struggle for recognition" is, and should be, at the center of social conflicts and social life.

Shouldn't the territory offer to people a space of recognition which is the key element of the regulation of any nonviolent social system and which lead to a balanced human development? This is the point. This recognition linkage, which develops in particular within the successive frameworks of the Love, the Right, or Solidarity, according to Honneth, couldn't it be also at the core of territorial assemblage? These arrangements should contribute to limit the feeling of contempt, of denial of recognition, or the feeling of reification shared by many citizens today. This recognition linkage is the bridge between human development and the various socialization arenas like families, social networks, or territorial communities. Topics of social cohesion, territorial resilience, or local development point are examples of the importance of the recognition.

A good place is a place where there are intersubjective recognition bonds through family, works, social network, friendship, etc. The formula could be: "*What (must be) common is the quality of interpersonal relationships, which should be good*".

In spite of human split and sometimes the radical post-humanism (Vandenberghe 2007) carried by the assemblage thinking—it exists finally enthusiastic experiences today around new kinds of territories, closed to assemblage perspective. We think about Smart City like Masdar or Songdo. We think at the Ecodistricts all around the world, or at Self-proclaimed autonomous towns (like Freetown Christiania in the Danish capital Copenhagen), or at the Multicultural "holy neighborhood" of the French city of Bussy-Saint-George. They are many reasons to consider that these examples of territorial assemblages are driven by some new political common grounds, certainly incomplete.

3 Summary

It is time to conclude. We summarize our main results here:

- Territory can be seen as an ongoing action (instead of an object), and doing so, as a story;
- "Stories are lived before they are told" said MacIntyre and this means that action has a story-like existence;

- In realistic (or pragmatics) theory of narration, story is an assemblage of heterogeneous entities or actants (problems, actors, plots or transformation, tools, setting, etc.) which are connected, therefore exist, and therefore perform together. Assemblage is an agency;
- By overstating the heterogeneity of entities and the "messy" character of connections between them, we come across assemblage thinking (instead of simply networked, structuralist, or systemic thinkings). Assemblage is a complex whole, but not a totality;
- But assemblage thinking is far from the common ground of actors, especially political people. So we need to understand what it means to inhabit an assemblage, and if it is a clean room for human;
- Doing so, we discuss about five political models, and give some examples of what it seems to be territorial assemblages.

References

1. Ricoeur P (1973) The model of the text: meaningful action considered as a text. New Literary History 5(1), What Is Literature? (Autumn, 1973), pp 91–117. Published by: Johns Hopkins University Press
2. Johnson J (2008) Hypernetworks in the science of complex systems (series on complexity science). Imperial College Press
3. Calvez P, Soulier E (2014) Sustainable Assemblage for Energy (SAE) inside Intelligent Urban Areas How massive heterogeneous data could help to reduce energy footprints and promote sustainable practices and an ecological transition. In: C4BD2014, Workshop on Complexity for Big Data, Workshop at IEEE BigData 2014, Washington, DC, USA, 27th, 28th, 29th, or 30th of October 2014
4. Soulier E (2014) ICCCI 2014, International Conference on Computational Collective Intelligence Technologies and Applications 2014, CompStory14, The First International Symposium on Computational Story, September 25, 2014, Seoul, Korea
5. Soulier E, Neffati H, Rousseaux F, Bugeaud F, Perez O, Saurel P, Calvez P, Leitzman M, Legrand J (2012) Towards a territorial intelligence based on territorial assemblages theory: urban community case of Cap Excellence in Guadeloupe. In: II INTI international conference La Plata 2012 territorial intelligence and globalization, tensions, transition and transformation, 17–20 October 2012, La Plata, 13 p
6. Lussault M, Stock M (2010) "Doing with space": towards a pragmatics of space. Soc Geogr 5:11–19. doi:10.5194/sg-5-11-2010, www.soc-geogr.net/5/11/2010/
7. Alasdair MacIntyre A (1981) After virtue: a study in moral theory. University of Notre Dame Press
8. Schapp W (1953) In Geschichten verstrickt (Entangled in Stories). Zum Sein von Mensch und Ding (trans: Heymann B). 2nd edn. 1976, Wiesbaden
9. Michael J, Sandel MJ (1984) The procedural republic and the unencumbered self. Polit Theory 12(1):81–96, Published by: Sage Publications, Inc
10. Latour B (2007) Reassembling the social: an introduction to actor-network-theory (Clarendon Lectures in Management Studies). Oxford University Press
11. Delanda M (2006) A new philosophy of society: assemblage theory and social complexity. Continuum, London & New York
12. Austin J (1955) In: Urmson JO, Sbisà M (eds) How to do things with words: the William James lectures delivered at Harvard University in 1955, 1962. Clarendon Press, Oxford

13. Czarniawska B (2004) On time, space, and action nets. Organization 11(6):773–791, ISSN 1350–5084, 2004 SAGE
14. Müller M (2015) Assemblages and actor-networks: rethinking socio-material power, politics and space. Geogr Compass 9(1):27–41. doi:10.1111/gec3.12192
15. Rouvroy A, Berns T (2013) Gouvernementalité algorithmique et perspectives d'émancipation. Le disparate comme condition d'individuation par la relation? Réseaux 177:163–196
16. Honneth A (1992) The struggle for recognition: the moral grammar of social conflicts. Polity Press, 1995
17. Mink LO (1970) History and fiction as modes of comprehension. New Lit Hist 1(3):541–558
18. Zizek S (2003) Organs without bodies: Deleuze and consequences. Routledge
19. Callon M (2008) What does it mean to say that economics is performative? In: MacKenzie D, Muniesa F, Siu L (eds) Do economists make markets? On the performativity of economics. Princeton University Press

Fractal Density and Singularity Analysis of Extreme Geo-Processes

Qiuming Cheng

1 Introduction

Since the principle of density was discovered by the Greek scientist Archimedes approximately 2000 years ago, density has become a fundamental property of mass or energy as a well-known physical concept with a variety of applications. Density as a scale independent property of material or energy has been treated as a fundamental physical parameter and variable of many physical models with applications in nearly all fields of study, ranging from physics to engineering, economics, and the social sciences. The density of a material or energy is defined as its mass or energy per unit volume. Therefore, density often is characterized as units of mass over volume (e.g., g/cm^3, kg/m^3) or energy over volume (J/cm^3, w/L^3). For example, continental crust, which consists mostly of granitic rock, has a density of about 2.7 g/cm^3 and the Earth's mantle of ultramafic rock has a density of about 3.3 g/cm^3. The density of seawater varies with temperature and salinity of the water. Although the density of seawater varies at different points in the ocean, a good estimate of its density at the ocean's surface is 1025 kg/m^3. Density of air is a temperature and pressure dependent parameter. For given temperature and pressure the density of air is independent of the volume of air. Analogy definition of concept of density has been commonly used in many other fields such as in social studies a population density can be defined as a number of people per unit region. Point

Q. Cheng (✉)
Department of Earth and Space Science and Engineering, York University, 4700 Keele Street, Toronto, ON, Canada M3K1P3

Department of Geography, York University, 4700 Keele Street, Toronto, ON, Canada M3K1P3

State Key Lab of Geological Processes and Mineral Resources, China University of Geosciences, Beijing 100083 and Wuhan 430074, China
e-mail: qiuming@yorku.ca; qiuming@cug.edu.cn; qiuming.cheng@iugs.org

P. Bourgine et al. (eds.), *First Complex Systems Digital Campus World E-Conference 2015*, Springer Proceedings in Complexity, DOI 10.1007/978-3-319-45901-1_41

density as number of points per unit area and road density as length of roads or a number of road segments per unit area are often seen in spatial analysis. According to the ordinary density the measure or mass within a volume can be estimated as the integral of the density multiplied by the volume. However, from multifractal point of view the mass or measure defined within a volume can show singularity in which the density measured as the mass over volume may approach to infinite as the volume becomes infinitely small [1]. The ratio of mass over volume does not converge; in this case the density does not exist according to the ordinary definition [2]. In the follow section the new form of density, a generalized form of the above ordinary density will be introduced which quantifies the density of complex objects.

We show that the end products for many types of singular processes can be characterized by fractal mass density or energy density. The concept and index of fractal density will be utilized in dynamic models involved in the modelling of extreme processes [2, 3].

2 Fractal Density and Density-Scale Power-Law Model

According to the concept of ordinary density, the mass density of an object (ρ) can be calculated by the following equation:

$$\rho = \frac{m(v)}{v}, \tag{1}$$

where $m(v)$ represents the mass contained in a volume and ρ is the average density of an object. If the density of the object is homogenous, then the density calculated in (1) becomes independent of volume. The unit of the density is determined by the ratio of the mass and volume, for example, g/cm^3. However, if the object has heterogeneous properties, the density may vary from place to place and the average density in (1) varies with different size of v. In this case a localized density must be calculated using the derivative of the mass over volume:

$$\rho = \frac{\mathrm{d}m(v)}{\mathrm{d}v} = \lim_{v \to 0} \frac{m(v)}{v}. \tag{2}$$

The preceding density exists only if the limit converges when the volume becomes infinitesimal. If the limit does not converge, then the density does exist. As a generalization of Eq. (2) I introduce a new Eq. (3) where there exists a parameter α (with positive value) so that the following limit converges:

$$\rho_\alpha = \lim_{v \to 0} \frac{m(v)}{v^{\frac{\alpha}{3}}}. \tag{3}$$

The value of ρ_α can be considered as generalized density because the ordinary density defined in Eq. (2) becomes a special case of Eq. (3) when $\alpha = 3$, the normal dimension of volume. This new density is named fractal density which is defined as mass or energy per unit "fractal set." The fractal density defined in Eq. (3) has as unit the ratio of mass to a fractal set of α dimensions; for example, g/cm$^\alpha$ or kg/m$^\alpha$. Similarly, the units of fractal energy density can be J/cm$^\alpha$ or w/L$^\alpha$. Combining Eqs. (2) and (3) yields the following relationship between the ordinary density and the fractal density:

$$\rho(v) = \rho_\alpha v^{-[1-\alpha/3]}. \tag{4}$$

Thus, the ordinary density obeys a power-law relationship with volume which has the following properties [4, 5]:

$$\lim_{v\to 0} \rho = \begin{cases} 0, & \text{if } \alpha > 3, \\ \infty, & \text{if } \alpha < 3, \\ \rho_\alpha, & \text{if } \alpha = 3. \end{cases} \tag{5}$$

In accordance with these properties, ordinary density becomes volume dependent when $\alpha \neq 3$ and it tends to either zero or infinity when the volume becomes infinitesimal. The notation of fractal density used in Eqs. (3) and (4) can be replaced by the following general model associating the fractal density and the ratio of mass and scale (ε—linear size of an E-dimensional set):

$$\rho(\varepsilon) = \rho_\alpha \varepsilon^{-[E-\alpha]}. \tag{6}$$

The power-law relation between the ordinary density and the scale is determined by two parameters: the fractal density ρ_α which is independent of scale and the exponent–singularity index α (fractal dimension), or $\Delta\alpha = E - \alpha$; the latter also is known as the co-dimension of fractal density. The singularity index ($\Delta\alpha$) measures the deviation of the fractal dimension from the dimension of normal density. These two parameters (ρ_α and $\Delta\alpha$) can be estimated from observed data by measuring the intercept and slope of a straight line on the log–log plot of m against ε.

3 Fractal Integral and Fractal Differential of Fractal Density

Integral and differential are two fundamental operations involved in modern calculus and in many other mathematical and physics subjects. The traditional integral and differentials are defined based on additive property of measure. When the measure no longer possesses additive property then the classical integral and differential may not exist. Therefore, the ordinary integral and differential operations do not applicable to the fractal density with singularity. Here I will extend the ordinary integral and differential operations to define fractal integral and differential operation as follows:

$$f'_{\alpha}(x) = \frac{df(x)}{dx^{\alpha}} = \lim_{\upsilon \to 0} \frac{\Delta f(x)}{\Delta x^{\alpha}}, \tag{7}$$

where $\Delta f(x)$ and Δx represent the increment of function $f(x)$ over an increment of Δx. The convergence of the limit of form (7) can be defined as α-fractal derivative of function $f(x)$. Similarly we can define the fractal integral of function $f(x)$ as follows:

$$\int f(x)dx^{\alpha} = \lim_{\Delta x \to 0} \sum f(x_i)\,(\Delta x)^{\alpha}, \tag{8}$$

where the $f(x_i)$ is the height of function $f(x)$ in the small range of $[x_i, x_i + \Delta x]$. If the limit (8) converges, then the limit can be named α-fractal integral of function $f(x)$. It must remind that the fractal derivative defined in this paper is different from the fractional derivative (fractional order) available in the literature $f^{(\mathrm{v})}(x)$, where v can be a non-integer order. The fractional derivative assumes that the normal integer order derivative $f^{(\mathrm{n})}(x)$ does exist. The fractal derivative is based on fractal dimension of the measure whereas the fractional derivative is based on fraction order of derivative defined on normal measure. As an example, let us take a power-law function to demonstrate the fractal derivative. Assume $f(x) = cx^b$, the ordinary derivative of the function $f'(x) = cbx^{b-1}$, which does not exist at $x = 0$ if $0 < b < 1$. The integral of the function is $\int f(x)\mathrm{d}x = c/\,(b+1)\,x^b + 1$, does not converge if $b < -1$ at $x = 0$. The fractal derivative at $x = 0$ exists and $f_{\alpha}{}'(x) = c$, if $\alpha = b$. Further discussion about fractal derivative and integral will be given in separate paper.

4 Fractal Density and Extreme Geo-Processes

In the remainder of this paper I demonstrate that fractal density ($\Delta\alpha \neq 0$) characterizes anomalous mass accumulation or energy release caused by extreme geo-processes occurred in the Earth's crust originated from cascade earth dynamics (e.g., mantle convection and plate tectonics) and self-organized criticality (e.g., avalanches of slab breakoffs or faults). The examples to be analyzed in the rest of the paper are river flow density caused by storms or energy density released by earthquakes and intensity of earthquakes [6]. Other examples introduced in the author's previous publications include heat flow over mid-ocean ridges [2], and element concentration anomalies in surface media (stream sediments and regolith) caused by hydrothermal mineralization [7].

4.1 Fractal Density of River Peak Flow Caused by Storms

In storm hydrology, an important phenomenon is the peak flow and recession described by the stream's discharge hydrograph, a graph showing the rate of flow (discharge) versus time past a specific point in a river. The rate of flow is typically expressed in cubic meters or cubic feet per second ($m^3\ s^{-1}$ or $f^3\ s^{-1}$). The stream rises to peak flow after each storm event and then falls in a slow recession. Extreme flow events may occur due to accumulation of anomalous water volume in a short period time and often cause floods when the river flow exceeds the capacity of river discharge. The author's early work shows that extreme flow events depict strong local singularity that can be estimated by the accumulative average river flow [6].

$$Q^*(<t) = \frac{1}{t}\sum_{i=1} Q_i, \tag{9}$$

where Q_i ($i = 1, 2, \ldots$) are the river flow observed at time t_i and Q^* stands for accumulative average river flow for a time period t. At the singular location of a flow series, the quantity Q^* follows a power-law relation with measuring unit t

$$Q^*(<t) = \rho_\alpha t^{-\Delta\alpha}, \tag{10}$$

where ρ_α (with unit m^3s^{-1}/day^α) is the fractal flow rate with α as the fractal dimension.

This model was applied to fit the peak flow data recorded at river gauging stations in the Oak Ridges Moraine (ORM) in southern Ontario, Canada. Several gauging stations in the study area were selected for this study. These stations have records of mean daily flow (m^3/s) and daily rainfall data since 1900 [8]. Figure 1a shows several gauging stations superimposed on a digital elevation model (DEM) [9]. Figure 1b shows the results obtained for the flood caused by storm after Hurricane Hazel hit the area in October 1954. The power-law model was applied to fit the observed data by least-squares method which yielded $Q^* = 23.55\ t^{-0.603}$, $\Delta\alpha = 0.603$, $R^2 = 1.000$. Similarly, the power-law model was applied to all 17 main peak flow data from the Cataract Station on the Credit River. The values of singularity index ($\Delta\alpha$) calculated from the 17 peak flow series range from 0.20 to 0.75. These positive values indicate that the peak flow series each show strong singularity around the flow peak implying that the water volume released during the short period time around the peak flow is non-linearly proportional to the time duration. The singularity index may be useful for characterizing the flow in different river systems and environments. As an example, we plot the singularities calculated for the 17 events that occurred during 1900–2000 at the Cataract Station in Fig. 1c. From this plot one can see that the singularity shows general decreasing trend for the time period studied. A linear regression can be fitted to the data by least-square method yielding $y = -0.003x + 6.67$, with $R^2 = 0.423$, and the student t-value $= 3.31$ (sample size $n = 17$). From this trend one can estimate the future time in about 223 years when the singularities dismiss. Of course this is based on the same decay trend and it must be kept in mind that the large error involved in the regression will lead to substantial variance of prediction. The cause of decay of singularity might be due to urban buildup and man-made discharge systems in the area, evolution of the river systems and drainage networks of the area, or attributed to climate change.

4.2 *Fractal Energy Density of Intensity Caused by Earthquakes*

Any earthquake is due to a typical self-organized criticality (SOC) extreme process and avalanches which cause anomalous energy release within narrow spatial–temporal intervals. Patterns of the spatial, temporal, and magnitude distribution of earthquakes have been the focus of many studies and the main attributes involved in models for prediction. Studies have demonstrated that the energy release of earthquakes follow power-law relations with accumulative time and scale of space. The energy released from earthquakes can be expressed as the Gutenberg–Richter equation [10]

$$E = E_0 \exp(bM), \tag{11}$$

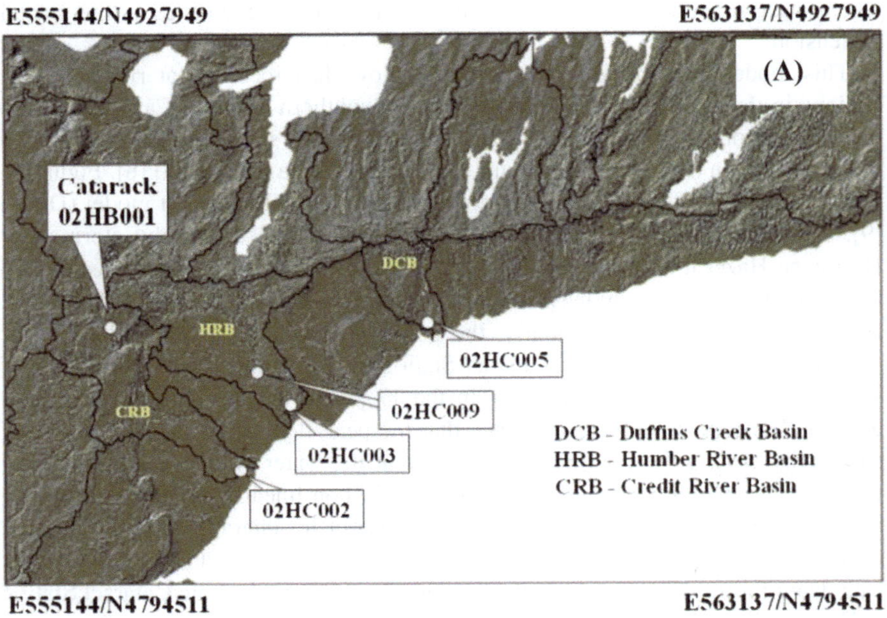

Fig. 1 (**a**) Shaded relief of digital elevation model of the ORM [9]. *Lines* represent major basins in the area. *White dots* represent the locations of river gauging stations chosen for the study. *Labels* in boxes are the IDs of gauging stations. (**b**) Plot shows the average flow (Q^*) as calculated for model (10). *Straight line* was fitted by least-squares method. (**c**) Plot showing a general descending trend of singularity values calculated from 17 events recorded at the Cataract station

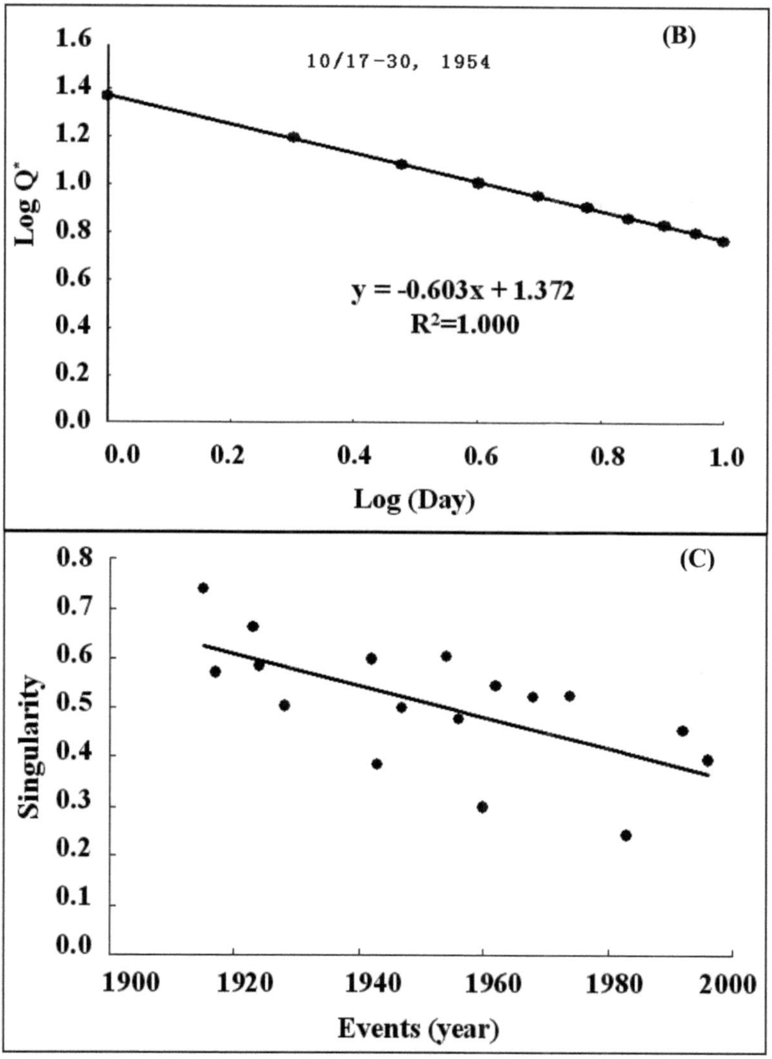

Fig. 1 (continued)

where E is the energy released in an earthquake, E_0 is a threshold energy, M is the earthquake magnitude, and b is a constant value. The Gutenberg–Richter equation indicates that the earthquake magnitude is a logarithmic transformation of energy release. It was proposed by Apostol [11] that the accumulative energy of earthquake and cumulative time follows a power-law relation with

$$t/t_0 = (E/E_0)^r, \tag{12}$$

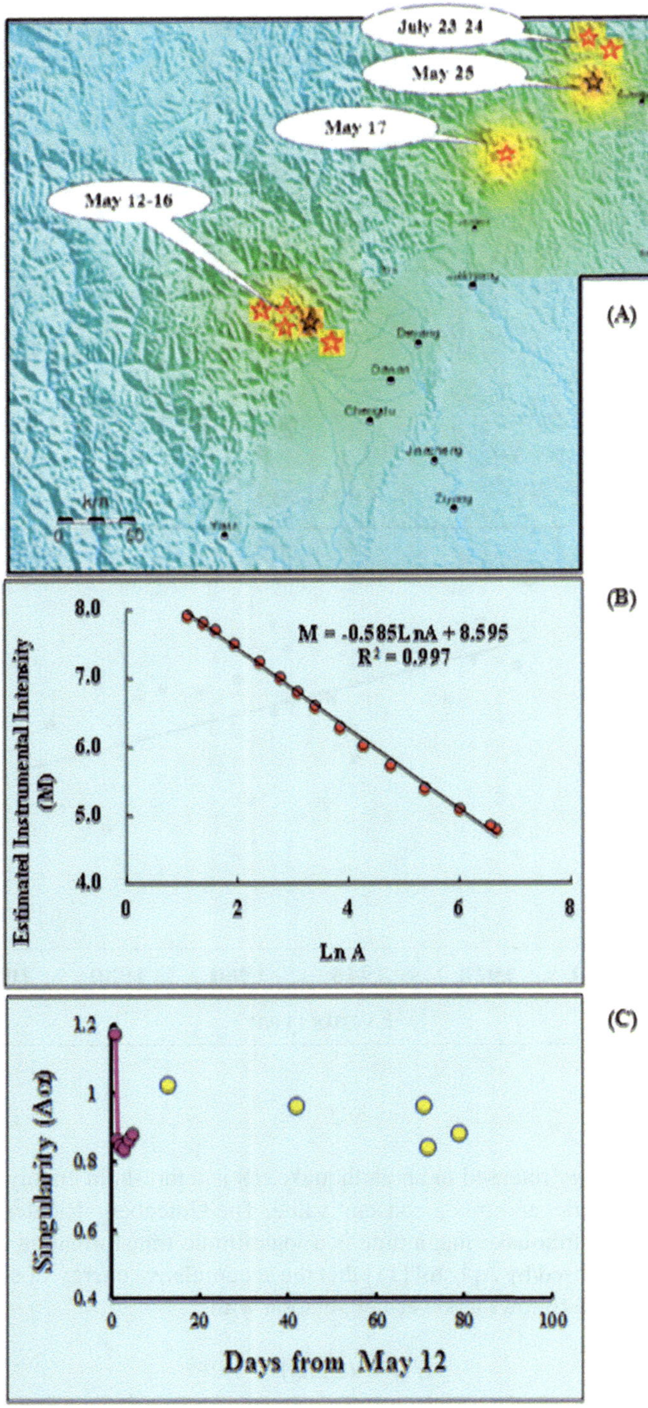
July 23 24
May 25
May 17
May 12-16
(A)
(B)
Estimated Instrumental Intensity (M)
M = -0.585LnA + 8.595
R² = 0.997
Ln A
(C)
Singularity (Aα)
Days from May 12

where t_0 is a threshold time and r is a parameter accounting for the geometrical and physical properties of earthquakes. An alternative relation was proposed by Turcotte [12] to associate the energy and length scale of focal zone as

$$E/E_0 \approx (R/R_0)^{1/r}, \tag{13}$$

where R and R_0 are scale lengths characterizing the focal zone. The exponent $1/r$ can be viewed as fractal dimension. This equation can be rewritten as a relation between energy and area

$$\begin{aligned} E(A) &= E_0/\left(\sqrt{A_0}\right)^{1/r}\left(\sqrt{A}\right)^{1/r} = \rho_a\left(\sqrt{A}\right)^{a} \\ \rho(A) &= \rho_a\left(\sqrt{A}\right)^{-(2-\alpha)}, \end{aligned} \tag{14}$$

where $\rho_\alpha = E_0/\left(\sqrt{A_0}\right)^{\alpha}$ is the fractal energy density with $1/r = \alpha$ as the fractal dimension. It can be seen that the unit of fractal energy density ρ_α is the ratio of energy and areal scale to the power of α and the energy density per area, $\rho(A)$, follows power-law relationship with area.

This function was fitted to data of intensity caused by the Wenchuan earthquakes which occurred in Wenchuan, Sichuan, China, May 2008. The Mercalli intensity scale used is a seismic scale for measuring the intensity of an earthquake. It measures the overall effect of an earthquake empirically from various observed effects. The intensity of an earthquake then is not totally determined by its magnitude but incorporates its effects on people, human structures, and the natural environment. The National Earthquake Information Center of the United States Geological Survey has compiled and maintained an online database providing intensity maps of major earthquakes that have occurred in the world (URL: http://earthquake.usgs.gov/earthquakes/shakemap/). Figure 2 shows the distribution of intensity maps of the earthquakes including 8 aftershocks with magnitude 6 or above as occurred in Wenchuan during May–July of 2008. The Wenchuan earthquake with 8.0 magnitude that occurred at about 2 AM on May 12 was the deadliest earthquake killing 69,197 people and leaving 18,222 missing. Strong aftershocks, some exceeding magnitude 6, continued to hit the area even months after the main quake, causing new casualties and damage.

Fig. 2 (**a**) Intensity maps showing earthquakes with magnitude above 6 that occurred in Wenchuan, Sichuan, China, in May of 2008. An earthquake of magnitude 8 occurred on May 12, and was followed by aftershock earthquakes in May. Data were taken from the website of National Earthquake Information Centre of the United States Geological Survey (URL: http://earthquake.usgs.gov/earthquakes/shakemap/). *Stars* represent location of earthquakes. *Color* indicates the level of intensity. (**b**) Plot showing power-law relationship between intensity and area for the main earthquake on May 12. *Red dots* represent observed data and the *black solid line* is for model fitted by the least-squares method. (**c**) Plot showing the singularity indexes ($\Delta\alpha$) estimated for the 9 earthquakes that occurred in May and June, 2008 in the same region. The results for the first 6 earthquakes are marked as *purple* and the other 5 as *yellow* for comparison

The data intensity and affected area for the main quake and all 8 aftershocks were fitted with power-law functions by the least-squares method. The results obtained for the main quake are estimated as, $\rho(A) = 5377.6(\sqrt{A})^{-1.17}$, with coefficient of determination $R^2 = 0.997$. Similarly, the data for 8 aftershocks were analyzed with the results shown in Fig. 2c. These results indicate that the main quake with singularity $\Delta\alpha = 1.17$, the first 5 aftershocks with singularity $\Delta\alpha \approx 0.83$–$0.88$, and the other 5 aftershocks, which occurred in July, with singularity $\Delta\alpha$ between 0.88 and 1.00. The result may imply different types of mechanisms were responsible for immediate aftershocks and more distant aftershocks.

5 Concluding Remarks

In this paper I have shown that the fractal density as a generalization of ordinary density can be used to characterize the singularity caused by extreme geo-processes such as those resulting in earthquakes, floods, heat flow at the mid-ocean ridges, and hydrothermal systems which occurred in the Earth's crust resulting in mineral and energy resources. The examples demonstrate that peak river flow caused by storm depicts fractal density following power-law relation with time after peak flow. Intensity caused by earthquakes depicts fractal density around epicenters with intensity decays following power-law relations for increase of area around the epicenter. Other examples shown in other papers include heat flow at mid-ocean ridges shows fractal density and the heat flow decays with distance from mid-ocean ridges following power-law relation. The element concentration in stream sediments caused by mineralization depicts fractal density and the concentration decays with distance from the center of anomalies following power-law relation. It can be generally concluded that the fractal density and local singularity analysis are applicable for most of the extreme geo-processes occurred in the Earth's crust originated from cascade earth dynamics (e.g., mantle convection) and self-organized criticality (e.g., slab breakoffs and faults as avalanches).

Acknowledgments This research has been supported by an NSERC Discovery Research "Research and development of multifractal methods and GIS technology for mineral exploration and environmental assessments" (ERC-OGP0183993).

References

1. Mandelbrot BB (1989) Multifractal measures, especially for the geophysicist. Pure Appl Geophys 131:5–42
2. Cheng Q (2016) Fractal density and singularity analysis of heat flow in oceanic ridges. Sci Rep. doi:10.1038/srep19167
3. Cheng Q (2015) Multifractal interpolation method for spatial data with singularity. J South Afr Inst Min Metal 115:1–6

4. Cheng Q (2007) Mapping singularities with stream sediment geochemical data for prediction of undiscovered mineral deposits in Gejiu, Yunnan Province, China. Ore Geol Rev 32:314–324
5. Cheng Q, Agterberg FP (2009) Singularity analysis of ore-mineral and toxic trace elements in stream sediments. Comput Geosci 35:234–244
6. Cheng Q (2012) Singularity theory and methods for mapping geochemical anomalies caused by buried sources and for predicting undiscovered mineral deposits in covered areas. J Geochem Explor 122:55–70
7. Cheng Q, Li L, Wang L (2009) Characterization of peak flow events with local singularity method. Nonlinear Process Geophys 16:503–513
8. HYDAT CD-ROM User's Manual: Surface Water and Sediment Data, Atmospheric Environment Program, Version 96 – 1.04 User's Manual, Environment Canada, p 95 (1996)
9. Kenny FM (1997) A chromostereo enhanced digital elevation model of the Oak Ridges Moraine Area, Southern Ontario and Lake Ontario Bathymetry. Geological Survey of Canada. Open file 3423, scale 1:200 000
10. Gutenberg B, Richter CF (1944) Frequency of earthquakes in California. Bull Seismol Soc Am 4:185–188
11. Apostol BF (2006) A model of seismic focus and related statistical distributions of earthquakes. Phys Lett A 357:462–466
12. Turcotte DL (1997) Fractals and chaos in geology and geophysics, 2nd edn. Cambridge University Press, Cambridge, 398 pp

Globalisation, Multilingualism and Multiculturalism in Mozambique

Sarita Monjane Henriksen

1 The Adoption of Portuguese as Official Language

As a background to the present study, it is worth mentioning that switching between two or more languages is part of the daily language practices of the average Mozambican, considering that the majority speaks an L1, usually a Mozambican Bantu Language, in the home domain, an L2, the country's official language, Portuguese, used mostly as the country's *Lingua Franca* and also for all formal and public administration domains and in a few instances, a foreign language, which in most cases will likely be English.

High levels of linguistic and cultural diversity and bi/multilingualism characterise Mozambique, a country, which as a result of the colonisation process, and similarly to other countries in the region, has adopted the ex-colonial language, Portuguese, as the only official language, at the time of its independence in 1975. The adoption of the former colonial languages as official languages has been common practice in many independent countries in Africa. Among the reasons that appear to be behind the choice of the ex-coloniser's languages at the expense of the local African languages is the fact that the newly independent countries did not really have any other alternative. It should be said that, in many countries, most of the African languages either did not have a written form or decisions had to be made concerning the language to select from amidst the mosaic of languages [1].

For Ruíz [2, p. 7], a large part of the work in the field of language planning has, in fact, been inspired by the "preponderance of problem-oriented language planning approaches", which seem to establish a link between language and language diversity with social problems and therefore multilingualism being perceived as

S.M. Henriksen (✉)
Universidade Pedagógica, Maputo, Mozambique
e-mail: shenriksen@up.ac.mz; sarita-henriksen@hotmail.com

P. Bourgine et al. (eds.), *First Complex Systems Digital Campus World E-Conference 2015*, Springer Proceedings in Complexity, DOI 10.1007/978-3-319-45901-1_42

ultimately leading to a lack of social cohesiveness[1]; with everyone speaking their own language, political and social consensus being *impossible* (p. 10).[2] In fact, for Tsonope [5], it is widely argued that promotion of various languages impairs national cohesion, increasing the risk of the emergence of conflicts between communities.[3]

Still in relation to the adoption of the ex-colonial language, at the expense of the Mozambican Bantu languages, Ganhão [8] admitted that although spoken by a very small percentage of Mozambicans, at the time of independence, Portuguese was certainly a politically neutral language, which would serve well the purposes of nation building, and surely a better option than any other national language for avoiding conflicts. The view of multilingualism or language diversity as a problem was certainly present in Mozambique at the time of independence. The vision of a multilingual Mozambique was out of the political agenda; the key goal at the time was to urgently build the Mozambican nation and make it work as such. The fact that there were so many languages spoken by the various ethnic groups constituting the new Mozambican state, and the fact that none of them was spoken nation-wide as a common language or *Lingua Franca*, would probably have constituted a hindrance to the birth of the new Mozambican nation.

2 Portuguese in Mozambique: Status and Attitudes

As a result of its adoption as the only official language, and thus its connection to jobs in the formal labour market, including both the public and private sectors, and its exclusive use in the education system,[4] the Portuguese language has enjoyed a very high status and prestige. It is a language, thus, seen as the key to academic,

[1]We have to look at the equation—multilingualism means problems—with caution, as a number of examples exist throughout the world of linguistically homogeneous areas (like Northern Ireland, Rwanda or Somalia), which also lack social cohesiveness; and if we compare them with Switzerland, which is extremely heterogeneous but still cohesive, then it becomes difficult to argue in favour of the equation.

[2]A very interesting critical review of the suggested correlation between high linguistic diversity and level of socioeconomic development or the relationship between multilingualism and social wealth of a country is presented by Coulmas [3], who draws on Pool [4, p. 222] who argues that "linguistically highly fragmented countries are always poor".

[3]See David Laitin [6] for an interesting discussion of the correlation between language policy and civil war and the evidence he presents "against claims that the elimination of minority grievances would be a sure fire way of lowering the incidences of civil war" (p. 178). See also François Grin [7] for the costs of maintaining cultural diversity.

[4]It should be highlighted that the *Plano Curricular do Ensino Básico* or the Curricular Plan for Basic Education [9, p. 17] stipulates that the medium of instruction in the whole education system is Portuguese, regardless of the fact that the large majority of children do not speak it when they start schooling.

social, professional and economic ascendance. It is a language linked to the top of the social hierarchy, and a language highly valued by the social and political elites and equally by the masses.[5]

The Portuguese language in Mozambique is no longer seen as the ex-coloniser's language, but a Mozambican language in its own right.[6] It certainly differs from the Portuguese of Portugal or the Portuguese of Brazil in many respects, and particularly at the lexical and phonological level. It is quite common, nowadays, to hear and read about the nativisation of the Portuguese language in Mozambique, which according to Firmino [10, p. 143] comprises not only a linguistic dimension, as new usages are being developed, but also a symbolic dimension, characterised by the emergence of new social attitudes and ideologies. Gonçalves [12, p. 61] points out that Mozambican Portuguese is characterised by the coining of new words, as well as extensive borrowings not only from the Bantu languages, but also from English; although it still draws on the European Portuguese, new semantic values and syntactic properties have developed.

Lopes [13, p. 475], on the other hand, argues that the Portuguese language is no longer foreign, as it "has been evolving as a naturalised variety to serve the needs of Mozambicans". The naturalisation and indigenisation processes have led to the acquisition of new features, adapting the language to the "local realities, including the journalistic and literary registers of use". In Couto's [14] view, alterations to the Portuguese language go beyond the linguistic domain and reveal a different perception of the world and life. "Mozambicans are in the process of transcending their role as simply users of the Portuguese language and assuming a status in which they are co-producers of this means of expression".

[5]Firmino's [10] study on attitudes to Portuguese in Maputo city appears to confirm that the Portuguese language is highly valued due to its pragmatic function, or in his own words, *"os sentimentos em relação ao Português é de que esta língua se tornou um importante instrumento linguístico em Maputo, em parte como resposta às exigências do Mercado linguístico e socioeconómico, mas também como resultado de uma consciência metapragmática da mudança do seu estatuto social"*. Essentially what Firmino states is that feelings in relation to Portuguese indicate that this is an important linguistic tool in Maputo, partly as a response to the demands of the linguistic and socioeconomic market, but also as a result of a metapragmatic awareness to change in its social statute. In her study of Trilingualism in Guinea-Bissau, Carol Benson came up with a similar conclusion as per the attitudes of ordinary people towards the Portuguese language. She argues that "although only a small percentage of Guineans claim to speak Portuguese (9 % total according to the 1991 census), there is a widespread, unquestioning belief in its value for future employment and other opportunities" [11, p. 170].

[6]Although in the official discourse and, most particularly, in the Constitution of the Republic, there is a reference to the *national languages*, meaning territorially and ethno-culturally Mozambican, and the Portuguese language (still seen as an exogenous language).

3 Multilingualism and Multiculturalism in Mozambique

As already mentioned above, Mozambique is a multiethnic, a multicultural and a multilingual country, and this heterogeneity is certainly a result of the encounter and contact, over the years, between different groups of people, among them, Islamic coastal traders, European colonisers and the indigenous populations. Sixteen main ethnic groups, mostly of Bantu origin, are recognised in Mozambique. In addition, we also find population groups of European descent, mixed Euro-Africans, Indians and Chinese.

As a result of ongoing globalisation, the levels of multilingualism and multiculturalism have surely increased in Mozambique, due to the presence of large numbers of Chinese nationals working mainly in the construction industry, Indian and Brazilian nationals working for the mine industry, just to mention a few. Globalisation, characterised by increasing levels of population movements and migration, is just one side of the coin. The other side is certainly the partition of Africa in the 1880s, which in my view, has also contributed greatly to increasing levels of ethno-linguistic and cultural diversity in the African territories and particularly in Mozambique. As Mazrui and Mazrui [15, p. 5] indicate, the "national boundaries of most African States lack the underpinning of any national linguistic identity"; in the same token, Kashoki [16, p. 186] argues that

> The largely arbitrary nature of the manner in which present-day African countries came into being as sovereign nation states is directly responsible for their present highly multi-ethnic, multicultural and multilingual 'national' character – sometimes, as in the case of Tanzania and Nigeria, containing as many as 100 or more 'languages' or 'dialects' within their borders.

Both Mazrui and Kashoki defend the view that the current linguistic diversity of many African countries is in part an artificial outcome of the colonisation process, resulting from the manner in which their borders were conceived by the colonial powers.

4 The Mozambican National Languages: Status and Attitudes

Until recently,[7] the use of the Mozambican National Languages had mostly been assigned to the home, family and other informal domains such as shops and market places. In addition, some of the Mozambican languages have also been used in radio broadcasting, particularly for news, and also for religious purposes. In regard to

[7]Most specifically, early 1990s, with the introduction of the national languages in literacy development and under the umbrella of the Mother Tongue Bilingual Education experimental project.

the status of the Mozambican National Languages, it would be quite prudent to argue that feelings and attitudes towards them have, throughout the years, been quite ambivalent; a mixture of positive and negative feelings.

Soon after independence, the use of these languages in the public sphere (particularly schools) was strongly discouraged by the FRELIMO government, the education authorities and also at the level of certain families (especially, urban and literate). In other words, there was no space, in the public sphere, for the use of the Mozambican National Languages. Among the reasons that appeared to have prevented their use in the public sphere were, the need to promote national unity by speaking a neutral language, the fact that these languages were seen as underdeveloped (particularly when it came to their readiness to be used as vehicles of instruction, issues related to their orthography or lack of it, etc.). One could perhaps attempt and say that, in fact, they seemed to be deprived of prestige, considering that during the colonial period and in the pre-independence years they were all considered dialects, with all the related connotations.[8]

Regarding the question of attitudes to indigenous African languages, Kamwanga-malu [17, p. 730] points out that

> A set of beliefs ... perpetuate the colonial myth that indigenous African languages do not have the linguistic complexity to be used in higher domains; and that these languages are good only to preserve African cultures and traditions. [Such a model, in his opinion, has] ideological implications that condemn languages to perennial status as underdeveloped. Consequently, in the post-colonies in Africa the position of the indigenous African languages in education and other higher domains has remained closely linked to the inherited colonizer's model, which perpetuates the hegemony of ex-colonial languages over the indigenous African languages.

Although the use of the Mozambican National Languages has been excluded from the formal contexts, they are strongly viewed as vehicles and symbols of the Mozambican national ethno-linguistic and cultural identity and, for that reason, their vitality is quite high.[9] They are still being transmitted from parent to child, from generation to generation, particularly in the rural areas. In fact, with their introduction in the education system, it is possible to argue, without a doubt, that there is a renewed vigour and even prestige in speaking and using these languages. Recent developments in Mozambican society point to the introduction of these languages also in the public television, introduction of *major* and *minor* courses in the Mozambican Bantu languages at university level (*Universidade Pedagógica*) and increasing research on these languages.

[8]In the popular use, the term dialect is commonly seen as a linguistic variety which is "inferior", less prestigious, and something "less" than a proper or a fully-fledged language.

[9]It is probably this strong identification with their mother tongues, as markers of ethno-cultural, linguistic and group identity that has contributed to the vitality and maintenance of these languages.

5 The English Language

Attitudes to the English language are very positive in Mozambique, due to the fact that it is the most spoken foreign language taught and learned in the Country, the main working language at the level of the Southern African Development Community (SADC), the Commonwealth of Nations, the African Union (AU), of which Mozambique is part of and also the main *Lingua Franca* used by other international organisations, such as the United Nations (UN).

Mozambique appears, geographically, like an island surrounded by English Speaking countries, such as South Africa, Swaziland, Zimbabwe, Zambia, Malawi and Tanzania. Although Portuguese is spoken in countries such as Angola, Cape Verde, Guinea-Bissau, Brazil, Portugal and to a small extent in Macau, in order to communicate with the rest of the world, Portuguese alone would not suffice. However, the findings of the study conducted by Henriksen [1] indicate that the English language would neither be enough. While admitting the importance of speaking English, the children in the urban school seem to look at multilingualism (including the use of Portuguese and the Mozambican National Languages) as part of the answer to the challenges of globalisation.

6 Challenges in Managing Multilingualism and Diversity

Managing multilingualism and diversity at society and school level raise a number of challenges that include, among others, the following:

(1) the need to manage people's attitudes to minority and majority languages or Languages of Wider Communication (LWCs), people's perceptions on the symbolic value of languages;
(2) the issue of the costs involved, for instance, in providing mother tongue education to everyone and the costs of defining a Language Policy that includes all languages and population groups and
(3) the need to eliminate ethnic and social cleavages.

In spite of the above-mentioned challenges, a number of successful examples are reported from different parts of the world. For instance, among others, we have in Europe the "European School Model"[10] in providing multilingual education,

[10]See Baetens Beardsmore [18] for details on the European School Experience, comprising provision of education in eight or nine different languages (majority European languages). Of course, I am not suggesting that this model would be easily replicated in a context such as Mozambique, because of the very favourable conditions of the European School Model, characterised by a whole wealth of resources (human, material, etc.); conditions which are not easily available in Mozambique.

and in India, there is the case of Tribal Multilingual Education.[11] As many have argued before me, although managing diversity is a very "arduous and expensive" endeavour,[12] it is worth it. Multilingualism and language diversity should be seen as "sources of knowledge and enrichment" [21], as "much more of an asset than a disadvantage" [22] or as very appropriately put by Jo Lo Bianco [23] as resources for the individual, the society and the economy. Linguistic Diversity enriches our world and our reality; because "all languages are depositories of knowledge and some of the endangered languages constitute the only possibility of access to valuable indigenous knowledge that reaches far back into the history of human species" [24, p. 3].

7 Concluding Remarks

Based on the above and informed by the best international practices, I advocate for the need to devise an inclusive and democratic language education policy for multilingual contexts such as Mozambique, in which there is room for smaller or *minority* languages and bigger or *majority* languages. This is a policy that is grounded on the discourse on Linguistic Human Rights (LHRs), it is a policy aimed at maintaining the country's linguistic diversity, while promoting its cultural and economic development, and at the same time fostering national integration, and helping the Mozambicans to keep abreast of world developments and participate actively in the so-called knowledge society.

References

1. Henriksen SM (2010) Language attitudes in a primary school: a bottom-up approach to language education policy in Mozambique. PhD Dissertation, Roskilde Universitet, Denmark
2. Ruíz R (1984) Orientations in language planning. In: Mackay SL, Wong SC (eds) Language diversity. Problem or resource? Newbury House Publishers, New York, pp 3–25
3. Coulmas F (ed) (1982) Linguistic minorities and literacy – language policy issues in developing countries. Mouton Publishers, Berlin, pp 29–37
4. Pool J (1972) National development and language diversity. In: Fishman J (ed) Advances in the sociology of language. The Hague, pp 213–230

[11] See Mishra [19] for an account of Tribal Multilingual Education in India.

[12] On the 2nd February 2008, Jutta Limbach [20], President of the Goethe-Institut, stated the following, in her paper entitled *Plurilingualism and Multilingualism—Obstacles on the Route towards a European Public:* "For some, the postulate of multilingualism appears to be an annoying national relic within the mosaic of the future European culture. However, this criticism misjudges the very special nature of European integration. The EU member states and their people do not want to follow the model of the nation state when shaping the European Union. When singing the praise of multilingualism, we must not forget a particularly weighty argument—the fact that language pluralism proves to be arduous and expensive."

5. Tsonope J (1995) Prospects of the indigenous languages of Botswana: implications of the Government White Paper No. 2 of 1994. Mosenodi 3(1&2):5–13
6. Laitin D (2004) Language policy and civil war. In: Van Parijs P (ed) Proceedings of the seventh Francqui Colloquium. De Boeck, Brussels. Retrievable at http://www.uclouvain.be/cps/ucl/doc/etes/documents/12.Laitin.pdf
7. Grin F (2004) On the costs of diversity. In: Van Parijs P (ed) Proceedings of the seventh Francqui Colloquium. De Boeck, Brussels. Retrievable at http://www.uclouvain.be/cps/ucl/doc/etes/documents/13.Grin.pdf
8. Ganhão F (1979) O Papel da Língua Portuguesa em Moçambique. 1 Seminário Nacional sobre o Ensino da Língua Portuguesa. Unpublished Paper. Ministry of Education and Culture, Mozambique
9. Ministério da Educação MINED (2003) Plano Curricular do Ensino Básico. Objectivos, Política, Estrutura, Plano de Estudos e Estratégias de Implementação. DINAME, Maputo
10. Firmino G (2005) A "Questão Linguística" na África Pós-Colonial. O Caso do Português e das Línguas Autóctones em Moçambique. Texto Editores, Maputo
11. Benson C (2004) Trilingualism in Guinea-Bissau and the question of instructional language. In: Hoffmann C, Ytsma J (eds) Trilingualism in family, school and community. Multilingual matters, Clevedon, pp 166–184
12. Gonçalves P (1996) Português de Moçambique: Uma Variedade em Formação. Livraria Universitária, Universidade Eduardo Mondlane, Maputo
13. Lopes AJ (1998) The language situation in Mozambique. J Multiling Multicult Dev 19(5 & 6):440–486
14. Couto M (1986) Uma Maneira Moçambicana de Contar Histórias Moçambicanas. In an interview to the Gazeta de Artes e Letras. Tempo 835
15. Mazrui A, Mazrui A (1998) The Power of Babel - language and governance in the African experience. James Carrey Ltd., Oxford
16. Kashoki ME (2003) Language policy formulation in multilingual Southern Africa. J Multiling Multicult Dev 24(3):184–194
17. Kamwangamalu N (2006) Bi-/Multilingualism in Southern Africa. In: Bhatia TK, Ritchie WC (eds) The handbook of bilingualism. Blackwell Publishing, Oxford, pp 725–741
18. Baetens Beardsmore H (1995) The European school experience in multilingual education. In: Skutnabb-Kangas T (ed) Multilingualism for all. Swets & Zeitlinger B.V., Lisse, pp 21–68
19. Mishra M (2004) Contextualizing classrooms in tribal area schools in Orissa. An experiential learning. In: Language in India, vol 4, 4 April 2004
20. Limbach J (2008) Plurilingualism and multilingualism – obstacles on the route towards a European public? Paper presented at the British Council Conference – Languages for Europe – held in Berlin, 1–2 February 2008
21. Aarts R, Extra G, Yagmur K (2004) Multilingualism in the Hague. In: Extra G, Yagmur K (eds) Urban multilingualism in Europe. Immigrant minority languages at home and school. Multilingual Matters, Clevedon, pp 193–220
22. Hélot C, Young A (2006) Imagining multilingual education in France: a language and cultural awareness project at primary level. In: García O, Skutnabb-Kangas T, Torres-Guzmán M (eds) Imagining multilingual schools. Languages in education and glocalization. Multilingual Matters Ltd., Clevedon, pp 69–90
23. Lo Bianco J (1987) National policy on languages. Australian Government Publishing Service, Canberra
24. Alexander N (2006) Communicating in local languages: a prerequisite for community access. In: Fourth session of the Intergovernmental Council for UNESCO's Information for All Programme (IFAP), Paris

Lagrangian Approach to Phytoplankton Mesoscale Biogeography in the Kerguelen Region

Alice Della Penna, Simon Wotherspoon, Thomas W. Trull, Silvia De Monte, Craig Johnson, and Francesco d'Ovidio

1 Context

Knowledge of phytoplanktonic biogeography is key [1, 2] in order to separate regions characterized by different biogeochemical processes [3–8], recognize biodiversity hotspots [9], identify potentially favourable foraging regions for higher trophic levels [10, 11] and design adaptive sampling strategies. Playing different biogeochemical roles, different phytoplanktonic groups respond differently to change both in the global climate perspective [12] and in local iron fertilization [13]. In the open ocean, the definition of biogeographical provinces (i.e. regions characterized by specific communities) at the sub- and mesoscale (1–100 km, few weeks to months) is challenging. Biogeochemical and ecological interactions

A. Della Penna (✉)
Quantitative Marine Sciences PhD Program, Institute for Marine and Antarctic Studies, University of Tasmania, and Commonwealth Scientific and Industrial Research Organisation, Hobart, TAS, Australia

Sorbonne Universités, Univ Paris Diderot Cité, UMR 7159, LOCEAN-IPSL, 75005 Paris, France
e-mail: alice.dellapenna@gmail.com

S. Wotherspoon • C. Johnson
Institute for Marine and Antarctic Studies, University of Tasmania, Hobart, TAS, Australia

T.W. Trull
Commonwealth Scientific and Industrial Research Organisation, Oceans and Atmosphere Flagship, Hobart, TAS, Australia

Antarctic Climate and Ecosystems Cooperative Research Centre, Hobart, TAS, Australia

S. De Monte
École Normale Supérieure, UMR 7625 Ecologie et Evolution, Paris, France

F. d'Ovidio
Sorbonne Universités, UPMC Univ Paris 06, UMR 7159, LOCEAN-IPSL, F-75005 Paris, France

P. Bourgine et al. (eds.), *First Complex Systems Digital Campus World E-Conference 2015*, Springer Proceedings in Complexity, DOI 10.1007/978-3-319-45901-1_43

between different groups of phytoplankton are embedded in a contrasted and highly dynamical environment characterized by oceanic eddies and fronts. Here we use a Lagrangian approach (i.e. based on observable measured along water parcel trajectories) to identify areas of expected diatom dominance in an iron-limited system, the Kerguelen region (Indian Sector of the Southern Ocean). The Kerguelen Islands arise from a large plateau that acts as a source for iron, that is otherwise a limiting factor for phytoplankton growth in the Southern Ocean [14]. Sediment resuspension from the Kerguelen Plateau is thought to enrich the water with iron and support one of the largest blooms of the Southern Ocean, that extends for more than 250,000 km^2 downstream [15]. Being an example of large scale natural iron fertilization the Kerguelen region is object of several investigations that are, however, limited by its remoteness. Relatively little is known about its phytoplankton biogeography and observations are sparse and clustered by the plateau, excluding most of the large downstream plume [16, 17].

2 Preliminary Results

Previous studies [18, 19] have shown that horizontal advection can be related with the maximum extent of the Chlorophyll plume developing East from the Kerguelen Plateau. Here, we suggest to use altimetry-derived oceanic currents (from Delayed Time Maps of Absolute Dynamic Heights (DT-MADT)—developed by CNES/CLS Aviso) to explain the spatial distribution of diatom dominance. Diatoms are the Southern Ocean's phytoplanktonic group that iron fertilization studies suggest to be mostly affected by iron scarcity [19]. In our "threshold" model, water parcels are enriched in iron by interactions of oceanic currents with the shallow bathymetry and they are advected downstream. As water parcels leave the plateau, iron is consumed by growing phytoplankton and scavenged by sinking particles resulting in a decrease in iron concentration that depends on how long before water parcels have been enriched. To quantify the iron content of water parcels, we use a Lagrangian diagnostic defined by d'Ovidio et al. [19]: the "water age"—how long ago a water parcel has been in contact with the shallow bathymetry of the plateau (shown in Fig. 1a). Once this quantity is measured, a proxy for iron content can be computed according to the equation:

$$[Fe] = [Fe_0]\, e^{\lambda_1 t_1 + \lambda_2 t_2} \qquad (1)$$

where $[Fe_0] = 150\,\mathrm{mol/m^2}$ is the iron concentration on the Kerguelen Plateau (measured during the field program KEOPS2 [19]) and $\lambda_1 = -0.041 d^{-1}$ and $\lambda_2 = -0.058 d^{-1}$ are coefficients for abiotic scavenging and abiotic scavenging and biological consumption combined. t_1 and t_2 represent the times since a specific water parcel has left the plateau during the winter and during the blooming season, respectively. Figure 1b displays a map of the proxy for iron.

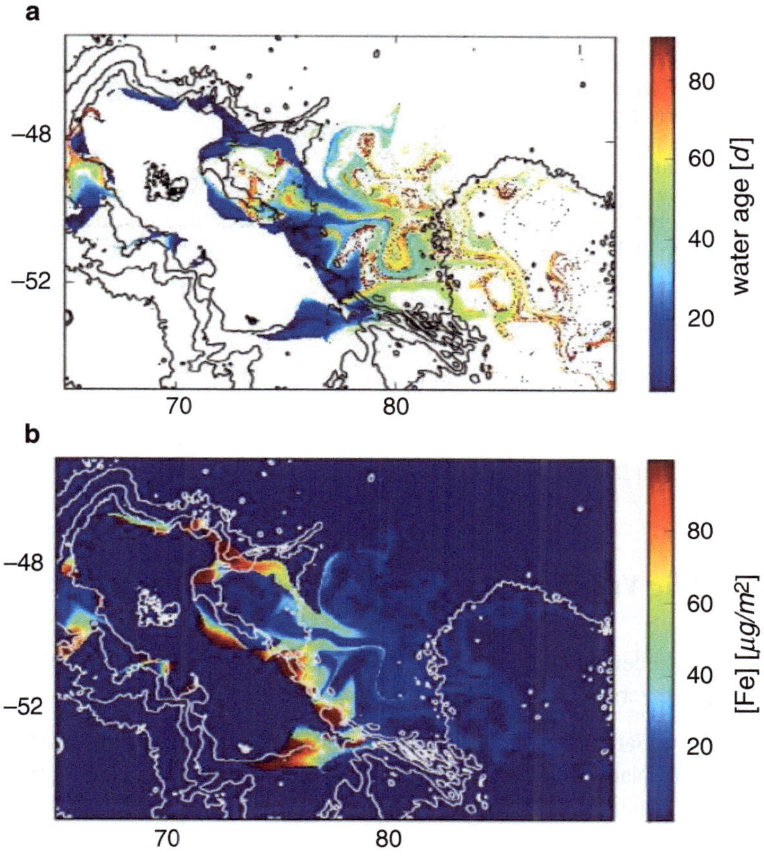

Fig. 1 Regional map of the water age diagnostic (**a**) and of the proxy for iron (**b**). Water age corresponds to the time has passed since a specific water parcel has been in touch with the Kerguelen Plateau—that we assume to be the main source of iron. The proxy for iron is computed by assuming constant abiotic scavenging and biological consumption rates

Assuming that iron is the main driver of community composition, we expect that for high iron concentrations (larger than $35\,\mathrm{mol/m^2}$) diatoms will dominate phytoplanktonic communities whereas water parcels-depleted water parcels will be dominated by other types of phytoplankton.

3 Perspectives

Preliminary statistical analyses suggest that the output of the threshold model relates significantly with the PHYSAT [20, 21] re-analysis, an empirical algorithm that allows to identify the dominant phytoplanktonic type from ocean colour images.

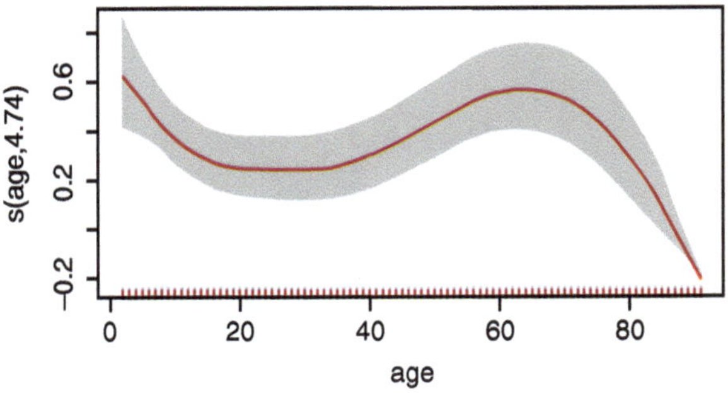

Fig. 2 Shape of the smooth function that best fits the relationship between water age and remotely sensed diatom dominance. The *y*-axis represents the log(odds)—directly proportional to the probability—of observing diatom dominance for a given water age

A pixel by pixel, day by day, General Additive Model [22] relating the remotely sensed maps and the threshold model suggests a significant relationship ($p-value < 2 * 10^{-16}$) between the increase of water age and the decrease of probability of observing diatom dominance (Fig. 2).

4 Scientific Validation

This paper has been unanimously validated in a collaborative review mode with the following reviewers:

- Leonardo A. Saravia, from Area Biologia y Bioinformitica, Instituto de Ciencias, Universidad Nacional de General Sarmiento, J.M. Gutierrez 1159 (1613) Los Polvorines, Buenos Aires Argentina
- Paul Bourgine, from CNRS
- Britt Stephens.

References

1. Longhurst AR (2010) Ecological Geography of the Sea. Alan R. Longhurst. Academic, London. ISBN 0124555594
2. Oliver, M. J., and A. J. Irwin (2008), Objective global ocean biogeographic provinces, Geophys. Res. Lett., 35 :L15601. doi:10.1029/2008GL034238
3. Dutkiewicz S, Follows MJ, Bragg JG (2009) Modeling the coupling of ocean ecology and biogeochemistry. Glob Biogeochem Cycles 23:GB4017. doi:10.1029/2008GB003405
4. Dutkiewicz, S., M. J. Follows, and J. G. Bragg (2009), Modeling the coupling of ocean ecology and biogeochemistry, Global Biogeochem. Cycles, 23, GB4017, doi:10.1029/2008GB003405.
5. Falkowski PG, Barber RT, Smetacek V (1998) Biogeochemical controls and feedbacks on ocean primary production. Science 281:200–206
6. Allen JT et al (2005) Diatom carbon export enhanced by silicate upwelling in the northeast Atlantic. Nature 437:728–732
7. Palmer TN, Lindstrom E (1991) Spatial variability in the sink for atmospheric carbon dioxide in the North Atlantic. Nature 350

8. Trull TW et al (2014) Chemometric perspectives on plankton community responses to natural iron fertilization over and downstream of the Kerguelen plateau in the southern ocean. Biogeosci Discuss 11:13841–13903
9. De Monte S, Soccodato A, Alvain S, d'Ovidio F (2013) Can we detect oceanic biodiversity hotspots from space&quest. ISME J 7:2054–2056
10. Moline MA, Claustre H, Frazer TK, Schofield O, Vernet M (2004) Alteration of the food web along the Antarctic peninsula in response to a regional warming trend. Glob Chang Biol 10:1973–1980
11. Kopczynska EE (1992) Dominance of microflagellates over diatoms in the Antarctic areas of deep vertical mixing and krill concentrations. J Plankton Res 14:1031–1054
12. Barton AD et al (2013) The biogeography of marine plankton traits. Ecol Lett 16:522–534
13. Boyd PW et al (2007) Mesoscale iron enrichment experiments 1993–2005: synthesis and future directions. Science 315:612–617
14. Boyd PW, Arrigo KR, Strzepek R, vanDijken GL (2012) Mapping phytoplankton iron utilization: insights into Southern Ocean supply mechanisms. J Geophys Res 117:C06009. doi: 10.1029/2011JC007726.
15. Mongin M, Molina E, Trull TW (2008) Seasonality and scale of the Kerguelen plateau phytoplankton bloom: a remote sensing and modeling analysis of the influence of natural iron fertilization in the southern ocean. Deep Sea Res II Top Stud Oceanogr 55:880–892
16. Lasbleiz M et al (2014) Pigments, elemental composition (c, n, p, and si), and stoichiometry of particulate matter in the naturally iron fertilized region of Kerguelen in the southern ocean. Biogeosciences 11:5931–5955
17. Uitz J et al (2009) A phytoplankton class-specific primary production model applied to the Kerguelen islands region (southern ocean). Deep Sea Res Part I Oceanogr Res Pap 56:541–560
18. Mongin MM, Abraham ER, Trull TW (2009) Winter advection of iron can explain the summer phytoplankton bloom that extends 1000 km downstream of the Kerguelen plateau in the southern ocean. J Mar Res 67:225–237
19. d'Ovidio F et al (2015) The biogeochemical structuring role of horizontal stirring: Lagrangian perspectives on iron delivery downstream of the Kerguelen plateau. Biogeosci Discuss 12:779–814
20. Alvain S, Moulin C, Dandonneau Y, Bréon FM (2005) Remote sensing of phytoplankton groups in case 1 waters from global SeaWiFS imagery. Deep Sea Res Part I Oceanogr Res Pap 52:1989–2004
21. Alvain S, Moulin C, Dandonneau Y, Loisel H (2008) Seasonal distribution and succession of dominant phytoplankton groups in the global ocean: a satellite view. Global Biogeochem Cycles 22:GB3001. doi:10.1029/2007GB003154
22. Hastie, Trevor J., and Robert J. Tibshirani. Generalized additive models. Vol. 43. CRC, New York

Erratum

First Complex Systems Digital Campus World E-Conference 2015

P. Bourgine et al. (eds.), *First Complex Systems Digital Campus World E-Conference 2015*, Springer Proceedings in Complexity, DOI 10.1007/978-3-319-45901-1

DOI 10.1007/978-3-319-45901-1_44

Paul Bourgine • Pierre Collet • Pierre Parrend

The original version of this book was inadvertently published with incorrect sequence of editors as given below

Pierre Parrend

Paul Bourgine

Pierre Collet

The correct sequence has been updated as given below.

Paul Bourgine

Pierre Collet

Pierre Parrend

The online version of the original book can be found under
http://dx.doi.org/10.1007/978-3-319-45901-1

P. Bourgine et al. (eds.), *First Complex Systems Digital Campus World E-Conference 2015*, Springer Proceedings in Complexity, DOI 10.1007/978-3-319-45901-1_44

Index

P. Bourgine et al. (eds.), *First Complex Systems Digital Campus World E-Conference 2015*, Springer Proceedings in Complexity, DOI 10.1007/978-3-319-45901-1

MIX
Papier aus verantwortungsvollen Quellen
Paper from responsible sources
FSC® C105338

If you have any concerns about our products,
you can contact us on
ProductSafety@springernature.com

In case Publisher is established outside the EU,
the EU authorized representative is:
Springer Nature Customer Service Center GmbH
Europaplatz 3, 69115 Heidelberg, Germany

Printed by Libri Plureos GmbH
in Hamburg, Germany